Molds and Mycotoxins

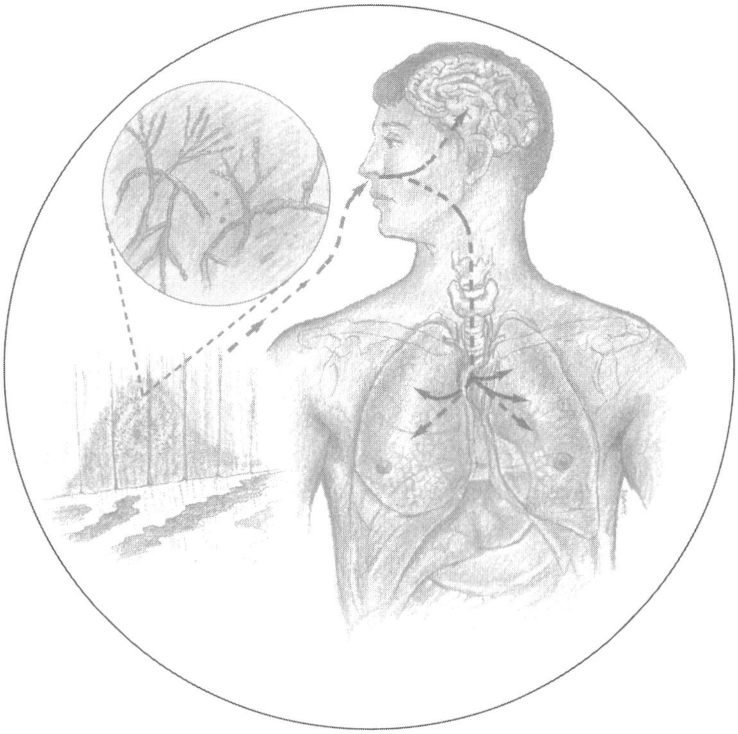

Molds and Mycotoxins

Papers from an International Symposium

Papers originally published in the
Journal Archives of Environmental Health
Vol. 58, Nos. 7 and 8

Kaye H. Kilburn, M.D., Editor
Ralph Edgington Chair of Medicine
Keck School of Medicine
University of Southern California

Heldref Publications
Washington, D.C.
2004

Molds and Mycotoxins

Heldref Publications
1319 Eighteenth Street, NW
Washington, DC 20036-1802
www.heldref.org
1-800-365-9753

ISBN: 0-916882-08-X

Printed in the United States of America

Cover illustration by Owen T. Davis

About the Editor
Kaye H. Kilburn received his M.D. from the University of Utah College of Medicine in 1954. He completed post-graduate studies at Case Western Reserve Medical School, University of Utah, Duke University, and University of London. He has served on several medical faculties across the country. He currently holds the Ralph Edgington Chair of Medicine at the University of Southern California's Keck School of Medicine.

CONTENTS

Mold Conference Objectives and Summary

Kaye H. Kilburn, M.D., and William J. Rea, M.D.

IMPAIRED FUNCTION AND DISEASE associated with living in a home in which mold is found seems new, but may have originated with the 1973 energy crisis. There are an estimated 40 million Americans exposed. As frequently occurs with "new" medical problems, public media, the legal profession, and insurance companies writing homeowners policies seem most aware of the problem. Physicians are consulted to render diagnoses, and to measure functional losses, assay symptoms, and opine as to whether molds found on surfaces, in the air, or evidenced by antibodies in the patient's blood are causing any or all of the manifestations. Patients hope that physicians can treat them to stop symptoms and reverse the abnormalities.

This book collects 18 papers that address these concerns and chemical sensitivity. The papers were presented at the 21st Annual International Symposium on Man and His Environment in Health and Disease, held in Dallas, Texas, in June 2003 under the joint sponsorship of the American Academy of Environmental Medicine and the American Environmental Health Foundation. The objective of the conference was to determine what circumstances have enabled molds to cause disease, why this has occurred relatively recently, and how it can be prevented. The participants included oceanographers studying coral reefs, chemists (of mycotoxin), microbiologists, physicians, otolaryngolgists, physiologists, immunologists, engineers, industrial hygienists, architects, and designers.

The 30 papers presented in 3-1/2 days stimulated much discussion. The organizers decided to publish 18 of the papers because concepts crystallized sufficiently to give a coherent story that explained much about mold/mycotoxin disorders. Most clear is the need to design houses and other buildings to breathe and ventilate out moisture and high humidity so as to take away the water needed for the growth of mold. Remediation of moldy buildings depends on those same principles.

Building boomed with the end of World War II to furnish 2 decades of pent-up demand from the Great Depression and the war. Let us review this scenario. To meet the demand, home building went to tracts, particularly in the sun states. Fast meant cheap and tight. As homes metamorphosed after 1950 to reduce ventilation of living spaces, attics, and walls, less air exchange increased humidity and decreased escape of evolved gases. Also, imagine that walls' materials shifted from plaster on cedar lath with a lime coat that was too alkaline to grow mold or bacteria, to drywall: gypsum with a neutral pH sandwiched between layers of paper. Consider that plumbing pipes changed from iron to copper to plastic, the latter likely to leak. Outer walls and shower stalls leaked into unventilated cul-de-sac wall spaces. You have pictured a modern house. Add air conditioning with ducts poorly insulated and preserving heat or cold inside but depending nearly exclusively on ducted air. The 1973 energy crisis dictated conservation of heat and cold that was accomplished by reduced air exchanges, thus maximal reuse of treated air.

Add to this the next diabolical scenario. Dust crosses continent-to-continent borne on winds that can move particles from deserts in Asia or Africa across the Atlantic in 4 to 5 days. In these dusts are spores of molds. One of these, *Aspergillus sydowii*, has decimated coral reefs in the Caribbean. Associated *Aspergillus–Penicillium* mixtures caused asthma epidemics in many states and inoculated molds into a new habitat on moist surfaces with neutral pH within the walls of homes.

A *Stachybotyrus* species that has been associated with lung hemorrhage in infants has been grown from several Cleveland homes. Here molds grew avidly, produced mycotoxins, and released these from surfaces or within walls. These mold particles and spores rich in mycotoxins were inhaled by residents. Spores and hyphi carried mycotoxins into human airways and lungs, causing inflammation that was evidenced by asthma, bleeding, and airway obstruction. Nasal sinuses became inflamed and obstructed. Mycotoxins crossed into the blood, and then through the blood-brain barrier. Hitting the brain, the toxins damaged neurons to impair balance, vision, reaction time, blink reflex, strength, cognition recall memory, concentration, concept juggling, and speed of physiological and mental processing. Dynamic brain (SPECT) scans demonstrated abnormalities in some patients that correlated with observed cognitive and memory impairments.

Avoidance is the principal medical and hygienic

strategy to prevent mold-induced disease. Dilute indoor air with abundant ventilation, and dry inner wall spaces and surfaces with drier outdoor air. This will eradicate the breeding places for mold and stop mycotoxin production and release. Buildings modified to be tight in the last 50 years must be made to breathe. Heat exchangers may be added to conserve energy, but indoor air must turn over with outdoor air.

These papers summarize the preliminary conclusions from studies of several hundred patients, present evidence for severe neurobehavioral impairment, nasal sinus and lung dysfunctions, and observations of immunological disturbance. How these tie together, their suggested treatments, and prognoses are works in progress. These papers introduce the reader to the issues.

Indoor Mold Exposure Associated with Neurobehavioral and Pulmonary Impairment: A Preliminary Report

KAYE H. KILBURN
University of Southern California
Keck School of Medicine
Environmental Sciences Laboratory
Alhambra, California

ABSTRACT. Recently, patients who have been exposed indoors to mixed molds, spores, and mycotoxins have reported asthma, airway irritation and bleeding, dizziness, and impaired memory and concentration, all of which suggest the presence of pulmonary and neurobehavioral problems. The author evaluated whether such patients had measurable pulmonary and neurobehavioral impairments by comparing consecutive cases in a series vs. a referent group. Sixty-five consecutive outpatients exposed to mold in their respective homes in Arizona, California, and Texas were compared with 202 community subjects who had no known mold or chemical exposures. Balance, choice reaction time, color discrimination, blink reflex, visual fields, grip, hearing, problem-solving, verbal recall, perceptual motor speed, and memory were measured. Medical histories, mood states, and symptom frequencies were recorded with checklists, and spirometry was used to measure various pulmonary volumes and flows. Neurobehavioral comparisons were made after individual measurements were adjusted for age, educational attainment, and sex. Significant differences between groups were assessed by analysis of variance; a *p* value of less than 0.05 was used for all statistical tests. The mold-exposed group exhibited decreased function for balance, reaction time, blink-reflex latency, color discrimination, visual fields, and grip, compared with referents. The exposed group's scores were reduced for the following tests: digit-symbol substitution, peg placement, trail making, verbal recall, and picture completion. Twenty-one of 26 functions tested were abnormal. Airway obstructions were found, and vital capacities were reduced. Mood state scores and symptom frequencies were elevated. The author concluded that indoor mold exposures were associated with neurobehavioral and pulmonary impairments that likely resulted from the presence of mycotoxins, such as trichothecenes.
<Key words: balance impairment, cognitive impairment, indoor air, mycotoxins, *Stachybotrys atra*, volatile organic chemicals>

THE ENERGY CRISIS OF THE 1970s, which resulted in decreased importation of Middle East crude oil to the United States and Europe, stimulated energy conservation in these 2 regions. After air leaks in buildings were plugged and air exchanges decreased, many people became sick indoors. Initial reports of Sick Building Syndrome (SBS) were subsequently doubted, and building occupants were considered hysterical[1–3] or to have suffered from "crowd disease."[3] Continuing reports of respiratory symptoms, increased sensitivity to chemicals, and memory and concentration defects spurred a search for causes.[4] Volatile organic chemicals[5] and formaldehyde[6] were considered as possible causes of SBS; however, their levels correlated poorly with symptoms.[7] Some individuals who moved out of mold-contaminated buildings soon after the onset of their symptoms improved,[8] but their symptoms recurred upon their return, whereas other individuals' symptoms persisted.

Asthma and flu-like symptoms have been associated with molds that grow indoors.[9,10] People complain of odors, memory loss, lack of concentration, sleep disturbances, and depression.[10,11] Mold growth has been associated with water condensation in walls; in air ducts; and under carpets, tiles, and flooring.[9] Water has leaked from roofs, windows, water pipes, shower stalls, bathtubs, and drains. Mold spores were observed microscopically and were grown in culture from both air and surface samples.[9,10] Patients' serum immunoglobulin (Ig)G, IgM, IgE, and IgA antibodies to molds were elevated, and their saliva contained elevated IgA antibodies.[10] Antibodies to aflatoxins, trichothecenes, and satratoxins occurred frequently.[11] Case recognition relied on symptoms, on molds verified microscopically, and on the basis of what grew on culture. Patients' respiratory symptoms, complaints of defective memory and concentration, impaired balance, and depression[10,11] led to measurements of lung and central nervous system functions.

Method

Study subjects. Sixty-five consecutive adult patients who consulted the author after being exposed to molds in their respective homes in California, Arizona, and Texas—a self-selected case series—were evaluated in 2001 and 2002. The patients' homes had mold growth on walls and floors, and the molds were identified microscopically. Cultures of indoor air samples frequently grew *Stachybotrys chartarum, Aspergillus, Penicillium, Chaetomium, Alternaria, Fusarium,* and *Rhizopus* at levels that exceeded concentrations found outdoors. Other fungal genera included *Actinomycetes* and *Cladosporium,* as well as the bacteria *Corynebacterium, Shigella, Agrobacterium,* and *Pseudomonas*. Each patient's serum was analyzed for antibodies to 14 categories of mold and to aflatoxin, trichothecenes, and satratoxin. Indoor air mycotoxins were not measured because methods for such measurements do not currently exist.

The 65 adult patients were compared by analysis of variance (ANOVA) with a referent group of 202 unexposed community-dwelling individuals who lived in Wickenburg, Arizona, and who were recruited and tested in 1993 and retested in 1996.[12] From this group and from comparable groups, equations for the calculation of predicted values were developed for each test.[12] The observed test values in unexposed and mold-exposed individuals were divided by the predicted values (i.e., [observed/predicted] × 100) that adjusted for sex, height, weight, and education (i.e., highest grade attained or level of higher education completed).

Reference subjects were selected at random from voter registration rolls, and individuals with occupational exposure to chemicals, and those who had medical or neurological diseases, were excluded via interview. Reference subjects comprised the largest of 3 groups, the measurements for which produced prediction equations.[12] Subjects were reimbursed for their time spent participating in this study, and all individuals provided informed consent after the procedures of the study were explained fully. The study protocol was approved by the Human Studies Research Committee of the University of Southern California, Keck School of Medicine.

Questionnaires. The subjects' questionnaires were checked by computer-guided card readings, and omissions were rectified by the subjects. The frequencies of occurrence of 35 common health complaints[13,14] were rated on a scale from *never* (1) to *daily* (11). Subjects answered the American Rheumatism Association's lupus erythematosus questions[15] and a standard respiratory questionnaire,[16] and gave histories of occupational and other exposures to chemicals, pesticides and herbicides, tobacco, alcohol, and drugs (prescription and illicit). Histories included recorded incidents of unconsciousness; anesthesia; head trauma; and medical, neurological, or psychiatric illnesses.[14,17] The questionnaires and neurophysiological and neuropsychological test batteries evolved through studies of histology technicians and firefighters, and through examinations of the effects of formaldehyde,[14,17,18] thermolysis products of polychlorinated biphenyls (PCBs),[19] and exposure to toluene-rich chemical waste.[20] These instruments also took into consideration several groups of unexposed reference subjects.[12,20]

Neurophysiological tests

Simple reaction time and visual 2-choice reaction time. Simple reaction time (SRT) and visual 2-choice reaction time (CRT) were the times from appearance to cancellation—by tapping a touch pad A or S—of a 10-cm block A for SRT and 10-cm blocks A and S for CRT, both of which were measured with a computerized instrument (Neurotest, Inc. [Pasadena, California]).[21] The lowest median score of the final 7 attempts in each of 2 trials of 20 attempts was accepted for SRT and for CRT.

Body balance. Body balance was measured while the subject was standing erect, with his or her feet aligned closely, side-by-side. The position of the head was tracked by 2 microphones from a sound-generating stylus on a headband (Neurotest, Inc. [Pasadena, California]). Results were processed by software (Graf/Bar Mark II [Scientific Accessories Corp. {Shelton, Connecticut}]) and expressed as the mean speed of sway in cm/sec.[22] The minimal sway speed was recorded for 3 consecutive 20-sec trials with eyes open, and 3 with eyes closed, in alternating sequence.

Blink reflex latency R-1. Blink reflex was measured electromyographically with surface electrodes placed on the lateral orbicularis oculi muscles bilaterally,[23,24] after the right and left supraorbital notches were tapped with a light hammer, which also triggered a recording computer (Neurotest, Inc. [Pasadena, California]). Ten R-1's were averaged for each side, and failures were recorded.[24]

Color confusion index. The color confusion index was measured with the desaturated Lanthony 15-hue test under conditions of constant illumination,[25] and the index was scored by Bowman's method.[26]

Hearing. Hearing was measured with standard programmed audiometers (Model ML-AM [Microaudiometrics {South Daytona, Florida}]) at stepped frequencies of 500–8,000 Hz. Results were expressed as the sum of hearing loss at each frequency for each ear.

Visual fields. For the threshold testing of visual fields, an automated, computerized perimeter recorder (Med Lab Technologies [North Wales, Pennsylvania]) that mapped the central 30° of the right and left eyes individually, was used. The performance of each eye was determined as the sum (in decibels) of the threshold of 80 points within the central 30°.

Neuropsychological testing. Immediate memory or recall was measured by memory or recall associated with information contained in 2 stories gleaned from Wechsler's Memory Scale–Revised (WMS-R).[27] Culture Fair (Scale 2 form A) tested for nonverbal, nonarithmetic intelligence, on the basis of the selection of 4 sets of designs[28,29]: logical series, difference, completion, and pattern definition and transfer. Culture Fair resembles Raven's progressive matrices.[30] The 46-word multiple-choice vocabulary test used originated from the multidimensional aptitude battery.[31] Digit symbol substitution from the Wechsler Adult Intelligence Scale–Revised (WAIS-R)[32] tested attention and integrative capacity. Information, picture completion, and similarities—all of which were also gleaned from the WAIS-R—tested for long-term (i.e., embedded or hold) memory. The time required to place 25 pegs in a grooved pegboard (from the Lafayette battery) to make trails A and B assessed dexterity, coordination, and decision-making; fingertip number writing measured peripheral perception and discrimination (from the Halstead-Reitan battery[33,34]). Subjects self-appraised their emotional status during the preceding week by using the Profile of Mood States (POMS)[35]—which comprises 65 words that describe tension, anxiety, depression, anger, vigor, fatigue, confusion—and the Limbic Check List.[36] Recall of the Rey 15 figures assessed malingering.[37]

Neurological examinations. The author conducted neurological examinations to assess cranial nerve function, muscle movement and strength, and cerebellar signs (e.g., past pointing, tremors, alternating movements, stance, and gait).

Respiratory flows and vital capacities. Respiratory flow and vital capacity were measured during a full inspiration while the subject, with a nose clip attached to his or her nose, stood upright and exhaled into a volume-displacement spirometer (Model 822 [Sensor Medics {Anaheim, California}]). The procedure was repeated until readings from 2 forced expirations comported within 5%.[38] Records were traced with a digitizer; measured with software (Graf/Bar Mark II [Scientific Accessories Corp. {Shelton, Connecticut}]), and compared with predicted values adjusted for height, age, sex, and years of cigarette smoking.[39]

Alcohol and carbon monoxide measurements. Fuel-cell analyzers (Alcohol Intoximeter [St. Louis, Missouri] and a Micro Smokerlyzer [Bedfont Scientific Ltd {Rochester, U.K.}] were used to measure alcohol and carbon monoxide, respectively, in air expired following a 20-sec breath hold.[20]

Statistical analyses. Scores and computed data were entered into an IBM-compatible microcomputer of our design. Descriptive and analytical computations—adjusted for differences in age, education, sex, height, and weight—were completed with stepwise linear-regression modeling in order to develop prediction equations for each test. Computations were completed with Stata Statistical Software, version 8 (Stata Corporation [College Station, Texas]). Prediction equations were developed from measurements of 293 subjects. Measurements had symmetrical distributions.[12] The observed measurements and scores for each patient were compared with his or her individual predicted values and expressed as percent predicted. The exposed group's percent predicted values were then compared with those of the control group, using ANOVA. Factors such as family income, hours of general anesthesia undergone, POMS score, and depression score were excluded because their coefficients were insignificant. Statistical significance was defined as $p < 0.05$. Abnormalities for the 65 patients and 202 referent subjects were counted (Table 1), after bilateral tests (e.g., hearing) were assigned a value of 0.5 per side—except for visual field performance which was assigned a value of 1 per side, and balance with eyes open and eyes closed which were scored at 2 each.

Results

All 65 patients (i.e., approximately 40 families) had observed mold growing on the walls in their kitchens, bathrooms and/or service areas, and sometimes on the bedroom and closet walls. Musty odors were traced to electrical plates and air-conditioning ducts that provided access to inner walls. Invariably, several mold genera grew in cultures of indoor and outdoor air samples obtained simultaneously; occasionally, endotoxin-producing bacteria were present. Variability of sampling and characterization methods made quantitative comparisons impractical; however, the data were correlated with the complaints, impairment profiles, and serum antibodies.

The mold-exposed group was not statistically significantly older than the unexposed group, but they had greater educational attainment. The exposed group's mean SRTs and CRTs were prolonged significantly, as evidenced by the percent predicted, compared with unexposed subjects (Table 1). With respect to balance, measured sway speeds were faster and significantly abnormal—especially with eyes closed—in the mold-exposed group. Blink reflex latency R-1 was delayed significantly. Color discrimination errors were increased.

Table 1.—Characteristics and Test Results for Mold-Exposed Subjects Compared with Referent (Unexposed) Subjects, as Percentage of Predicted

	Exposed (n = 65)		Unexposed (n = 202)		
Subject characteristic/test	$\bar{x}$	*SD*	$\bar{x}$	*SD*	p*
Subject characteristic					
Age (yr)	47.3	13.7	46.6	20.6	0.624
Education level (yr)	14.6	2.5	12.9	2.3	0.0001
Neurophysiological test					
Reaction time (ms)					
Simple	103.8	6.6	99.9	3.7	0.0001
Choice	102.4	4.4	100.0	2.5	0.0001
Balance sway speed (cm/sec)					
Eyes open	137.0	44.6	100.2	20.0	0.0001
Eyes closed	167.4	95.3	103.1	26.7	0.0001
Blink reflex latency R-1 (ms)					
Right	114.9	13.2	99.4	14.6	0.0001
Left	115.4	12.7	96.4	13.2	0.0001
Hearing loss					
Right	105.7	38.3	101.5	24.6	0.436
Left	105.3	35.4	98.3	23.2	0.213
Color discrimination errors					
Right	72.0	42.6	102.6	51.1	0.0001
Left	74.0	49.0	102.6	51.1	0.0001
Visual field performance					
Right	87.4	21.7	100.0	22.8	0.006
Left	89.6	19.1	101.1	21.7	0.008
Grip strength					
Right	90.1	18.8	99.3	17.5	0.0001
Left	86.8	22.1	99.1	17.5	0.0001
Neuropsychological test					
Culture Fair	95.8	21.3	101.2	20.0	0.063
Digit symbol	93.5	10.4	101.5	9.2	0.0001
Vocabulary	88.6	28.6	99.2	20.8	0.015
Verbal rtecall					
Immediate	79.5	24.6	99.8	31.1	0.0001
Delayed	71.7	38.0	99.9	41.3	0.0001
Pegboard	92.2	15.7	101.8	25.7	0.001
Trails A	103.7	9.2	100.3	8.3	0.006
Trails B	102.9	9.1	100.4	7.5	0.026
Finger writing errors					
Right	99.4	7.3	100.0	7.5	0.647
Left	101.0	9.7	100.0	7.8	0.513
Information	82.6	36.1	101.5	39.4	0.0007
Picture completion	75.2	29.6	99.3	32.2	0.0001
Similarities	93.4	29.3	98.1	41.2	0.396
Total abnormalities	9.9	5.0	2.3	2.4	0.0001

Notes: $\bar{x}$ = mean, and *SD* = standard deviation.
*p values by analysis of variance.

Errors were modeled as the reciprocal of quadratic power, therefore abnormal observed values decreased compared with predicted scores. Visual field performance was reduced significantly in both eyes in the exposed group. Grip strength was also significantly lower in the exposed group. Hearing losses were not significantly different between the 2 groups.

The results of psychological tests revealed significantly lower scores in the mold-exposed group for digit symbol substitution and vocabulary; however, Culture Fair (i.e., problem-solving) scores were not significantly different between the 2 groups. Immediate verbal recall (of stories) was significantly lower in the exposed group, and decreased further following a 30-min delay.

Times for peg placement trail-making A and B were all significantly prolonged in the exposed group, thus reflecting perceptual motor impairment. Fingertip number writing perception was not impaired in either group.

Information and picture completion, but not similarities, were significantly different in the exposed group, despite the fact that these tests are usually unaffected by chemical exposures.[40]

In the mold-exposed group, 12 of 14 physiological functions, and 9 of 13 psychological tests, were affected adversely by molds and mycotoxins, for a total of 21 of 27 neurobehavioral tests. Mold-exposed subjects averaged 9.9 abnormal tests, compared with 2.3 among the referents ($p < 0.0001$ [Table 1]).

The mood states (i.e., POMS) scores were 3 times higher in the exposed group than in the unexposed group—a significant difference. Vigor was reduced, and tension, depression, anger, fatigue, and confusion were higher in the exposed group (Table 2). The POMS score equals the sum of the 5 adverse moods, minus vigor. Mean frequencies of 35 symptoms averaged 5.0 ± 1.9 (mean ± standard deviation) in the exposed group, compared with the controls (2.6 ± 1.1), which was a significant difference ($p < 0.0001$).

Vital capacity was reduced significantly in the mold-exposed group, as was forced expiratory volume in 1 sec ($FEV_{1.0}$)—results evidencing airway obstruction (Table 3). The decrease in vital capacity reduced the slowly emptying lung space; therefore, mid-flow forced expiratory flow (i.e., FEF_{25-75}), terminal flow (FEF_{75-85}), and the $FEV_{1.0}$-to-forced vital capacity (FVC) ratio were all higher in the exposed than the unexposed group.

Fifty-two percent of mold-exposed patients had current symptoms of peripheral neuropathy, and 62% of these patients had a history of peripheral neuropathy. This percentage compared with 14% of unexposed subjects.

Sixty-eight percent of patients had developed excessive sensitivity to chemicals (i.e., chemical triggers) following mold exposure, and 57% reacted to more than 2 chemicals. However, regression analysis for the influence of numbers of chemical triggers, age, POMS score, and other factors on total abnormalities showed that only POMS had a significant correlation coefficient ($r = .031698$, $p < 0.03$), but with little effect ($r^2 = 5.7\%$). Regression analysis for the influence of peripheral neuropathy score, age, and other factors on total abnormalities showed a correlation coefficient for peripheral neuropathy score of 0.2709596 ($p < 0.029$; $r^2 = 6.2\%$)—again, with little effect.

Table 2.—Profile of Mood States (POMS) Test Results for Mold-Exposed Subjects Compared with Unexposed Subjects

	Exposed ($n = 65$)		Unexposed ($n = 202$)		
POMS symptom	$\bar{x}$	*SD*	$\bar{x}$	*SD*	*p*
Score*	64.0	46.1	21.0	31.6	0.0001
Tension	17.0	8.8	9.1	5.8	0.0001
Depression	15.7	13.2	8.1	9.3	0.0001
Anger	14.0	11.1	8.3	7.9	0.0003
Fatigue	16.2	8.2	7.6	6.1	0.0001
Vigor	11.1	7.1	18.3	6.3	0.0001
Confusion	13.6	7.0	6.1	4.5	0.0001
Symptom frequency	5.0	1.9	2.6	1.1	0.0001

Notes: $\bar{x}$ = mean, and *SD* = standard deviation.
*The POMS score is the sum of tension, depression, anger, fatigue, and confusion, minus vigor.

Table 3.—Pulmonary Function Test Results for Mold-Exposed Subjects Compared with Unexposed Subjects, as Percentage of Predicted

	Exposed ($n = 65$)		Unexposed ($n = 202$)		
Pulmonary function test	$\bar{x}$	*SD*	$\bar{x}$	*SD*	*p**
FVC	91.9	11.1	101.6	15.1	0.0001
$FEV_{1.0}$	88.4	12.7	93.6	15.8	0.016
FEF_{25-75}	93.6	28.9	88.1	35.0	0.256
FEF_{75-85}	91.6	51.1	78.1	52.7	0.070
$FEV_{1.0}$/FVC	77.2	7.1	72.9	9.5	0.0007

Notes: $\bar{x}$ = mean, *SD* = standard deviation, FVC = forced vital capacity, $FEV_{1.0}$ = forced expiratory volume in 1 sec, FEF_{25-75} = mid pulmonary forced expiratory flow, and FEF_{75-85} = terminal pulmonary flow.
**p* values by analysis of variance.

The neurological abnormalities seen most frequently in mold-exposed persons were past-pointing in 48 subjects, decreased or absent vibration sensation in the feet in 35, ataxia in 32, slow alternating movements in 21, dysmetria in 16, and tremors in 12. No one had clonus or cogwheel rigidity. One third of the exposed patients had more than 4 abnormal cerebellar signs and abnormally fast sway speed with eyes closed.

Satratoxin antibodies were increased in exposed subjects by 178%, but antibodies to aflatoxin and trichothecenes were decreased by 55% and 22%, respectively. Serum antibody titers to mold showed that IgM increased by 114%, IgE by 47%, and IgG by 42% among the exposed subjects (Table 4). Serum antibody titers to molds or mycotoxins were not predictive of neurobehavioral abnormality, as evidenced by the results of regression analyses. Pulmonary impairment, as shown by the results of spirometric tests, was not correlated with mold concentrations in air, or with antibody titers.

Impairment in 6 patients exposed to additional contaminants of sewer gas, construction materials (including organic solvents), or pesticides averaged below the group mean. Excluding these individuals from analysis did not affect the interpretation; therefore, they were retained in the analysis. No patient had a neurological or psychiatric disease. Severe depression subsequent to mold exposure was diagnosed and treated in 4 patients.

Eight mold-exposed patients were tested 11–20 mo following their initial evaluations, after they had left the homes in which exposure occurred. Repeat testing showed that no individual had experienced improved functions; in fact, 7 of the 8 patients (88%) had worsened and 1 individual's symptoms were unchanged. These observations, which allow for comparison of functions across time and enable each individual to function as his or her own control, confirm comparisons of groups after the application of prediction equations. Absent additional mold exposure, functions in 88% (n = 7) of these patients had deteriorated during the course of a year.

Table 4.—Differences in Prevalence of Abnormal Titers (in %) of Antibodies to 14 Molds and 3 Mycotoxins in Mold-Exposed Patients with Symptoms and Impairments, Compared with Community (Unexposed) Subjects

Antibodies measured	Exposed (n = 54)	Unexposed (n = 48)	Difference between groups
Mold*			
IgG	74.0	52.0	> 42
IgM	94.0	44.0	> 114
IgA	33.0	79.0	< 42
IgE	117.0	79.0	> 47
Mycotoxin			
Aflatoxin	9.2	16.7	< 55
Satratoxin	40.7	22.9	> 178
Trichothecene	13.0	16.7	< 22

Note: Ig = immunoglobulin.
*The following mold antibodies were measured: *Alternaria tenus, Aspergillus fumigatus, Aspergillus niger, Aspergillus versicolor, Chaetomium globosum, Cladosporium herbarum, Epicoccum nigrum, Geotrichium candidum, Penicillium notatum, Phoma herbarum, Pullullaria pullulans, Rhizopus nigricans, Rhodotorula glutinis, Stachybotrys chartarum.*

Discussion

Many neurobehavioral measurements in 65 subjects exposed to molds and mycotoxins at home were abnormal, compared with their individual predicted values. Mold exposure was associated with impaired balance, color discrimination, visual field performance, and grip strength, as well as slowed reaction times—but *not* with decreased hearing. Verbal recall was decreased, perceptual motor functions were impaired for peg placement and trail making, and problem solving was reduced in digit symbol substitution. Vocabulary and long-term memory for information and picture completion were impaired. Excessive cerebellar abnormalities—evidenced by the results of neurological examinations—correlated with measured balance as excessive sway speed.

These mold-exposed patients complained of persistent flu-like illnesses, onset of asthma (not verified by exam), beginning of severe fatigue, impaired memory and concentration, frequent dizziness, and unsteady balance. These symptoms were accompanied by the patients noticing musty odors and black mold growth on baseboards and lower walls (particularly in the bathrooms, adjoining bedrooms, and closets) and on outer walls beneath window sashes and where walls joined ceilings. Concurrent with symptoms, water leaks were usually found from roofs, windows, outer walls, or within the walls. Microscopic examination of touch preparations showed the spores and hyphae of many different molds. Patients' serum and saliva antibodies for many molds exceeded laboratory reference values, but only satratoxin antibodies were elevated compared with an unexposed group.

Clinical judgment attributed these impairments to mold exposure, after other possibilities had been considered, inasmuch as these patients were without preexisting medical or neurological diseases or traumatic brain damage. In addition, mold-exposed patients had no previous chemical exposures and they had rarely used pesticides.

Pulmonary function abnormalities indicated severe airways obstruction that reduced vital capacity and $FEV_{1.0}$, and their reduced ratio was characteristic of bronchiolitis.[41,42] There is no accepted method for the measurement of mycotoxins in air or for the quantification of molds in air (except by spore count or culture), and no serum assay or biomarker is available; therefore, we lack the doses for a dose-response curve of adverse effects on the brain.

Mycotoxins are recognized by bioassays for toxic activity on cells or enzymes, and trichothecene activity can be estimated in air or particles.[43–46] Dust samples from a Montreal, Canada, office building were assayed by thin-layer chromatography, using 4-(*p*-nitrobenzyl) pyridine to identify the 12,13-epoxy group of trichothecenes.[44] Methanol extractions of water-damaged building materials and gypsum board, assayed in feline skin and lung cells, were 200 times as toxic as board extract that was not water damaged.[44] Spores and fragments of 3 molds (i.e., *Aspergillus versicolor, Penicillium melinii,* and *Cladosporium cladosporioides*) share antigens and immunological reactivity.[45] A trichothecene assay (i.e., an inhibition of protein synthesis by using firefly luciferase translation in rabbit reticulocytes) correlated with fungal counts in indoor air.[47]

Trichothecenes identified by 3-[4,5-dimethylthiazol-2-yl]-2,5-diphenyltetrazolium bromide (MTT) cleavage, and shown chemically by high-performance liquid chromatography with diode array ultraviolet detection and gas chromatography-mass spectrometry, were cytotoxic for swine kidney cell cultures.[46] A chymotrypsin-like serin proteinase from *Stachybotrys chartarum* was isolated and purified from an infant with pulmonary hemosiderosis. It cleaved lung proteinase inhibitors, bioactive peptides, and collagen,[48] all of which suggested that it could destroy lung tissue.

The consecutive mold-exposed patients in the present study were referred by physicians or attorneys, or were self-referred. A representative study of a random sample of individuals would necessitate motivating individuals who had no complaints to submit to testing. Such a study might be feasible when mold problems are better understood and a biological marker is available.

Patients' exposures to toxins at home are a difficult epidemiological problem. They are not a cohort with shared space and exposure, as is the case in a workplace. Age, wellness, and hours per day of exposure vary more for people at home than for individuals at work. Self-selection for testing (bias) may ultimately result in an increase in symptoms, but it does not affect neurobehavioral measurements.[49,50] Measurements of effects are independent, even if symptoms are exaggerated. Clinical judgment has favored molds as causes, although homes—like schools and offices[8,9,42]—often contain other toxic chemicals.[6,7]

Impairment from mold exposure resembles that associated with formaldehyde exposure and the Indoor Air Syndrome—except for the lengthened blink reflex latency[6] that is typically associated with exposure to chlorinated solvents and arsenic,[49–51] but not with other chemicals.[52–54]

Antibody titers to trichothecenes, satratoxin, and aflatoxin did not predict impairment of lung or brain. An important relationship to consider in future studies is that which may exist between total and specific neurobehavioral abnormalities, and trichothecene concentrations—especially T-2 and satratoxin—in indoor air[47,55] and in patients' serum and fat.

Observations of cognitive and memory impairment in 20 previous patients exposed to *Stachybotrys atra,*[56] as well as abnormalities of balance, reaction time, recall, memory, and trail making in our preliminary group[11] are now confirmed.

Inhaled mycotoxins, liberated from indoor mold growth, caused brain impairment and neurological symptoms. Trichothecenes are epoxides that covalently adduct deoxyribonucleic acid, ribonucleic acid, protein, and microtubules of nerve axons,[56,57] providing mechanisms for lung, brain, and immune system toxicity.

Most therapeutic antibiotics, including penicillin and aminoglycosides, are products of molds[58] Other drugs produced by molds include ergot alkaloids, coumadins, and adriamycin C—used to treat cancer.[59] Asthma, cancer, wasting, hemorrhage, inanition, and death have been linked to mold exposure since ancient times.[10,59] But why have people developed sickness from molds in their homes during the most-recent decade?

Consider that after World War II construction of the interior walls of homes shifted from wood or metal lath, plaster, and lime coat to plasterboard (gypsum board)—a change that reduced cost. The inner paper (cellulose) layer of plasterboard encourages mold growth[60] when inner walls become damp from inadequate venting of moisture or from leaks from walls, roofs, or plumbing.[61] Variations in moisture levels stress molds to produce spores and toxins, such as trichothecenes and satratoxin,[62,63] that escape into living and working spaces.[43] Prior to the use of plasterboard, alkaline plaster or lime coating on walls discouraged fungal growth.

There are no antidotes for mold toxins.[59,62] They are likely to permanently impair brain function. Only the avoidance of exposure stops added damage; however, impairment may progress *absent additional exposure*—as was the case in 7 of 8 patients we tested twice—suggesting that *absorbed dose continues to damage neurons.*

Prevention

Limiting the damaging health effects of mycotoxin exposure requires primary prevention. An initial step is the development of wallboard that does not facilitate the growth of *Stachybotrys* and other toxic molds. Perhaps boric acid or sodium borate can be added to the paper layer or slurry of gypsum during the making of wallboard. Other agents that provide an acidic milieu when applied to surfaces may prevent mold growth; methoxymethane (dimethylether) removes aflatoxins from peanuts.[64] It is important, however, that any agent developed for the lessening of mold exposure not be toxic itself. Recognition by the medical profession and by the public that mold causes neurological and/or respiratory damage is important because these illnesses are often progressive and irreversible. The social, personal, and economic consequences are immense. Prevention by avoidance of exposure is imperative.

* * * * * * * * * *

The author thanks Drs. Ann Spurgeon, Nachman Brautbar, and Janette Sherman for their valuable suggestions and advice regarding the content of this report.

Remuneration for examination and testing was obtained from patients; their medical insurance companies; and, for several patients, from law firms that undertook litigation on the patients' behalf.

Submitted for publication September 20, 2003; revised; accepted for publication October 30, 2003

Requests for reprints should be sent to Dr. Kaye H. Kilburn, University of Southern California, Keck School of Medicine, Environmental Sciences Laboratory, Bldg. A7 #7401, 1000 S. Fremont Ave, Unit 2, Alhambra, CA 91803.

E-mail: Kilburn@usc.edu

* * * * * * * * * *

References

1. Landrigan PJ, Miller B. The Arjenyattah epidemic. Home interview data and toxicological aspects. Lancet 1983; 2 (8365–66):1474–76.
2. Hefez A. The role of the press and the medical community in the epidemic of "mysterious gas poisoning" in the Jordan West bank. Am J Psychiatry 1985; 142:833–37.
3. Small GW, Borus JF. Outbreak of illness in a school chorus. Toxic poisoning or hysteria? N Engl J Med 1983; 308: 632–35.
4. Kreiss K. The sick building syndrome: where is the epidemiological basis? Am J Public Health 1990; 80: 1172–73.
5. Burge S, Hedge A, Wilson S, et al. Sick Building Syndrome: A study of 4373 office workers. Ann Occup Hyg 1987; 31:493–504.
6. Kilburn KH. Indoor air effects after building renovation and manufactured homes. Am J Med Sci 2000; 320 (4):249–54.
7. Dally KA, Hanrahan LP, Woodbury MA, et al. Formaldehyde exposure in nonoccupational environments. Arch Environ Health 1981; 36:277–84.
8. Menzies R, Tamblyn R, Farant JP, et al. The effect of varying levels of outdoor air supply on the symptoms of sick building syndrome. N Engl J Med 1993; 328:821–27.
9. Burge PS, Finnegan M, Horsfield N, et al. Occupational asthma in a factory with a contaminated humidifier. Thorax 1985; 40:248–54.
10. Johanning E, Morey P, Jarvis B, et al. Health and immunology study following exposure to toxigenic fungi *Stachybotrys chartarum* in a water damaged office environment. Int Arch Occup Environ Health 1996; 68: 207–18.
11. Kilburn KH. Inhalation of moulds and mycotoxins. Eur J Oncol 2002; 7(3):197–202.
12. Kilburn KH, Thornton JC, Hanscom BE. Population-based prediction equations for neurobehavioral tests. Arch Environ Health, 1998; 53:257–63.
13. Kilburn KH, Warshaw R, Boylen CT, et al. Pulmonary and neurobehavioral effects of formaldehyde exposure. Arch Environ Health 1985; 40:254–60.
14. Kilburn KH, Warshaw R, Thornton JC. Formaldehyde impairs memory, equilibrium, and dexterity in histology technicians: effects which persist for days after exposure. Arch Environ Health 1987; 42:117–20.
15. Levin RE, Weinstein A, Peterson M, et al. A comparison of the sensitivity of the 1971 and 1982 American Rheumatism Association criteria for the classification of lupus erythematosus. Arthritis Rheum 1984; 27:530–38.
16. Ferris BG Jr. Epidemiology standardization project. Am Rev Respir Dis 1978; 118:7–54.
17. Kilburn KH, Warshaw RH. Neurobehavioral effects of formaldehyde and solvents on histology technicians: repeated testing across time. Environ Res 1992; 58:134–46.
18. Kilburn KH. Neurobehavioral impairment and seizures from formaldehyde. Arch Environ Health 1994; 49:37–44.
19. Kilburn KH, Warshaw RH, Shields MG. Neurobehavioral dysfunction in firemen exposed to polychlorinated biphenyls (PCBs): possible improvement after detoxification. Arch Environ Health 1989; 44(6):345–50.
20. Kilburn KH, Warshaw RH. Neurotoxic effects from residential exposure to chemicals from an oil reprocessing facility and Superfund site. Neurotoxicol Teratol 1995; 17: 89–102.
21. Miller JA, Cohen GS, Warshaw R, et al. Choice (CRT) and simple reaction times (SRT) compared in laboratory technicians: factors influencing reaction times and a predictive model. Am J Ind Med 1989; 15:687–97.
22. Kilburn KH, Warshaw RH. Balance measured by head (and trunk) tracking and a force platform in chemically (PCB and TCE) exposed and referent subjects. Occup Environ Med 1994; 51:381–85.
23. Shahani BT, Young RR. Human orbicularis oculi reflexes. Neurology 1972; 22:149–54.
24. Kilburn KH, Thornton JC, Hanscom B. A field method for blink reflex latency (BRL R-1) and prediction equations for adults and children. Electromyogr Clin Neurophysiol 1998b; 38:25–31.
25. Lanthony P. The desaturated panel D-15. Doc Ophthalmol 1978; 46:185–89.
26. Bowman KJ. A method for quantitative scoring of the Farnsworth panel D-15. Acta Ophthalmol 1982; 60: 907–16.
27. Wechsler D. A standardized memory scale for clinical use. J Psychol 1945; 19:87–95. Wechsler Memory Scale–Revised Manual. New York: The Psychological Corporation, 1987.
28. Cattell RB. Classical and standard score IQ standardization of the IPAT: culture-free intelligence scale 2. J Consult Psychol 1951; 15:154–59.
29. Cattell RB, Feingold SN, Sarason SB. A culture-free intelligence test II evaluation of cultural influences on test performance. J Educ Psychol 1941; 32:81–100.
30. Raven JC, Court JH, Raven J. Standard Progressive Matrices. Oxford, U.K.: Oxford Psychologists Press, 1988.
31. Jackson DN. Multidimensional aptitude battery. Port Huron, MI: Sigma Assessments Systems, 1985.
32. Wechsler D. Adult Intelligence Scale Manual–Revised. New York: The Psychological Corporation, 1981.
33. Reitan RM. A research program on the psychological effects of brain lesions in human beings. In: Ellis NR (Ed). International Review of Research in Mental Retardation. New York: Academic Press, 1966; pp 153–216.

34. Reitan RM. Validity of the trail-making test as an indicator of organic brain damage. Percept Mot Skills 1958; 8: 271–76.
35. McNair DM, Lorr M, Droppleman LF. Profile of Mood States. San Diego, CA: Educational and Industrial Testing Service, 1971/1981.
36. Teicher MH, Glod CA, Surrey J, et al. Early childhood abuse and limbic system ratings in adult psychiatric outpatients. J Neuropsychiatry Clin Neurosci 1993; 5(3): 301–06.
37. Rey A. L'examen Clinique en Psychologie. Paris: Presses Universitaires de France, 1964.
38. American Thoracic Society. Standardization of spirometry—1987 update. Statement of the American Thoracic Society. Am Rev Respir Dis 1987; 136:1285–98.
39. Miller A, Thornton JC, Warshaw R, et al. Mean and instantaneous expiratory flows, FVC and FEV_1: prediction equations from a probability sample of Michigan, a large industrial state. Bull Eur Physiopathol Respir 1986; 22: 589–97.
40. Ryan CM, Morrow LA, Bromet EJ, et al.. Assessment of neuropsychological dysfunction in the workplace: normative data from the Pittsburgh Occupational Exposures Test Battery. J Clin Exp Neuropsychol 1987; 9:665–79.
41. Northup SC, Kilburn KH. The role of mycotoxins in human pulmonary disease. In: Wyllie TD, Morehouse LG (Eds). Mycotoxic Fungi, Mycotoxins, Mycotoxicoses. Vol. 3. New York: Marcel Dekker, 1977; pp 91–108.
42. Savilathi R, Uitti J, Laippala P, et al. Respiratory morbidity among children following renovation of water damaged school. Arch Environ Health 2000; 55:405–10.
43. McLaughlin CS, Vaughan MH, Campbell IM, et al. Inhibition of protein synthesis by trichothecenes. In: Rodricks JV, Hesseltine CW, Mehlman MA (Eds). Mycotoxins in Human and Animal Health. Park Forest South, IL: Pathotox, 1977.
44. Smoragiewicz W, Cossette B, Boutard A, et al. Trichothecene mycotoxins in the dust of ventilation systems in office buildings. Int Arch Occup Environ Health 1993; 65:113–17.
45. Gorny RL, Reponen T, Willeke K, et al. Fungal fragments as indoor air biocontaminants. Appl Environ Microbiol 2002; 68(7):3522–31.
46. Johanning E, Gareis M, Nielsen K, et al. Airborne mycotoxin sampling and screening analysis. Proceedings of the 9th International Conference on Indoor Air Quality and Climate (Indoor Air 2002), June 30–July 15, 2002. Santa Cruz, CA: Indoor Air 2002 Conference Secretariat.
47. Yike I, Allan T, Sorenson WG, et al. Highly sensitive protein translation assay for trichothecene toxicity in airborne particulates: comparison with cytotoxicity assays. Appl Environ Microbiol 1999; 65:88–94
48. Kordula T, Banbula A, Macomson J, et al. Isolation and properties of stachyrase A, a chymotrypsin-like serine proteinase from *Stachybotrys chartarum*. Infect Immun 2002; 70(1):419–21.
49. Kilburn KH, Warshaw RH. Effects on neurobehavioral performance of chronic exposure to chemically contaminated well water. Toxicol Ind Health 1993; 9:391–404.
50. Kilburn KH. Chlorine-induced damage documented by neurophysiological, neuropsychological, and pulmonary testing. Arch Environ Health 2000; 55:31–37.
51. Kilburn KH. Neurobehavioral impairment from long-term residential arsenic exposure. In: Arsenic. Exposure and Health Effects. New York: Chapman and Hall 1997; pp 159–75.
52. Kilburn KH. Exposure to reduced sulfur gases impairs neurobehavioral function. So Med J 1997; 90:997–1006.
53. Kilburn KH. Evidence for chronic neurobehavioral impairment from chlorpyrifos. Environ Epidemiol Toxicol 1999; 1:153–62.
54. Kilburn KH. Visual and neurobehavioral impairment associated with polychlorinated biphenyls. Toxicology 2000; 21(4):489–500.
55. Andersson MA, Nikulin M, Köljalg U, et al. Bacteria, molds and toxins in water-damaged building materials. Appl Environ Microbiol 1997; 63:387–93.
56. Gordon WA, Johanning E, Haddad L. Cognitive impairment associated with exposure to toxigenic fungi. In: Johanning E (Ed). Bioaerosols, Fungi and Mycotoxins: Health Effects Assessment Prevention and Control. New York: Mt. Sinai School of Medicine, 2001.
57. Hayes JR, Campbell TC. Food additives and contaminants. In: Casarett and Doull's Toxicology. The Basic Science of Poisons, 3rd ed. New York: Macmillan, 1986; pp 771–800.
58. Waksman SA. Antibiotics of actinomycetes that have found important practical applications. In: Waksman SA (Ed). The Actinomycetes. A Summary of Current Knowledge. New York: Ronald Press, 1967; pp 208–26.
59. Rall TW, Schleifer LS. Oxytocin, prostaglandins, ergot alkaloids and other drugs; tocolytic agents. In: Gilman AG, Goodman LS, Rall TW, Murad F (Eds). The Pharmacological Basis of Therapeutics, 7th ed. New York: Macmillan, 1985; pp 926–45.
60. Hintikka EL. Human stachybotryotoxicosis. In: Wyllie TD, Morehouse LG (Eds). Mycotoxic Fungi, Mycotoxins, Mycotoxicoses. Vol. 3. New York: Marcel Dekker, 1977; pp 87–89.
61. Nielsen KF, Hansen MO, Larsen TO, et al. Production of mycotoxins on artificially inoculated building materials. Int Biodeterioration Biodegradation 1998; 42:9–16.
62. Dales R, Burnett R, Zwanenburg H. Adverse health effects among adults exposed to home dampness and mold. Am Rev Respir Dis 1991; 143:505–09.
63. Mirocha CJ, Pathre SV, Christensen CM. Chemistry of *Fusarium* and *Stachybotrys* mycotoxins. In: Wyllie TD, Morehouse LG (Eds). Mycotoxic Fungi, Mycotoxins, Mycotoxicoses. Vol. 1. New York: Marcel Dekker, 1977; pp 365–420.
64. Aibara K, Yano N. New approach to aflatoxin removal. In: Rodricks JV, Hesseltine CW, Mehlman, MA (Eds). Mycotoxins in Human and Animal Health. Park Forest South, IL: Pathotox, 1977.

Effects of Toxic Exposure to Molds and Mycotoxins in Building-Related Illnesses

WILLIAM J. REA
Environmental Health Center–Dallas
Dallas, Texas
NANCY DIDRIKSEN
Behavioral Medicine Associates
Richardson, Texas
THEODORE R. SIMON
Functional Imaging of Texas
Dallas, Texas
YAQIN PAN
ERVIN J. FENYVES
BERTIE GRIFFITHS
Environmental Health Center–Dallas
Dallas, Texas

ABSTRACT. The authors studied 100 patients who had been exposed to toxic molds in their homes. The predominant molds identified were *Alternaria, Cladosporium, Aspergillus, Penicillium, Stachybotrys, Curvularia,* Basidiomycetes, Myxomycetes, smuts, *Epicoccus, Fusarium, Bipolaris,* and *Rhizopus.* A variety of tests were performed on all, or on subgroups of, these patients. Sensitivities and exposures were confirmed in all patients by intradermal skin testing for individual molds (44–98% positive), and by measurement of serum antibodies. Abnormalities in T and B cells, and subsets, were found in more than 80% of the patients. The findings of trichothecene toxin and breakdown products in the urine, serum antibodies to molds, and positive intradermal skin tests confirmed mycotoxin exposure. Respiratory signs (e.g., rhinorrhea, sinus tenderness, wheezing) were found in 64% of all patients, and physical signs and symptoms of neurological dysfunction (e.g., inability to stand on the toes or to walk a straight line with eyes closed, as well as short-term memory loss) were identified in 70% of all patients. Objective abnormal autonomic nervous system tests were positive in all 100 patients tested. Brain scans, conducted using triple-head single photon emission computed tomography, were abnormal in 26 (86%) of 30 (subgroup of the 100) patients tested. Objective neuropsychological evaluations of 46 of the patients who exhibited symptoms of neurological impairment showed typical abnormalities in short-term memory, executive function/judgment, concentration, and hand/eye coordination.
<Key words: immunological, mold, mycotoxin, neurological>

TOXIC EXPOSURE to molds and mycotoxins in public buildings, office buildings, and homes is becoming commonplace. In the current study, 2 adults in each of 31 families (n = 62), plus an additional 38 "solo" individuals who lived alone (n = 2) or had no other family members affected (n = 36), had been exposed to mold in their own homes or apartments. These patients reported symptoms of respiratory, neurological, and immune dysfunction. (Children were not included in this study.)

Materials and Method

A total of 100 patients (61 females, 39 males; age range 21–78 yr; mean age 44.3 yr) were studied to determine the health effects of the presence of toxic molds in their homes. Their houses were analyzed by an independent professional mold evaluator to determine the type and duration of mold exposure that had occurred. All patients had to vacate their homes, at least during the period of remediation, and 10 left their homes permanently because of health concerns.

Signs and symptoms were recorded for each organ system by an examining physician. Mold and mycotoxin sensitivity were confirmed by positive skin whealing with intradermal serial dilution titration and/or the intradermal provocation method.[1–3] Immune abnormalities were confirmed by individual T- and B-cell counts, which were performed by flow cytometry.[4] Immune globulin was evaluated by the mold antibody assay (i.e., immunoglobulin [Ig]G, IgM, IgA, and IgE) method of Vojdani et al.,[5] and urine trichothecenes were measured by the method of Croft.[6] Delayed recall antigens were measured 48 hr postinjection to assess cell-mediated immunity,[7] using testing kits (Multi-Test® II) supplied by Lincoln Diagnostics, Inc. (Decatur, Illinois). The autonomic nervous system was evaluated by the pupillography method of Ishikawa et al.[8] and by the heart rate variability method of Riftine.[9,10]

Neuropsychological tests, performed in accordance with the method of Butler et al.,[11] were conducted by Dr. Didriksen. Brain scans were performed by Dr. Simon, using single-photon emission computed tomography in conjunction with a triple-head gamma camera (tripleSPECT), which provides imaging of the tracer distribution in 3 dimensions.[12] Scans were analyzed in accordance with the method of Simon.[13]

Table 1.—Indoor Mold Cultures Taken from 69 Homes*

Mold type	No. of homes	Percentage of homes
Alternaria	68	100
Aspergillus/Penicillium	68	100
Cladosporium	68	100
Stachybotrys	55	80
Epicoccus	55	80
Curvularia	54	79
Basidiomycetes	52	76
Myxomycetes	51	75
Smut	50	73
Fusarium	50	73
Bipolaris	46	67
Rhizopus	45	65

Source: Environmental Health Center–Dallas, Dallas, Texas, 2003, Bertie Griffiths, Ph.D.
*Two adults per house in 31 houses (= 62 individuals) + 1 person in each of 38 homes (not every mold found in each home). Average clean house = 0–4 fungal colonies per culture.

Results

Signs and symptoms. The predominant signs and symptoms observed in our subjects were as follows: (a) respiratory symptoms (i.e., sneezing, rhinorrhea, nasal stuffiness, dyspnea, and wheezing) in 64 patients; (b) neurological symptoms (short-term memory loss, imbalance, and dizziness) in 70 patients; (c) immunological symptoms (hypersensitivity to molds, foods, and chemicals) in 100 patients; (d) gastrointestinal symptoms (i.e., bloating, gas, and cramps) in 24 patients; (e) musculoskeletal symptoms (i.e., muscle and joint aches and tenderness) in 29 patients; and (f) cardiovascular symptoms (bruising, hemoptysis, and petechiae) in 10 patients.

Indoor mold cultures. The common molds found in the 69 homes analyzed represented a wide spectrum of molds. Results of the indoor mold cultures are given in Table 1.

Intradermal sensitivity testing. Each individual's mold and mycotoxin sensitivity and exposure were confirmed by intradermal injection with preservative-free antigens. Negative controls for saline were nonreactive, except in 5% of the patients, whereas positive controls for histamine were positive in all patients (Table 2). Intradermal skin test results for individual molds in the 100 patients showed reactions that varied from 44% to 98% positive results (Table 2). All patients reacted to at least 4 molds; therefore, 100% of the patients reacted to skin testing. The remaining 45 tests were also positive for the Ascomycetes molds *Leptosphaeria* (98%) and *Phaeosphaeria* (98%). A comparison group of 30 patients, 21–60 yr of age, who were selected for moderate sensitivity to molds and who did not receive massive exposures, showed 20% positive intradermal tests for molds.

For subgroups of the studied patients, intradermal tests for aflatoxin were positive for 21 of 27 (79%); 25 of 25 (100%) tested positive for trichothecene; and 22 of 24 (91%) tested positive for *Fusarium*. These intradermal tests were compared with those for the aforementioned 30 normal controls, of whom 1% were positive. (This intradermal test was developed relatively recently and thus was unavailable for the first 75–78 patients studied because its efficacy and safety were not verified until later in the study period.)

Immune parameters. White blood cell counts were abnormal in 11% of the 100 patients, and the total lymphocyte count was decreased in 23% of patients (Table 3). CD_{11} counts (i.e., total T_{11}-cells) were low in 29 patients, and high in 4, of the 100 patients studied. CD_4 counts (i.e., T_4-helper cells) were low in 32% of the patients and high in 40% of the patients. CD_8 counts (i.e., T_8-suppressor cells) were low in 22% of patients and high in 11%. B_4 lymphocyte counts were abnormal in 14% of the patients: elevated in 11% and depressed in 3%.

Cell-mediated immunity for delayed recall antigens (*Proteus*, tuberculin, *Candida*, *Streptococcus*, *Staphylococcus*, diphtheria, and tetanus) was positive for 3 or fewer antigens in 59 (71%) of 83 patients measured (Table 3). A cell-mediated immunity of 3 or less was considered abnormal. Nineteen patients were not mea-

Table 2.—Intradermal Mold Provocation/Neutralization Testing (*N* = 100 Patients)

Mold, mycotoxin, toxic metabolite, or algae	Patients			Comparison group: Avg. mold-sensitive		Comparison group: Nonsensitive	
	Positive	Tested	Percentage positive	Positive	Tested	Positive	Tested
Rhizopus	89	100	89	30	100	1	30
Aspergillus mix	81	84	96	20	100	2	30
Penicillium	89	100	89	5	100	1	30
Trichoderma	79	90	88	20	100	1	30
Cladosporium herbarum	75	79	95	25	100	2	30
Stachybotrys	75	87	86	30	100	3	100
Curvularia spp.	73	77	95	10	100	4	30
Sporobolmyces	71	84	85	5	100	1	30
Streptomyces	68	70	97	10	100	0	30
Monilia situ	68	75	91	5	100	0	30
Drechslera	68	77	88	5	100	1	30
Cladosporium fulvum	63	72	88	2	100	0	30
Leptosphaeria	61	62	98	2	100	1	30
Phaesophaeria	44	45	98	2	100	3	30
Lake algae	83	84	99	3	100	0	30
Aflatoxin	21	27	79	4	30	1	30
Fusarium	22	24	91	0	30	0	30
Trichothecene	25	25	100	6	30	1	30

Source: Environmental Health Center–Dallas, Dallas, Texas, 2003.

Table 3.—Major Immune Parameters (*N* = 100 Patients)

Cell type	Control range (n = 60)	Abnormal (%): Increase	Abnormal (%): Decrease
White blood cells	4,000–11,0000	2	9
Lymphocytes	1,200–3,300	7	23
T_{11}	1,000–4,000	4	29
T_4	657–1,770	40	32
T_8	325–1,050	11	22
T_4/T_8	1–2.7	21	5
B_4	82–477	11	3
Cell-mediated immune response†		71%	59/83

Source: Environmental Health Center–Dallas, Texas; 2003.
*More than 1 abnormality per patient.
†3 or less was considered abnormal. *Proteus*, tuberculin, *Candida, Streptococcus, Staphlycoccus*, diphtheria, tetanus, and delayed recall at 48 hr. Nineteen percent of the patients did not have delayed recall antigens measured because of unavailability of the testing solution.

sured because the laboratory was unable to provide recall antigens.

Serum mold antibody assays. Table 4 contains the results obtained for 100 patients for mold and mycotoxin antibodies in serum, including IgA, IgG, IgM, IgE, and Vojdani's reference ranges.[5] These serum mold antibody results were correlated with those for the positive intradermal skin tests, and were also compared with Vojdani's controls.[5] The number of patients with massive mold exposures exceeded controls for at least 4 molds, and often for as many as 10.

Trichothecene urine testing. Tests for trichothecene and its byproducts in urine were completed in accordance with the method of Croft.[6] Urine was collected and analyzed approximately 6 mo (range = 1–12 mo) after patients evacuated their homes. Not all of our subjects underwent this testing because the assay was not available at the beginning of the study. Of the 78 patients tested, 9 were in the "present but low" (3–4 units) category, 30 were in the "moderate" (5–8) category, and 39 were in the "high positive" (9+) category. These findings were compared with values determined for recovered, treated patients (0–2).

Comparison between positive urine mycotoxins and positive trichothecene serum antibodies revealed a 98% correlation in the 78 patients measured. One patient who had anaphylaxis had mycotoxins in urine; however, this patient also had negative serum antibodies. All of the urine specimens that were positive for mycotoxins (i.e., in the high, moderate, or low categories) correlated with positive mycotoxin skin tests; however, only 27 patients received mycotoxin skin tests, given the more recent development of the tests. Of the other patients who had positive tests for serum mycotoxin antibodies, 8 were positive for IgA to molds, 8 were positive for IgG to molds, and 6 were positive for IgM to molds. These 22 patients had no urine mycotoxins measured because the test was unavailable, inasmuch as the laboratory experienced difficulties in handling the volume of tests. Three patients were positive for IgE to molds, but none of these had urine measurements done.

Table 4.—Blood Mold Antibody Assay (N = 100 Patients)*

Mold or mycotoxin	IgA			IgG			IgM			IgE			Nonsensitive	Controls
	Elevated (*n*)	Tested (*n*)	%	Elevated (*n*)	Tested (*n*)	%	Elevated (*n*)	Tested (*n*)	%	Elevated (*n*)	Tested (*n*)	%		
Aspergillus fumigatus	4	21	19	9	21	43	5	21	24	4	21	19	0	1,600
Aspergillus niger	23	82	28	22	82	27	25	82	30	10	32	12	0	1,600
Aspergillus versicolor	6	17	35	6	17	35	2	17	12	2	17	12	0	1,600
Alternaria tenuis	4	48	8	19	48	40	14	48	29	6	48	13	0	1,600
Chaetomium globosum	11	42	26	16	42	38	15	42	36	7	42	17	0	1,600
Chrysagem	0	1	0	0	1	0	0	1	0	0	1	0	0	50
Cladosporium herbarum	12	67	18	25	67	37	20	67	30	8	67	12	0	1,600
Epicoccum nigrum	7	26	27	25	67	37	20	67	30	8	67	12	0	1,600
Fumigatus	0	2	0	0	2	0	0	2	0	0	2	0	0	1,600
Fusarium nachilform	0	2	0	0	2	0	0	2	0	0	2	0	0	1,600
Geotrichum candidiolum	0	6	0	2	6	33	1	6	17	0	6	0	0	1,600
Penicillium notatum	10	84	12	35	84	42	19	84	23	9	84	11	0	1,600
Rhizopus nigrans	2	9	22	7	9	78	6	9	67	1	9	11	0	1,600
Stachybotrys chartarum	16	85	19	35	85	41	28	85	33	13	85	15	0	1,600
Aflatoxins	26	92	28	26	92	28	35	92	38	15	92	16	0	1,600
Satratoxins	21	88	24	27	88	31	26	88	30	13	88	15	0	1,600
Trichothecene	32	88	36	38	88	43	27	88	31	10	88	11	0	1,600

Source: Environmental Health Center–Dallas, Dallas, Texas, 2003.
Notes: Immunoglobulin (Ig) E was significantly lower than IgG, IgM, IgG, and IgM. Ig levels determined by Vodjani, Immunosciences Laboratories, Beverly Hills, California.
*Not all individuals had the same assay, given the differences in findings in each home.

Autonomic nervous system measurements. Evaluation of autonomic nervous system function via eye pupillography[8] was performed in 58 patients and revealed abnormalities in 51, whereas 7 patients tested normal (Table 5A). These results revealed significant changes in the pupillary area (A_1), the constriction ratio (CR) of the pupillary area before and after a light challenge of 400 lux (A_3/A_1), and the speed of contraction (T_2) after a light challenge of 400 lux. Recovery time (T_5) showed no significant change.

Autonomic nervous system measurements via heart rate variability[9,10] were performed on 67 mold/mycotoxin-exposed patients, compared with 65 controls. Abnormalities were revealed in all patients (Table 5B). A comprehensive fitness score—including sympathetic–parasympathetic function with frequency recordings, the amount of microcirculation vasoconstriction, and the chronophoric reaction of the heart—was markedly elevated in all 67 patients measured (score = 10.16) compared with 65 controls (score = 6.06). The patients' scores were significantly abnormal ($p < 0.001$) and indicated autonomic nervous system impairment.

There was overlap in the 2 measurement methods in that 25 patients were administered both tests. The 7 patients who had a normal pupillography also had a heart rate variability change, and we considered them "positive" for the study.

Neurocognitive data. Neurocognitive data for 46 (65%) of the 70 neurotoxic patients (mean age = 47.95 yr, mean educational level = 15.32 yr) who reported massive mold/mycotoxin exposures were examined in the present study. Because of financial limitations, in some cases multiple familial neurocognitive recordings were not performed on members of the same family (i.e., husband and wife combining time in the same household), and it was this that contributed to the discrepancy in numbers of untested patients. Neuropsychological test batteries of varying comprehensiveness (including the Halstead-Reitan Neuropsychological Test Battery) were compared with normative data. Deficits were found primarily on measures of executive function, psychomotor problem-solving, and incidental memory. Scores on the Wechsler Memory Scale-III were within normal limits overall, with greatest impairment

Table 5.—Autonomic Nervous System Testing*

A. Pupillography†

Pupillary change parameters	Patients (n = 58) $\bar{x}$	Patients (n = 58) *SD*	Controls (n = 181) $\bar{x}$	Controls (n = 181) *SD*	Difference‡
A_1 (mm^2)	26.91	1.62	35.10	1.59	$p < 0.001$
CR (A_3/A_1) (mm^2)	0.29	0.01	0.46	0.03	$p < 0.001$
T_2 (ms)	269.6	5.6	199.0	5.2	$p < 0.001$
T_5 (ms)	1,616	83	1,414	70	$p < 0.05$

B. Heart Rate Variability§

	Patients (n = 67)	Controls (n = 65)	Difference
Fitness score	10.16	6.06	$p < 0.001$

Notes: $\bar{x}$ = mean, *SD* = standard deviation, A_1 = pupillary area, CR (A_3/A_1) = constriction ratio of the pupillary area before and after a light challenge of 400 lux, T_2 = speed of contraction after a light challenge of 400 lux, and T_5 = recovery time.
*Some patients were administered both tests (n = 25).
†Method: Ishikawa, Hamamatsu Photonics, Hamamatsu City, Japan.[8]
‡p values < 0.001 were significant.
§Method: Riftine, Heart Rhythm Instruments, Inc., Metuchen, New Jersey.[10]

on measures of visual memory. Fifty-seven percent of the 70 patients demonstrated mild to moderate impairment on the Halstead-Reitan Battery. Forty-two percent demonstrated mild to severe impairment on the Comprehensive Neuropsychological Screen, and an additional 17 percent scored in a low-normal range. Scores on measures of specific neuropsychological abilities showed some impairment of sensory and motor functions. Intelligent Quotient scores generally fell within expected ranges.

Brain scans. TripleSPECT brain scans of 30 patients, selected randomly for physical signs of imbalance, revealed a neurotoxicity pattern in 26 (86%). All 100 patients could not be evaluated because of cost constraints. Of the 26 with a neurotoxicity pattern, 15 individuals displayed mild neurotoxicity and 11 showed moderate neurotoxicity. In the glutathione-dependent areas of the cerebral hemispheres, the patterns showed a decrease in flow and function, temporal lobe asymmetry, and a flow of dye into soft tissue with areas of high dye uptake ("hot" areas) or no dye uptake ("cold" areas). Further details of tripleSPECT brain scans are contained in a study by Simon and Rea.[12]

Discussion

All patients received the same skin, blood, and autonomic nervous system tests, whereas , for financial reasons, a selected subgroup underwent the tripleSPECT brain scan, urine mycotoxin, and intradermal mycotoxin tests. Subgroup testing did not appear to bias the study, but confirmed the initial findings because of the high correlation. Clearly, toxic mold in our patients' homes was associated with adverse health effects in this series of patients. The 62 individuals of the 31 families appeared to have a more severe exposure than did the 38 solo patients inasmuch as multiple individuals in the same home were made ill. The individuals who lived alone (n = 2) or had no family members affected (n = 36) had histories similar to the families who were affected, but for an unknown reason only 1 member of this group was ill. However, in these 38 solo patients, the intensity of exposure (i.e., countable colonies vs. too numerous to count) appeared to be less than in the 62 family members. These findings could be related to the different lengths of time (3 mo to 2 yr) spent in the home, as well as the intensity of growth of the mold cultures. Given that mold counts are crude and have large variations in reported mold quantity, abnormalities in counts may have occurred and may have resulted in more exposure—even if patients were exposed to the same antigen. Even if the mold counts were the same, perhaps the less-damaged patients did not spend as much time in their homes (thus less exposure) as did the more severely damaged patients, or perhaps these individuals had more resistance to mold exposures. Alternatively, perhaps the mold/mycotoxin growth was not as vigorous or virulent as in the homes in which more than 1 patient was made ill. Technology to evaluate quantitative mycotoxin exposure and virulence is unavailable, making these parameters difficult to evaluate.

Water leakage that leads to mold growth involves leaking roofs and windows, and other water-related defects in structures (e.g., defective construction with leaking showers, sewer pipes, water pipes). The multiplicity of molds found in homes appeared to cause a total body load (burden) dyshomeostasis—as evidenced by the multiplicity of positive skin tests, pan immune sensitivity, and autonomic dysfunction which occurred alternatively or simultaneously in any given patient. Not only did the total burden of the mold load appear significant, but damage from a *specific* mold—up to 10 separate categories—also appeared to occur. One mold (*Stachybotrys*) and 1 mycotoxin (trichothecene) seemed particularly virulent, damaging all 3 systems: immune, respiratory, and neurological. A larger spectrum of many other molds not listed or considered pathogenic was found in the serum antibodies and skin tests of most patients, as well as in several individual homes and cases. These additional molds may also create a greater total body burden, which may aid in disrupting the immune and detoxification systems, thus augmenting the damage caused by the predominant molds. This additive or synergistic effect of multiple molds might also have caused the more severe disease seen in the massively exposed patients.

Another reason that most patients developed a near pan-sensitivity could have been the frequency of exposure. The molds commonly found in patients' homes were those typically found outdoors, like *Alternaria*, *Aspergillus*, and *Cladosporium*, whereas *Stachybotrys* is usually found indoors. These particular molds could cause specific sensitivity and toxicity, as well as trigger various aspects of the immune and detoxification systems—regardless of the patient's location. This phenomenon would tend to exacerbate one's illness once sensitization occurred. Clearing of symptoms would be difficult inasmuch as there would be minimal avoidance time from exposure to the triggering agent because exposure events occurred both indoors and outdoors.

Judging by the mycotoxins or breakdown products found in urine (in all 78 patients tested), antibodies (found in 99 of 100 patients tested in this category), and positive intradermal skin tests of a small subgroup of 27 patients tested for mycotoxins (i.e., aflatoxin [78%], fusarium [92%], trichothecene [100%]), there appears to be strong evidence of mycotoxin exposure. These patients did not respond clinically as the average mold-sensitive patient would, or as controls with no mold sensitivity would—the symptoms of the massively exposed group were much more widespread in our study. In addition to nasal, sinus, and bronchial symptoms, the exposed group had more neurological (70 of 100 patients) and immune system (100% vs. 20% controls) involvement. Not only were their laboratory findings more frequently and severely abnormal, but clinically these mold/mycotoxin-exposed patients were much sicker. Many became incapacitated and were unable to work. This contrasted with the average mold-sensitive patient (without a massive exposure) who typically can continue to work and has the usual symptoms related only to the respiratory system (e.g., rhinosinusitis, asthma, cough).

Parameters suggestive of more severe problems than just an average mold sensitivity included alterations in white blood cell count (in 19% of patients), alterations in T-lymphocyte subsets of helper cells (72%), changes in T_8 lymphocytes (33%), and changes in B lymphocytes (24%). These findings further substantiate that the immune system was significantly altered in these patients with massive mold/mycotoxin exposure. Cell-mediated immunity decreased in 71% of 83 patients measured, again pointing to abnormal T-cell function. It appears that the mycotoxin overload significantly disrupted the immune system of these severely exposed patients.

A high correlation was seen between the number of molds to which a patient was exposed (4+ per patient) and (a) their total immunoglobulin changes (100% change) and (b) the number of positive skin tests (44–98% [Table 2]). When tested intradermally, patients with massive mold/mycotoxin exposure demonstrated almost a pan sensitivity to these molds found on culture, and to other molds not found on culture. This finding contrasted with the average mold-sensitive patient who has sensitivities to only 20% of the molds tested. The finding of pan mold sensitivity suggested a severe alteration in both the immunological and neurological systems. Such a severe immunological reaction would likely trigger the autonomic system because of the close anatomical proximity and physiologic communication of the autonomic nerves. These alterations of both the nerves and immune cells may have resulted in a vulnerability to exposure from other molds in the ambient outdoor air and indoor air—not necessarily just the molds found in their immediate home exposures. Recognition sites on cell membranes might also have been damaged as a result of the severe reactions—or the harmonious function of the T-helper and T-suppressor cells might have been disrupted—thus allowing pan sensitivity to occur in these patients.

Neurological involvement from mold/mycotoxin exposures was evidenced not only by abnormal physical findings (i.e., inability to stand on toes with the eyes closed or to walk a straight line), but also by the abnormal results in objective measurements of autonomic nervous system dysfunction and by tripleSPECT brain scan results. An imbalance in the autonomic nervous system certainly emphasizes that there is a proven communication between the central nervous system and the immune system. Autonomic abnormalities occurred in virtually all of our patients, which suggests that mold/mycotoxin sensitivity and overload initially caused a basic change in the nervous system, which could then propagate some of the sympathetic and parasympathetic effects we saw in these patients. Alternatively, the autonomic nervous system change might have occurred later in the course of illness (no patients were measured until 3 mo following exposure) as a result of long-term exposure and ongoing reactions. The comparison groups with mild, average mold sensitivity exhibited very few autonomic dysfunctions.

Positive tripleSPECT brain scans (n = 26), combined with neuropsychological profiles, helped to objectively differentiate aberrations in cognition, recall, and executive function. The neurological and psychological tests correlated well with the tripleSPECT scans and autonomic tests to demonstrate a highly significant organic damage pattern. Although the number of tripleSPECT scans performed was low, the data were sufficient to allow correlation of outcomes with the autonomic and immune data. These findings suggest that direct toxic damage resulted from mold or mycotoxin exposure, or from an immunological overlay, as discussed earlier.

Our study (along with those of other investigators) has shown that systems other than respiratory system are affected in patients with massive mold/mycotoxin exposure. Gray et al.,[14] Kilburn,[15] and Campbell et al.,[16] independently, have demonstrated both immunological and neurological involvement secondary to large mold/mycotoxin exposures. Although none of these investigators conducted skin testing, their findings in regard to the home environment, clinical signs and symptoms, and serum antibodies were similar to those of our study. Correlations were found with a "toxic" building, high levels of mold cultures, and their patients' illnesses. Incapacitating factors of the mycotoxin- and mold-induced symptoms in their and our studies included short-term memory loss and inability to concentrate. The neurological involvement seen in our series, which caused patients to miss work or school and often prevented them from performing even manual tasks, appeared to be the result of mold- and mycotoxin-induced chronic fatigue. Patients found it difficult to move from their homes because of this fatigue, even though evacuation was necessary for their survival.

In summary, 100 patients exposed to toxic molds and mycotoxins in their homes were evaluated by various means (with subgroups receiving additional skin and urine testing for mycotoxins). Signs and symptoms reported were related primarily to the respiratory tract and to the neurologic and immune systems; other systems were also involved, although to a lesser degree. Serum antibodies, direct urine measurement, and intradermal skin tests for mold sensitivity demonstrated a high correlation with exposure and with alteration of the immune system. Objective alteration of the neurologic system was found by autonomic nervous system testing through the eye and heart nerves, by brain tripleSPECT scans that confirmed clinical neurological involvement, and by neuropsychological tests for memory loss, executive dysfunction, and hand/eye coordination.

* * * * * * * * * *

Submitted for publication September 20, 2003; revised; accepted for publication December 9, 2003.

Requests for reprints should be sent to William J. Rea, M.D., Environmental Health Center—Dallas, Suite 220, Dallas, TX 75231-4262.

E-mail: wjr@ehcd.com

* * * * * * * * * *

References

1. Rinkel HJ. Inhalant Allergy. Pt I: The whealing response of the skin to serial dilution testing. Ann Allergy, 1949; 7: 625–30.
2. Lee CH. Allergy Neutralization: The Lee Method. St. Joseph, MO: Tri Sigma Press, 1987.
3. Miller JB. Food Allergy: Provocative Testing and Injection Therapy. Springfield, IL: Charles C. Thomas, 1972.
4. Givan AL (Ed). Flow Cytometry—First Principles. New York: Wiley-Liss, 1992.
5. Vojdani A, Campbell AW, Kashanian A, et al. Antibodies against molds and mycotoxins following exposure to toxigenic fungi in a water-damaged building. Arch Environ Health 2003; 58(6):324–36.
6. Croft WA, Jastromski BM, Croft AL, et al. Clinical confirmation of trichothecene mycotoxicosis in patient urine. J Environ Biol 2002; 23(3):301–20.
7. Popa V, Holgate ST, Salvi S. Cell-mediated immunity in asthma. Am J Respir Crit Care Med 2002; 166:1607.
8. Ishikawa S, Natio M, Inabe KI. A new videopupillography. Opthalmologica 1970; 160:248.
9. Gu HG, Ren W, Lu QS, et al. Application of degree of complexity in heart rate variability analysis during orthostatic (correction of orthostatic) standing [in Chinese]. Space Med Med Eng (Beijing) 2001; 14(3):192–95.
10. Riftine A. Quantitative Assessment of the Autonomic Nervous System Based on Heart Rate Variability Analysis. Heart Rhythm Instruments, Inc., 173 Essex Ave., Metuchen, NJ 08840.
11. Harrell EH, Butler JR, Didriksen NA, et al. Harrell-Butler Comprehensive Neuropsychological Screen [unpublished assessment instrument]. Health Psychology/Behavioral Medicine Associates, 100 N Cottonwood Dr., Ste. 106, Richardson, TX 75080, 1986.
12. Simon TR, Rea WJ. Use of functional brain imaging in the evaluation of exposure to mycotoxins and toxins encountered in Desert Storm/Desert Shield. Arch Environ Health 2003; 58(7):406–09.
13. Ross GH, Rea WJ, Johnson AR, et al. Neurotoxicity in single photon emission computed tomography brain scans of patients reporting chemical sensitivities. Toxicol Ind Health 1999; 15:415–20.
14. Gray MR, Thrasher JD, Crago R, et al. Mixed mold mycotoxicosis: immunological changes in humans following exposure in water-damaged buildings. Arch Environ Health 2003; 58(7):410–20.
15. Kilburn KH. Indoor mold exposure associated with neurobehavioral and pulmonary impairment: a preliminary report. Arch Environ Health 2003; 58(7):390–98.
16. Campbell AW, Thrasher JD, Madison RA, et al. Autoantibodies and neurophysiologic abnormalities in patients exposed to molds in water-damaged buildings. Arch Environ Health 2003; 58(8). Forthcoming.

Use of Functional Brain Imaging in the Evaluation of Exposure to Mycotoxins and Toxins Encountered in Desert Storm/Desert Shield

THEODORE R. SIMON
Functional Imaging of Texas, P.A.
Dallas, Texas
WILLIAM J. REA
Environmental Health Center—Dallas
Dallas, Texas

ABSTRACT. In this retrospective analysis the authors compared brain scintigrams, performed using triple-head single-photon emission computed tomography (tripleSPECT), of subjects who were judged clinically impaired from exposure to toxins during the Desert Storm/Desert Shield military action, and of subjects exposed to mycotoxins, with those of normal controls. The scintigrams for both exposed groups exhibited similar patterns of abnormalities, which were consistent with neurotoxic impairment. The authors conclude that further study is needed to determine whether mycotoxin exposure may be a cause of abnormalities seen in tripleSPECT images.
<Key words: Desert Storm/Desert Shield, mycotoxin, SPECT>

TECHNETIUM-99M—labeled d,l-hexamethylenepropyleneamine oxime ([Tc-99m]HMPAO), and also known as [Tc-99m] exametazime (Amersham Corporation [Arlington Heights, Ilinois])—is an established tracer for scintigraphic examination of the brain. It is especially useful in conjunction with a triple-head gamma camera (Trionix Corporation [Cleveland Ohio]) and single-photon emission computed tomography technique (tripleSPECT), which provides imaging of the tracer distribution in 3 dimensions. Previous investigators have described regional cerebral blood flow in the brains of subjects exposed to neurologically active chemicals.[1–3] These studies, based on early observations by Anderson,[4] illustrated the functional distribution of glutathione. Inasmuch as glutathione function involves the clearing of toxins, it was hypothesized that the functional distribution of glutathione might be affected by exposure to toxins. This hypothesis was tested by using tripleSPECT techniques to examine veterans of the Desert Storm/Desert Shield military operation in Kuwait and Iraq, as well as individuals who had exposures to organic solvents or pesticides.[5–7] The scintigraphic pattern often seen in patients with neurotoxic exposure has been described as (a) shunting from the brain to the soft tissues, (b) asymmetric temporal lobar activity, (c) multiple hot and cold foci distributed throughout the cortex without regard to lobar distribution (referred to as "salt/pepper distribution"), and (d) mismatch of brain activity between the early (predominately blood flow) phase images and the late (predominately functional) phase images.

Exposure to mycotoxins can result in psychological, allergic, and toxic reactions.[8–11] We hypothesized that toxic reactions to mycotoxin would also affect the functional distribution of glutathione in the brain. We initially addressed this hypothesis with a retrospective analysis of tripleSPECT examinations of the brain conducted using [Tc-99m]HMPAO.[5]

Method

Consecutive patients who had experienced toxic reactions to mycotoxin exposure, and who were seen in the Environmental Health Center—Dallas and examined with tripleSPECT using [Tc-99m]HMPAO, were entered retrospectively into our pilot study. This population is discussed in detail in an article by Rea et al.[12] in this issue. Thirty subjects whose tripleSPECT examinations were performed between November 22, 2000, and December 5, 2002, were compared with subjects evaluated in our previous study of chemical sensitivities in normal controls vs. Desert Storm/Desert Shield participants.[5] Thus, we compared 3 groups: (1) subjects with mycotoxin exposure, (2) clinically impaired veterans of Desert Storm/Desert Shield, and (3) normal controls.

Imaging techniques, described earlier,[6] were identical among the groups. Briefly, the quality of the [Tc-99m]HMPAO tracer was ensured and the subject was invited into a private imaging room controlled for lighting, noise, and radio frequency emissions. The technologist explained the procedure in a stylized manner,[6] while an intravenous port was established. The subject was then positioned comfortably on the imaging table, within the imaging geometry, and the gamma camera was adjusted for acquisition. Next, the subject was asked to imagine a beach scene, which again was described in a stylized presentation that provided uniform cues regarding the various elements of the scene.[6] When the subject appeared relaxed, the technologist activated a Harvard pump which delivered a standard [Tc-99m]HMPAO dose at a standard rate over 2 min. After a prescribed delay, but still during the injection, a tripleSPECT acquisition was begun over the brain and continued for 2 min. After a 30-min delay, a 2nd tripleSPECT image was acquired over a period of 30 min. Both sets of data were computer-processed, digitally filtered using a phantom tested technique,[13] and back-projected onto 3 planes orthogonal to the canthomeatal plane. Images were tested, using standardized methods and software (Trionix Corporation [Cleveland, Ohio]), to exclude motion artifact, and displayed in a manner that provided uniform brightness for normal examinations between the early and late acquisition sets.

The neurotoxic subjects were grouped qualitatively according to degree of clinical impairment—mild, moderate, or severe—by the treating physician (WJR), who had no access to the imaging results. Initial image interpretations were performed by either of 2 physicians who were Board Certified in Nuclear Medicine and highly experienced in SPECT imaging of the brain using [Tc-99m]HMPAO. These initial interpretations, and an equal number of unrelated interpretations, were reviewed by TRS, who was unaware of the subject's clinical history or presentation. If the original interpretation lacked a comment regarding a component of the neurotoxic pattern, that component was graded during the review. Each component was graded qualitatively as normal, mild, moderate, or severe.

Table 1.—TripleSPECT and Clinical Findings for Mycotoxin-Exposed Subjects

				Neurotoxic pattern components					
Subject	Date of tripleSPECT procedure	Sex	Age (yr)	Salt/ pepper	Early/late mismatch	Temporal assymetry	Soft tissue shunting	Overall tripleSPECT assessment	Clinical assessment
1	01-03-01	M	67	Mild	Mild	Normal	Mild	Mild	Mild
2	01-03-01	F	50	Mild	Normal	Mild	Mild	Normal	Mild
3	04-25-02	F	39	Moderate	Mild	Mild	Mild	Moderate	Mild
4	03-19-02	F	59	Mild	Normal	Mild	Moderate	Mild	Severe
5	10-03-01	M	39	Moderate	Normal	Mild	Moderate	Mild	Severe
6	07-11-02	M	42	Moderate	Moderate	Mild	Moderate	Moderate	Moderate
7	06-26-02	F	43	Mild	Mild	Mild	Mild	Moderate	Mild
8	10-31-02	F	46	Mild	Mild	Mild	Mild	Mild	Severe
9	02-26-02	F	30	Mild	Normal	Mild	Mild	Mild	Mild
10	10-10-02	F	56	Moderate	Mild	Moderate	Mild	Normal	Severe
11	09-05-02	F	46	Moderate	Normal	Mild	Mild	Moderate	Severe
12	04-04-02	F	45	Moderate	Moderate	Mild	Mild	Moderate	Severe
13	08-30-02	M	5	Normal	Mild	Mild	Normal	Mild	Severe
14	11-26-02	F	33	Moderate	Normal	Normal	Normal	Mild	Severe
15	12-05-02	M	45	Moderate	Mild	Normal	Moderate	Moderate	Severe
16	02-12-01	F	38	Mild	Mild	Mild	Moderate	Moderate	Mild
17	06-26-02	F	59	Mild	Mild	Mild	Severe	Mild	Severe
18	03-12-02	M	41	Moderate	Normal	Mild	Normal	Mild	Severe
19	07-12-02	M	47	Mild	Moderate	Mild	Moderate	Moderate	Mild
20	08-15-02	M	20	Mild	Normal	Mild	Mild	Mild	Severe
21	05-10-02	F	49	Mild	Normal	Mild	Mild	Mild	Moderate
22	10-02-02	F	42	Mild	Normal	Mild	Normal	Normal	Severe
23	04-16-02	F	50	Mild	Normal	Normal	Normal	Normal	Severe
24	03-12-02	F	42	Mild	Mild	Mild	Normal	Mild	Severe
25	07-18-02	M	31	Mild	Mild	Mild	Moderate	Moderate	Moderate
26	06-05-01	F	50	Severe	Severe	Mild	Mild	Moderate	Moderate
27	08-08-02	F	37	Mild	Normal	Mild	Mild	Mild	Severe
28	03-08-02	F	47	Mild	Mild	Mild	Mild	Mild	Severe
29	11-22-00	M	53	Moderate	Mild	Mild	Mild	Moderate	Severe
30	09-25-02	F	49	Mild	Mild	Mild	Mild	Mild	Moderate

Notes: tripleSPECT = triple-head single-photon emission computed tomography, M = male, and F = female.

Results

The mycotoxin-exposed patients comprised 10 males and 20 females, 5–67 yr of age (average ± standard error of the mean [*SEM*] = 43.3 ± 11.9 yr). The 6 normal controls and 6 Desert Storm/Shield subjects were males who ranged in age from 23 to 54 yr and 25 to 46 yr, respectively.

Clinical and tripleSPECT assessments for the mycotoxin-exposed subjects are given in Table 1. Similar data for the Desert Storm/Shield and normal control groups were published previously.[5] Comparative images are shown in Figures 1–3. Abnormalities were detected in 26 of 30 mycotoxin-exposed patients, for an overall sensitivity of 87%. Sensitivities for individual components of the neurotoxic pattern were as follows: salt/pepper distribution, 97%; mismatched early vs. late phases, 60%; temporal asymmetry, 87%; and soft tissue shunting, 80%. All overall abnormal SPECT assessments were rated as mild or moderate, compared with only 40% of clinical assessments. Even considering the individual tripleSPECT components, only 2 patients exhibited severe abnormalities.

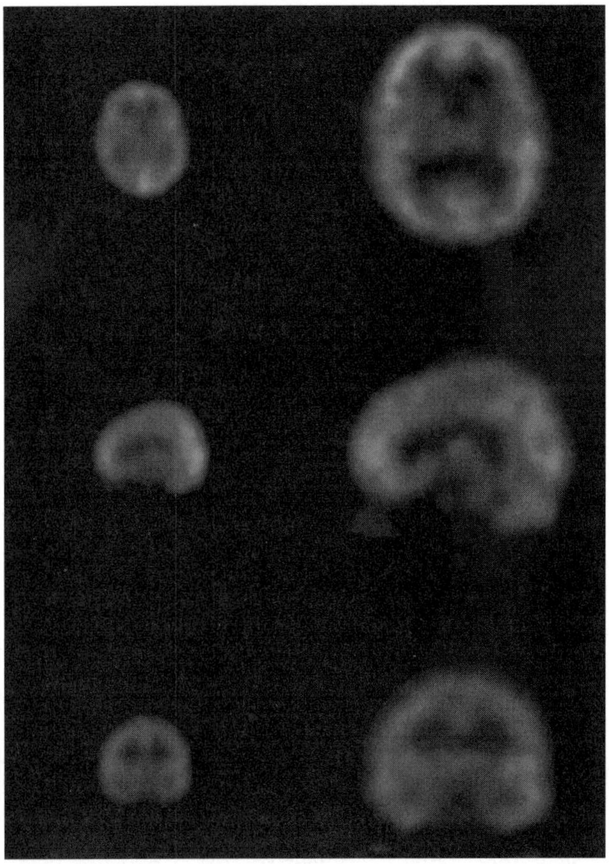

Fig. 1. TripleSPECT images of a normal control subject. These 6 images are representative of the complete examination. The left 3 images are taken from the early, predominately blood-flow, phase of glutathione detoxification; the right 3 images represent the late, predominately functional phase. The top 2 images are parallel to the canthomeatal plane; the middle 2 are in the sagittal plane; the bottom 2 are orthogonal to the other images, on a plane roughly corresponding to the coronal plane. The brighter areas indicate more tracer uptake. Images are in standard orientations: canthomeatal—anterior is up and left brain is on the right; sagittal—rostral is up and anterior is on the left; and coronal—rostral is up and left brain is on the right).

Discussion

In this retrospective pilot study, we observed similar abnormalities in subjects with clinically significant toxic mycotoxin exposure and those with toxic exposure sustained during the Desert Storm/Desert Shield action. These abnormalities were similar to those described elsewhere in conjunction with other toxic agents.[5,6] The pattern for these 2 groups differs from findings for normal controls evaluated with identical brain tripleSPECT techniques. In our mycotoxin-exposed group, the pattern tended to understate the degree of clinical presentation. None of the subjects with

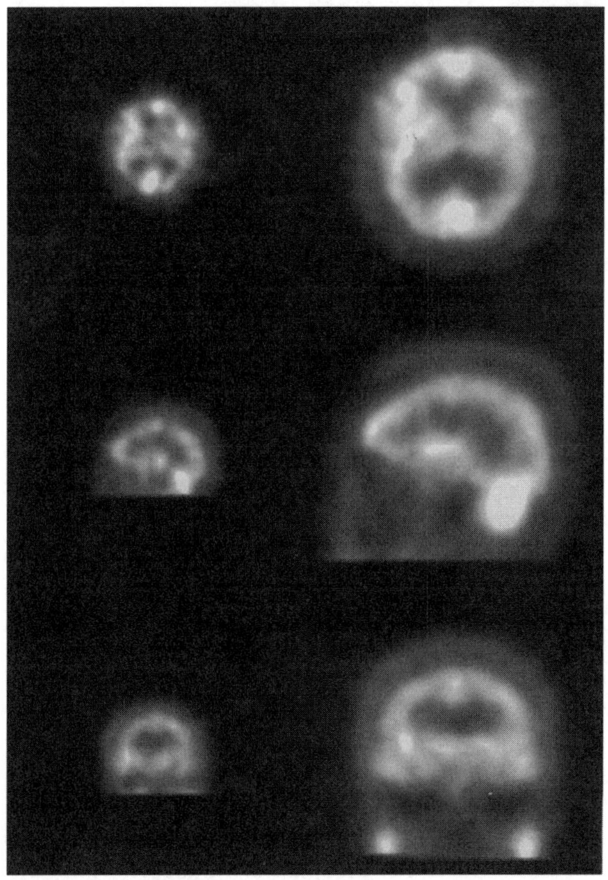

Fig. 2. TripleSPECT images of a mycotoxin-exposed subject, displayed as in Figure 1. Note the prominent shunting to the soft tissues, which appears as diffuse activity. Such shunting causes the brain structures to lose definition when viewed, with the extensive activity emitted from tracer that has escaped localization within the brain.

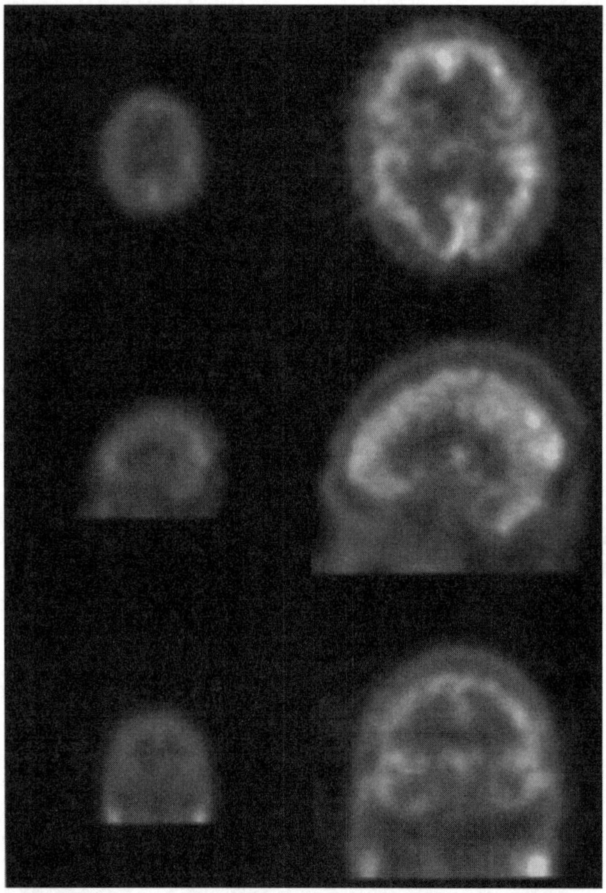

Fig. 3. TripleSPECT images of a Desert Storm/Desert Shield subject, displayed as in Figure 1. Note the well-defined neurotoxic pattern, with greater soft tissue activity (compared with brain activity) than seen in Figure 1, temporal lobe asymmetry, multiple hot and cold foci distributed throughout the cortex, and mismatched early- vs. late-phase brain activity.

clinically severe impairments had tripleSPECT scans rated as severely abnormal in the overall assessment. Further work is needed to understand the relative contributions of each component of the neurotoxic pattern, in order to improve accuracy in the overall grading of abnormal tripleSPECT examinations. An additional benefit may be the ability to discern quantitatively the distinct signatures of various causative agents on the basis of nuances in the pattern. Such work should include both clinical assessments and objective neuropsychological testing.

In this pilot study, the control group was neither age- nor sex-matched to the group exposed to mycotoxin. Clinical grading was assessed subjectively by a single treating physician, and image interpretations were reviewed by a single observer who was blinded to the clinical presentation. We believe that the results of our study justify a formal examination of mycotoxin exposure as a cause of abnormalities seen in tripleSPECT imaging of the brain using [Tc-99m]HMPAO.

* * * * * * * * * * *

Submitted for publication September 20, 2003; revised; accepted for publication October 30, 2003

Requests for reprints should be sent to Theodore R. Simon, M.D., Functional Imaging of Texas, P.A., 4429 Southern Avenue, Dallas, TX 75202.

E-mail: theodorersimon@yahoo.com

* * * * * * * * * * *

References

1. Erbas B, Bekdik C, Erengi F, et al. Regional cerebral blood flow changes in chronic alcoholism using Tc-99m HMPAO SPECT: comparison on CT parameters. Clin Nucl Med 1992; 17:123–27.
2. Holman BL, Arvalho PA, Mendelson J, et al. Brain perfusion is abnormal in cocaine-dependent polydrug users: a study using technetium-99m-HMPAO and ASPECT. J Nucl Med 1991; 32:1206–10.
3. Tumeh SS, Nagel JS, English RJ, et al. Cerebral abnormalities in cocaine abusers: demonstration by SPECT perfusion brain scintigraphy. Work in progress. Radiology 1990; 176:821–24.
4. Anderson AR. Tc-99m-D,L-hexamethylene-propylene-amine oxime (Tc-99m-HMPAO): basic kinetic studies of a tracer of cerebral blood flow. Cerebrovasc Brain Metab Rev 1989; 1:288–318.
5. Simon TR, Hickey DC, Fincher CE, et al. Single photon emission computed tomography of the brain in patients with chemical sensitivities. Toxicol Ind Health 1994; 10: 573–77.
6. Fincher CE, Chang TS, Harrell EH, et al. A comparison of single photon emission computed tomography findings in cases of healthy adults and solvent exposed adults. Am J Ind Med 1997; 31:4–14.
7. Ross GH, Rea WJ, Johnson AR, et al. Neurotoxicity in single photon emission computed tomography brain scans of patients reporting chemical sensitivities. Toxicol Ind Health 1999; 15(3–4):415–20.
8. Baldo JV and Ruff AL. Neuropsychological performance of patients following mold exposure. Appl Neuropsychol 2002; 9(4):193–202.
9. Sobotka TJ, Brodie RE, Spaid SL. Neurobehavioral studies of tremorgenic mycotoxins verruculogen and penitrem A. Pharmacology 1978; 16(5):287–94.
10. American College of Occupational and Environmental Medicine (ACOEM). Adverse Human Health Effects Associated with Molds in the Indoor Environment. Evidence-based Statement. Arlington Heights, IL: ACOEM, 27 October 2002.
11. New York City Department of Health and Mental Hygiene, Bureau of Environmental and Occupational Disease Epidemiology. Guidelines on Assessment and Remediation of Fungi in Indoor Environments. New York: New York City Department of Health and Mental Hygiene, January 2002.
12. Rea WJ, Didricksen N, Simon TR, et al. Effects of toxic exposure to molds and mycotoxins in building-related illnesses. Arch Environ Health, 2003;58(7):399–405.
13. Simon TR, Walker BS, Triebel JG. Target detection using single photon emission computed tomography. [Abstract] Proceedings of the 33rd Annual Meeting of the Society of Nuclear Medicine, June 1986, Washington, D.C. J Nucl Med 1986; 27:1006.

Mixed Mold Mycotoxicosis: Immunological Changes in Humans Following Exposure in Water-Damaged Buildings

MICHAEL R. GRAY
Progressive Healthcare Group
Benson, Arizona
JACK D. THRASHER
Sam-1 Trust
Alto, New Mexico
ROBERT CRAGO
Neurobehavioral Health Services
Tucson, Arizona
ROBERTA A. MADISON
Department of Health Sciences
California State University, Northridge
LINDA ARNOLD
ImmunTox LLC
Benson, Arizona
ANDREW W. CAMPBELL
Medical Center for Immune and Toxic Disorders
Spring, Texas
ARISTO VOJDANI
Immunosciences Lab, Inc.
Beverly Hills, California

ABSTRACT. The study described was part of a larger multicenter investigation of patients with multiple health complaints attributable to confirmed exposure to mixed-molds infestation in water-damaged buildings. The authors present data on symptoms; clinical chemistries; abnormalities in pulmonary function; alterations in T, B, and natural killer (NK) cells; the presence of autoantibodies (i.e., antinuclear autoantibodies [ANA], autoantibodies against smooth muscle [ASM], and autoantibodies against central nervous system [CNS] and peripheral nervous system [PNS] myelins). A total of 209 adults, 42.7 ± 16 yr of age (mean ± standard deviation), were examined and tested with (a) self-administered weighted health history and symptom questionnaires; (b) standardized physical examinations; (c) complete blood counts and blood and urine chemistries; (d) urine and fecal cultures; (e) thyroid function tests (T4, free T3); (f) pulmonary function tests (forced vital capacity [FVC], forced expiratory volume in 1 sec [$FEV_{1.0}$], and forced expiratory flow at 25%, 50%, 75%, and 25–75% of FVC [FEF_{25}, FEF_{50}, FEF_{75}, and FEF_{25-75}]); (g) peripheral lymphocyte phenotypes (T, B, and NK cells) and mitogenesis determinations; and (h) a 13-item autoimmune panel. The molds-exposed patients reported a greater frequency and intensity of symptoms, particularly neurological and inflammatory symptoms, when compared with controls. The percentages of exposed individuals with increased lymphocyte phenotypes were: B cells (CD20+), 75.6%; CD5+CD25+, 68.9%; CD3+CD26+, 91.2%; CD8+HLR–DR+, 62%; and CD8+CD38+, 56.6%; whereas other phenotypes were decreased: CD8+CD11b+, 15.6% and CD3–CD16+CD56+, 38.5%. Mitogenesis to phytohemagglutinin was decreased in 26.2% of the exposed patients, but only 5.9 % had decreased response to concanavalin A. Abnormally high levels of ANA, ASM, and CNS myelin (immunoglobulins [Ig]G, IgM, IgA) and PNS myelin (IgG, IgM, IgA) were found; odds ratios for each were significant at 95% confidence intervals, showing an increased risk for autoimmunity. The authors conclude that exposure to mixed molds and their associated mycotoxins in water-damaged buildings leads to multiple health problems involving the CNS and the immune system, in addition to pulmonary effects and allergies. Mold exposure also initiates inflammatory processes. The authors propose the term "mixed mold mycotoxicosis" for the multisystem illness observed in these patients.

<Key words: immune hyperactivation, immunotoxicity, mitogenesis, molds, mycotoxicosis, mycotoxins, proinflammatory immune toxicity, toxic encephalopathy>

THE POTENTIAL HARMFUL EFFECTS of exposure to mixed molds in inhabited buildings were recognized in early Biblical times. In the Old Testament,[1] Leviticus put forth a detailed protocol for the remediation of mold-contaminated structures, including the destruction of dwellings and personal belongings if remediation failed. Today, it is recognized that water intrusion into buildings leads to amplification of molds,[2–7] often requiring remediation.

Potentially toxic and immunogenic byproducts of fungi and molds include mycotoxins; 1,3-alpha-D-glucans; extracellular polysaccharides (EPS); enzymes; and solvents.[8–19] Occupants of affected structures can develop symptoms in multiple organ systems, including the upper and lower respiratory systems, central and peripheral nervous systems, skin, gastrointestinal tract, urinary tract, connective tissue, and the musculoskeletal system.[3,6,20–26] Human illness can result from 1 or all of the following: mycotic infections, or mycoses[27–32]; immunoglobulin (Ig)E-mediated sensitivity and asthma[33–35]; hypersensitivity pneumonitis and related inflammatory pulmonary diseases[36–39]; cytotoxicity[40–44]; immune suppression/modulation[45–47]; mitochondrial toxicity[48–51]; carcinogenicity[52,53]; nephrotoxicity[54]; and the formation of nuclear and mitochondrial deoxyribonucleic acid adducts.[55–58] Finally, in the infectious state, molds secrete exodigestive enzymes (EES) that cause tissue destruction, angioinvasion, thrombosis, infarction, and other manifestations of mycosis.[27,59–63]

We have evaluated and treated more than 209 patients who presented with multiorgan system symptoms resulting from exposure to molds in their homes, schools, or workplaces. Recognizing the complexity of health problems associated with mixed mold exposure, we undertook a multicenter investigation of patients with chronic health complaints attributed to exposure to mixed colonies of indoor fungi and molds. We employed detailed health and environmental history-gathering questionnaires, environmental monitoring data, physical examination, pulmonary function testing protocols, routine clinical chemistries, neurocognitive testing, and 16-channel quantitative electroencephalograms (QEEGs). In addition, we measured lymphocyte phenotypic markers (on T, B, and natural killer [NK] cells), antibodies to molds and mycotoxins, neuronal antigen antibodies, and leukocyte apoptosis. Herein we present data on symptoms, alterations in peripheral lymphocyte phenotypes, and autoantibodies observed in adult patients. Future communications will detail pulmonary abnormalities, antibodies to molds and mycotoxins, and neurobehavioral and QEEG changes observed in these patients, and will report the statistically significant multisystem correlations observed. Currently, we refer to the illness of these individuals as a "mixed mold mycotoxicosis" involving the immune system, the lungs, and the central and peripheral nervous systems, as well as a generalized inflammatory and irritant response to exposure to spores, hyphal fragments, mycotoxins, solvents, and other byproducts (e.g., EPS and EES).

Materials and Method

Patients. Two hundred nine adult patients with a history of exposure to mixed colonies of molds resulting from structural water intrusion in residential, workplace, or school-based settings were included in this study. Adults were considered to be males older than 12 yr of age and females older than 11 yr of age. The patients, 42.7 ± 16 yr of age (mean ± standard deviation), were evaluated from early 1994 through June 2003 and comprised 126 females (43.1 ± 15.2 yr) and 83 males (42.3 ± 17.1 yr). Patients involved in litigation numbered 71 (40.1 ± 16.7 yr) and nonlitigants numbered 1,368 (44.5 ± 15.3 yr). Litigation status was uncertain for 4 adult patients. Females under age 11 and males under age 12 were not considered as adults with respect to immune parameters and symptoms, and were therefore excluded from the data presented in this report, although many of these children were clinically ill.

Questionnaires. We asked the patients to complete 2 self-administered questionnaires developed by 1 of the authors (MRG), seeking information from the following areas of concern: (a) medical history, (b) occupational and general environmental history, (c) lifestyle and habits, and (d) a review of systems. The symptom frequency review entailed questions on 58 specific symptoms. In accordance with methods provided by Ziem,[64] we report on the 38 most frequently experienced symptoms. In brief, the symptoms were scored by the patient as occurring: 1 = daily to almost daily, 2 = several times a week, 3 = weekly, 4 = several times a month, 5 = monthly, 6 = rarely, if ever (considered a negative response), and 7 = unsure. For statistical purposes, the scores were reversed to reflect the ascending frequency of the reported symptoms when tabulated. The mean value and standard deviation for the frequency score for each symptom were determined for the whole group, for males alone, for females alone, and for litigant and nonlitigant patients.

Controls for the historical questionnaires were obtained by auditing responses to the same questionnaires administered to 28 consecutive adult patients presenting to our general medical clinic for initial "database" comprehensive physical examinations. The results from the first 28 consecutive Ziem symptom audits were used as controls for comparison with the mold-exposed patients' responses.

Physical examinations. Each patient underwent a thorough physical examination, performed by MRG. A standardized form for entering relevant physical findings was used to facilitate uniformity of the exam and recording of findings. Clinical laboratory samples were collected by certified phlebotomists at the community hospital located adjacent to the clinic in which the exams were conducted. All samples were either processed at the local hospital's American College of Pathology (ACP)-accredited clinical laboratory facilities, or forwarded to the appropriate ACP-accredited reference laboratories. Guidelines of the U.S. Centers for Disease Control and Prevention (CDC) were followed for the handling of all lymphocyte tissue cultures tested.

Clinical laboratory tests. The following standard diagnostic laboratory tests were performed by Clinical Laboratory Improvement Amendment and Medicare-certified national reference and specialty laboratories: complete blood count (CBC); comprehensive metabolic panel (CMP); urinalysis; urine culture (if indicated); stool (occult blood, fungal, and mold) culture; erythrocyte sedimentation rate (ESR); C-reactive protein (CRP); thyroid profile (thyroid-stimulating hormone [TSH], free T4, and free T3); antinuclear autoantibodies (ANA); and rheumatoid factor (RF). All clinical laboratory tests were compared with the laboratories' published expected reference ranges, according to generally accepted procedures and practices. Serology was performed for the following herpes viruses: herpes I, II, VI, and varicella; Epstein Barr; and cytomegalovirus.

Peripheral lymphocyte phenotype determination. Whole venous blood was sent in laboratory-provided silicon-treated, sodium-heparinized glass e-vac tubes by overnight courier to Antibody Assay Laboratories (AAL), Santa Ana, California, following procedures prescribed by the lab to ensure that the cells were viable in accordance with CDC requirements. AAL performed mitogen tests with phytohemagglutinin (PHA), and with pokeweed and concanavalin A (Con A). Tuberculin purified protein derivative (PPD), tetanus toxoid, and *Candida albicans* mannoprotein were used as control mitogens. Any blood samples that failed to meet CDC requirements were discarded and redrawn. Quality assurance was performed using negative and positive controls. Mononuclear cells were isolated using Ficoll-isopaque density gradient centrifugation.[65] Not all of the 209 adults had all of the tests performed; therefore, *n* varies slightly for some of the parameters, as noted in the relevant tables.

Peripheral white blood cells (i.e., total white cell count and total lymphocyte count) were enumerated by AAL. In addition, the percentages of the following lymphocyte phenotypes were determined: B cells (CD20+); T cells (CD3+); T-helper (CD4+) and T-suppressor (CD8+) cells; interleukin (IL)-2 receptor-bearing T cells (CD5+CD25+); activated T cells (CD3+CD26+ and CD3+HLA–DR+); activated T-suppressor cells (CD8+CD38+ and CD8+HLA–DR+); complement receptor-bearing T-suppressor cells (CD8+CD11b+); and NK cells (CD3–CD16+CD56+). Monoclonal antibodies to CD antigens were purchased from Becton Dickinson (Los Angeles, California), except for CD26 (Beckman Coulter [Miami, Florida]). Flow cytometry was performed using a Coulter Epic XL MCL flow cytometer (Beckman Coulter), in accordance with the manufacturer's instructions.

Mitogenesis. Mitogenesis responses to PHA and Con A were evaluated on peripheral lymphocytes of all 209 patients, using the colorimetric MTT [3-(4,5-dimethylthiazol-2-yl)-2,5-diphenyltetrazolium bromide] assay.[66] Viable cells activate the MTT, which is measured colormetrically at 570 nm with an EAR 400 microplate reader (SLT Labinstruments [Salzburg, Austria]). Mononuclear cells were isolated and suspended in 0.1 ml RPMI 1640 medium at 10^6 cells/ml. They were cultured in RPMI 1640 supplemented with 10% fetal calf serum and antibiotics (i.e., penicillin and streptomycin) (Irvine Scientific [Santa Ana, California]). Cells from each individual were tested at 3 different concentrations of the mitogens for optimum stimulation. The tests were performed in triplicate and were reported as the average of 3 concentrations.

NK cell function testing. NK cell function was evaluated by AAL, using K562 cells (immortal cell line) (Coriell Institute for Medical Research [Camden, New Jersey]) as target cells. In brief, the patient's NK cells were incubated with K562 cells and a fluorescein derivative that is hydrolyzed by live K562 cells. The percentage of cells that retained the dye following killing was measured with an Epics Flow cytometer. The result is expressed as a percentage of kill.

Autoantibody determinations. Autoantibodies against smooth muscle (ASM), brush border (ABB), parietal cells (APC), mitochondria (AMIT), and nucleic acids/nucleoproteins (ANA) were determined with standard indirect immunofluoresence.[67,68] The controls for ASM and ANA were reported previously.[69] Immunoglobulin (Ig)A, IgM, and IgG antibodies to myelin sheath were detected with frozen monkey spinal cord (CNS) and sciatic nerve (PNS) as substrates, employing an indirect immunofluorescent technique for which antihuman immunoglobulin conjugated to fluorescein was used.[68,69] The controls for antimyelin antibody testing consisted of 32 chiropractic students (20 males and 12 females), 29 ± 9 yr of age, as reported previously.[69] Antibodies against thyroglobulin and thyroid peroxidase were measured by automated assays performed using chemoluminescence on DPC Immulite 2000 (Diagnostic Products Corp. [Los Angeles, California]), with DPC reagents.

Statistical analysis. All statistical analyses were performed using Statistica 10.0 for Windows (StatSoft, Inc. [Tulsa, Oklahoma]). Statistical tests included basic descriptive statistics, critical *t* tests, critical *z* tests, odds ratios, and analysis of variance (ANOVA).

Results

Physical examinations and routine diagnostic tests. Physical exams revealed (a) nasal mucosal hyperproliferation and inflammatory changes, (b) relative alopecia, (c) cough and wheezing, (d) frequent balance problems (Romberg positive), and (e) increased dermal fluorescence on ultraviolet inspection. The CBC, CMP, ESR, CRP, RF, thyroid panel, urine analyses, urine cultures, and stool analyses and stool cultures did not vary from expected values, except for a low mean total bilirubin of 0.40 (reference range: 0.4–1.0).

Serology tests for all viruses were negative for either active or reactivation infections. Lymphadenopathy indicative of acute or reactivation herpes viral infections was absent in all cases.

Symptoms. Initially, litigants were compared with nonlitigants for each symptom. Critical *t* tests for each symptom revealed no difference between the 2 groups (data not shown). The responses of males vs. females for each symptom were also compared (data not shown). Critical *t* tests showed that females had a greater frequency of the following symptoms: excessive fatigue, headache, memory problems, "spaciness"/disorientation, lightheadedness, slurred speech, weak voice, spasms, coordination problems, vision changes, rash, cold intolerance, heat intolerance, chest discomfort, excessive thirst, swallowing problems, flushing skin, rapid pulse, palpitations, bruising, and swelling ankles.

Results for males and females were grouped together, and the frequency of each symptom expressed by the mold-exposed patients was compared with frequencies reported by the 28 controls (Table 1). As indicated in the table, exposed patients had an increased frequency of expression for the following symptoms: excessive fatigue, headache, nasal symptoms, memory problems, spaciness and disorientation, sinus discomfort, coughing, watery eyes, throat discomfort, slurred speech, lightheadedness, dizziness, weakness, bloating, insomnia, spasms, coordination problems, vision changes, rash, chest tightness, and wheezing.

Peripheral lymphocytes. The percentage of lymphocyte phenotypes measured in the peripheral blood of the patients was compared for litigants vs. nonlitigants. Critical *t* tests revealed no significant difference between the 2 groups for each lymphocyte phenotype (data not shown). Table 2 summarizes our observations following a comparison of males vs. females with respect to each phenotype. Critical *t* tests revealed no difference between males and females. As a result of these observations, all data were grouped into an "all patients" category. The mean percentages of total lymphocytes for all patients exceeded the expected laboratory ranges (95% confidence intervals [CIs]). The percentages of complement-receptor–bearing suppressor cells and NK cells were within expected laboratory ranges, but were on the low side of the 95% CI for each cell type. The percentages of individuals with results outside of the expected laboratory ranges for each phenotype also are presented in Table 2.

Mitogenesis. The results of mitogenic stimulation with PHA and Con A are summarized in Table 2, along with laboratory expected ranges at 95% CI. The average mitogen responses to PHA and Con A were on the low side of the expected ranges. The percentages of all patients with mitogenesis below the expected ranges were 26.2% for PHA and 5.9% for Con A.

NK cell activity. NK cell (CD3–CD16+CD56+) activity was normal for all study subjects.

Autoantibodies. IgG, IgM, and IgA antimyelin antibodies against CNS and PNS myelin were compared for litigants vs. nonlitigants and for males vs. females. Critical *z* tests revealed no difference between males and females (data not shown). Similarly, no difference was observed in myelin autoantibodies when litigants were compared with nonlitigants, except for PNS IgA (< 0.05 and > 0.02, respectively). However, given the number of tests applied (6), this difference was considered insignificant. Thus, antimyelin antibodies were grouped for all patients and compared with controls (Table 3). Observations for antibodies (IgG, IgM, and IgA) against neurofilament antigen in the mold-exposed patients, compared with controls, are presented in Table 4.

The percentages of patients with increased ANA and ASM autoantibodies were compared for males vs. females and for litigants vs. nonlitigants (data not shown). The percentages of males and females with elevated ANA autoantibodies were 25.9% and 31.7%, respectively, and with elevated ASM autoantibodies were 34.6% and 30.3%, respectively. Similar values for litigants vs. nonlitigants were 27.5% and 30.6%, respectively, for ANA, and 27.5% and 25%, respectively, for ASM. Critical *z* tests revealed no differences in the percentages for males vs. females or for litigants vs. nonlitigants. Results for male and female patients were then combined and compared with the controls on the basis of odds ratios. The controls' values for ANA and ASM were 1.8% and 14.6%, respectively (n = 55). The ORs at 95% CI for ANA and ASM were 11.00 and 2.58, respectively, both being significant (Table 5).

Discussion

We made the following observations with respect to the 38 most frequently reported symptoms in our study (Table 1): (a) The mixed-mold–exposed patients expressed symptoms at a greater frequency than the controls; (b) Exposure to mixed molds caused significant morbidity, leading affected individuals to seek medical assistance; (c) The difference in symptoms between litigants and nonlitigants was not significant; (d) A statistically significant increase was seen in the frequency of

Table 1.—Frequency of Occurrence of the 38 Most Frequently Reported Symptoms, in Mold-Exposed Patients vs. Controls

Symptom*	Mold-exposed patients (*n* = 209)		Controls (*n* = 28)			
	$\bar{x}$	*SD*	$\bar{x}$	*SD*	*p*	*p* females
Excessive fatigue	5.8	1.9	4.3	2.1	0.0001	0.001
Headache	5.2	1.9	4.1	2.0	0.005	< 0.006
Nasal symptoms	5.1	2.2	4.1	2.0	0.02	
Memory problems	5.1	2.1	3.3	1.6	0.0002	0.005
"Spaciness"/disorientation	4.8	2.3	3.2	1.8	0.0007	< 0.01
Sinus discomfort	4.7	2.2	3.6	1.8	0.01	
Coughing	4.6	2.2	3.2	1.6	0.001	
Watery eyes	4.6	2.1	3.4	1.7	0.004	
Throat discomfort	4.5	2.1	3.4	1.7	0.008	
Slurred speech	4.5	2.3	3.1	2.0	0.002	< 0.02
Lightheadedness	4.4	2.2	3.2	1.4	0.006	0.01
Joint discomfort	4.4	2.3	3.7	2.1	NS	
Dizziness	4.3	2.1	3.1	1.4	0.005	
Weakness	4.2	2.3	3.0	1.7	0.008	
Bloating	4.2	2.2	3.2	1.6	0.02	
Insomnia	4.1	2.2	3.8	2.0	NS	
Weak voice	4.1	2.2	2.8	1.4	0.003	0.02
Spasms	4.0	2.2	3.8	2.1	NS	0.04
Coordination problems	4.0	2.2	2.9	1.4	0.01	0.009
Vision changes	3.9	2.3	2.9	1.4	0.02	0.008
Rash	3.9	2.2	2.9	1.7	0.02	0.02
Numbness	3.9	2.2	3.4	1.7	NS	
Cold intolerance	3.9	2.4	3.1	1.8	NS	0.002
Heat intolerance	3.8	2.4	3.6	2.0	NS	0.003
Chest tightness	3.8	2.2	2.6	1.3	0.006	
Chest discomfort	3.7	2.2	3.0	1.3	NS	0.02
Frequent urination	3.7	2.3	3.8	2.1	NS	
Excessive thirst	3.6	2.3	3.4	2.0	NS	0.01
Ringing ears	3.6	2.2	4.4	2.4	NS	
Wheezing	3.6	2.0	2.6	1.3	0.02	
Swallowing problems	3.2	2.0	3.0	1.7	NS	0.008
Skin flushing	3.1	2.1	2.8	1.6	NS	0.04
Bladder control problems	3.1	2.0	2.8	1.4	NS	
Rapid pulse	3.0	2.0	2.6	0.9	NS	0.04
Palpitations	2.8	1.9	2.4	0.8	NS	0.003
Bruising	2.8	1.7	2.4	0.9	NS	0.003
Swelling ankles	2.7	1.8	2.6	1.5	NS	0.02
Hearing changes	2.7	1.8	2.6	1.5	NS	

Notes: $\bar{x}$ = mean, *SD* = standard deviation, and NS = not significant.
*Symptoms were compared for females vs. males.

symptoms among women compared with men; and (e) The most frequently reported symptoms were neurological (i.e., headache, memory difficulty, slurred speech, spaciness, lightheadedness, dizziness, weakness, coordination problems, and changes in vision), state of well-being (excessive fatigue, bloating, rash, discomfort, and muscle spasms), and ophthalmic and upper/lower respiratory (nasal symptoms, sinus discomfort, coughing, watery eyes, throat discomfort, weak voice, chest tightness, and wheezing). Overall, the symptom complex we observed was consistent with observations reported by others.[3,6,12,20–26] The preponderance of symptoms involving the CNS and state of well-being are reflective of injury to the CNS, as reported by Kilburn[25] and Anyanwu et al.[26] The increased frequency of symptoms in females is consistent with their greater representation in several other clinical conditions (e.g., fibromyalgia and related disorders,[70,71] autoimmune diseases,[72–74] and exposure to molds[6]). The greater representation of females with respect to symptoms may suggest that xenobiotics, estrogenic solvents, and/or mycoestrogens in their mold-contaminated environs play a role in their illnesses.[75,76] And, finally, the absence of a difference in symptoms between litigants and nonlitigants supports the assertion that individuals who exercise their legal rights through litigation do not exaggerate their symptoms, nor are they prone to malingering.[77–79]

The lymphocytes measured in our mold-exposed pa-

Table 2.—Lymphocyte Phenotypes Observed in Males vs. Females, and in Combined Sexes (All Patients)

Cell type	Designation	Expected range (95% CI)	% variation from expected	All patients* (n = 206) $\bar{x}$	All patients* SD	All patients* % abnormal	Males (n = 83) $\bar{x}$	Males SD	Males % abnormal	Females (n = 123) $\bar{x}$	Females SD	Females % abnormal
B	CD20+	5, 15	> 15	17.9	6.1	75.6	18.1	6.9	75.6	17.7	5.5	75.6
IL-2 receptor-bearing T	CD5+CD25+	0, 8	> 8	9.2	4.7	68.9	8.9	4.7	70.7	9.4	4.7	67.5
Activated T	CD3+CD26+	0, 30	> 30	45.9	11.8	91.2	44	12.2	91.5	47	11.3	91.0
	CD3+HLA-DR+	0, 7	> 7	11.8	5.9	95.1	11.9	7.5	92.7	11.7	4.5	96.7
Activated and suppressor	CD8+CD38+	0, 8	> 8	15.4	6	56.6	15.6	6.4	54.9	15.3	5.8	57.8
	CD8+HLA-DR+	0, 3	> 3	4.7	5	62	5.2	6.4	61	4.4	3.7	63.6
Complement-receptor-bearing suppressor	CD8+CD11b+	5, 45	< 5	5.7	3.7†	15.6	6.3	4§	18.3	5.3	3.5**	13.1
Natural killer	CD3–CD16+CD56+	5, 20	< 5	7.6	11	38.5	9.8	16.1	48.8	6.1	3	31.7
Phytohemagglutinin	PHA	96, 195	< 96	104	23.5‡	26.2	104	21.4#	25	104	25**	27
Concanavalin A	Con A	94, 354	< 94	108	27.2‡	5.9	108	23.9#	6.3	108	29.2**	5.7

Notes: CI = confidence interval, $\bar{x}$ = mean, and SD = standard deviation. Not all tests were performed on all individuals, as noted.
*t and z tests of males and females revealed no significant differences; therefore, the 2 sexes were combined to form All Patients.
†n = 204.
‡n = 202.
§n = 82.
#n = 80.
**n = 123.

Table 3.—Percentage of Individuals with Antibody Titers > 1:4 against Central Nervous System (CNS) Myelin and Peripheral Nervous System (PNS) Myelin for Each Isotype, in Mold-Exposed Patients vs. Controls

	CNS myelin antibodies						PNS myelin antibodies					
	IgG		IgM		IgA		IgG		IgM		IgA	
Subject	*n*	%	*n*	%	*n*	%	*n*	%	*n*	%	*n*	%
All patients	199	67.3	201	43.3	201	55.7	201	61.7	201	45.8	201	60.2
Controls*	32	12.5	32	12.5	32	12.5	32	12.5	32	12.5	32	6.25
Odds ratio	14.40		5.34		10.59		14.40		5.39		19.10	
95% CI	4.85, 42.9		1.8, 15.8		3,58, 31.3		4.86, 42.87		1.82, 15.93		4.44, 82.6	

Notes: Ig = immunoglobulin, and CI = confidence interval.
*The percentage of controls that exceeded 1:4 had antimyelin titers of 1:8.

Table 4.—Percentage of Individuals with Abnormally High Titers to Neurofilament Antigen, in Mold-Exposed Patients vs. Controls

	% abnormal		
Subject	IgG	IgM	IgA
Patients (*n* = 93)	6.45	31.6	36.2
Controls (*n* = 100)	0.0	2.0	2.1
Odds ratio	Infinity	22.615	27.767
95% CI		5.224, 99.900	6.437, 119.784

Notes: Ig = immunoglobulin, and CI = confidence interval.

Table 5.—Odds Ratios (ORs) and 95% Confidence Intervals (CIs) for Antinuclear Autoantibodies (ANA) and Autoantibodies against Smooth Muscle (ASM)

Autoantibody	OR	95% CI
ANA	11.00	2.56, 46.50
ASM	2.58	1.26, 6.26

Note: Autoantibodies for males and females were combined and compared with combined controls A and B.

tients demonstrated increased expression of various activation markers when compared with expected laboratory ranges (Table 2), as follows: CD3+CD26+ (activated T cell), 91.2% of patients; CD3+HLA–DR+ (class II major histocompatibility molecule [MHC] and marker of activation), 95.1%; CD5+CD25+ (IL-2 receptor-bearing T cell), 68.9%; CD20+ (mature antigen-producing B cells), 75.6%; and CD8+HLA–DR+, 62%. In light of these observations, the functional role of each of these activation markers should be considered. Expression of CD26, reflecting cellular activation, is diagnostic of (or prognostic for) a variety of nonallergic clinical conditions (e.g., autoimmune disorders, various tumors, hematological malignancies, and inflammatory conditions)[80] and is also increased in individuals ill from exposure to other xenobiotics.[69,81] HLA–DR, a class II MHC, recognizes either allogenic (self) MHC molecules or foreign protein, which means that it has recognized foreign antigens bound to self class II MHC molecules.[82] HLA–DR is expressed on immune cells in inflammation,[83] asthma,[84] autoimmune diseases,[85] and neurological disorders.[86] CD25 is considered to be a natural regulatory marker of T cells and plays a major role in IL-10 production and in controlling the immune response to self and foreign antigens.[87,88] CD20 (B) cells produce antibodies and are the source of immunoglobulins directed against foreign and self-antigens; they also play a central role in autoimmunity. Currently, these cells are the focus of anti-CD20 compounds directed toward immunotherapy in B-cell malignancies and autoimmune diseases.[89,90] Collectively, the increased presence of activated T cells, and increased B cells, implies a *proinflammatory* state. Moreover, the reduction of CD8+CD11b+ (complement-receptor–bearing suppressor) cells is commensurate with increased expression of activation markers. This level of overexpression of activation markers is related to dramatic antigenic stimulation in patients with a history of mixed mold exposure. In addition, the increase in HLA–DR+ expression reflects the presence of increased autoimmunity. In the aggregate, this situation represents a proinflammatory, immune toxic state.

The effects of mold exposure on the human immune system have been reported previously. Johanning et al.[6] found a significant decrease in CD3 T cells, along with a slight decrease in mitogenesis to both Con A and PHA, following mold exposure in a water-damaged building. However, activation markers were not studied. In children exposed to high levels of residential mold contamination (vs. control children from a low-contamination environment) there was a significant increase in CD3CD45RO (memory T cell) expression, with a concomitant decrease in the helper/suppressor ratio that persisted for 12 mo.[91] Finally, animal feed production workers exposed to mixed mycotoxins, with aflatoxin concentrations of 1.55 to 6.25 ng/m^3, had an increase in tumor necrosis factor-alpha (TNF-α).[92] In addition, the animal feed workers had a shift in lactic dehydrogenase (LDH) isoenzymes, with a significant increase in LDH1 (spleen) and LDH3 (lungs). These observations corroborate the immune changes reported herein and support the conclusion that exposure to mixed molds and their byproducts causes the expression of immune markers of activation, as well as at least 1 inflammatory cytokine—TNF-α. Furthermore, the decreases we observed in the percentage of peripheral blood NK cells and response to PHA further support the concept that immune dysregulation is occurring, and represents a "promoter" state for the expression and development of malignancies.

The subjects in our study exhibited a high risk for producing autoantibodies to nuclei, smooth muscle, CNS and PNS myelin, and neurofilament (Tables 3–5). The presence of autoantibodies ANA, ASM, and CNS and PNS myelin has been reported following exposure to other xenobiotics.[69,81,93] High titers of ANA are associated with various types of connective tissue injury and/or connective tissue diseases.[94] ASM antibodies are nonspecific, occurring in a variety diseases, including autoimmune hepatitis,[95,96] vascular events,[97] rheumatoid arthritis,[98] *Mycoplasma pneumoniae*,[99] bronchial suppuration,[100] autoimmunity,[101] and asthmatic bronchitis.[102] The antimyelin autoantibodies—initially identified in Guillian-Barre syndrome[68]—are now recognized to represent several different neuronal antigens, including various gangliosides, tubulin, chondroitin sulfate, and sulfatide, found in neuropathies.[103–108] Thus, we have recently incorporated neuron neurofilament antigen into our protocol and have found increased neurofilament antibodies in these patients (Table 5). In summary, individuals exposed to mixed molds produce several different autoantibodies. Work is in progress to determine the significance of these antibodies in conditions such as lupus erythematosus, autoimmune neuropathy, and a multiple-sclerosis–type syndrome.

A systemic shift in the Th1/Th2 balance to a Th2 immune profile (e.g., TNF-α, IL-4, and IL-10 cytokines) has been reported for Gulf War veterans and chronic fatigue patients, as well as in asthma and lupus erythematosus.[109–112] Shift to Th2 profile leads to an increase in various diseases that are exacerbated by decreased Th1. Therefore, future research into immunological alterations should include testing for Th2 profile and cytokines, particularly because CD5+CD25+ (IL-2) and multiple autoantibodies were present in the patients in our study.

The single limitation of our study needs to be addressed. The immune profiles (Table 3) were compared with expected laboratory ranges, rather than with healthy control subjects. However, the absence of controls should not be considered excessively limiting in the evaluation of immune profiles of these mold-exposed patients. First, the percentage of individuals with increased activation expression on T cells greatly exceeded the maximum expected range as published by the testing laboratory. For example, the maximum percentages of total recorded lymphocytes for controls in regard to CD5+CD25+, CD3+CD26+, CD3+HLA–DR+, CD8+ CD38+, and CD8+HLA–DR+ were 8%, 30%, 7%, 8%, and 3%, respectively. The percentages of mold-exposed patients that exceeded these values were 68.9%, 91.2%, 95.1%, 56.6%, and 62.0%, respectively, for each of the phenotypes. In addition, the mean percentages for CD20+, CD5+CD25+, CD3+CD26+, CD3+HLA–DR+, CD8+CD38+, and CD8+HLA–DR+ cell counts exceeded the laboratory's 95% CI, which further supports the preceding observation. This represents a greatly increased frequency of activation markers when compared with maximum expected ranges. Second, expression of the autoantibodies ANA, ASM, CNS and PNS myelin, and neurofilament was significant in the mold-exposed patients vs. controls. Thus, the presence of autoantibodies is commensurate with immune activation, and, finally, the antineuronal antigen (neurofilament)-specific antibodies are strongly associated with a wide array of degenerative neurological disorders of undetermined origin.

In this study, we have shown that individuals exposed to mixed colonies of molds in water-damaged buildings have several abnormalities among their immune parameters. These include (a) immune activation markers, with elevated CD26+, HLR–DR+, CD25+, and CD38+ phenotypes in the peripheral blood; (b) the presence of autoantibodies (ANA, ASM, CNS and PNS myelin, and neurofilament); and (c) decreased complement-receptor–bearing T-suppressor (CD11b+) cells. Future research should be directed toward clarifying the Th1/Th2 profile, and accompanying cytokines, in humans affected adversely by mixed mold exposure. Also, efforts should be made to correlate the abnormal immune parameters with other measured abnormalities found in individuals exposed to mixed colonies of structural fungi and molds, and their associated mycotoxins, extracellular polysaccharides, exodigestive enzymes, hyphae fragments, and spores.

* * * * * * * * * * *

Submitted for publication September 16, 2003; revised; accepted for publication November 24, 2003.

Requests for reprints should be sent to Michael Gray, M.D., M.P.H., C.I.M.E., 300 S. Ocotillo Road, Benson, AZ 85602.
E-mail: docmike007@aol.com

* * * * * * * * * * *

References

1. The Bible. Approved (King James) Version. Oxford 1888 ed. Leviticus 14:34–47.
2. Gravesen S, Nielsen PA, Iverson R, et al. Microfungal

contamination of damp buildings—examples of constructions and risk materials. Environ Health Perspect 1999; 107(suppl 3):505–08.
3. Croft WA, Jarvis BB, Yatawara CS. Airborne outbreak of trichothecene toxicosis. Atmos Environ 1986; 20:549–52.
4. Peltola J, Andersson MA, Haahtela T, et al. Toxic-metabolite-producing bacteria and fungus in an indoor environment. Appl Environ Microbiol 2001; 67:3269–74.
5. Shelton BF, Kirkland KH, Flanders WE, et al. Profiles of airborne fungi in buildings and outdoor environments in the United States. Appl Environ Microbiol 2002; 68: 1743–53.
6. Johanning E, Biagini R, Hull DL, et al. Health and immunology study following exposure to toxigenic fungi (*Stachybotrys chartarum*) in a water-damaged office environment. Int Arch Occup Environ Health 1996; 68: 207–18.
7. Andersson MA, Nikulin M, Kooljalg U, et al. Bacteria, molds, and toxins in water-damaged building materials. Appl Environ Microbiol 1997; 63:387–93.
8 Nielsen KF, Gravesen S, Neilsen PA. Production of mycotoxins on artificially and naturally infested building materials. Mycopathologia 1999; 145:43–56.
9. Claeson AS, Levin JO, Blomquist G, et al. Volatile metabolites from microorganisms grown on humid building materials and synthetic media. J Environ Monit 2002; 4:667–72.
10. Tuomi T, Reijula K, Johnsson T, et al. Mycotoxins in crude building materials from water-damaged buildings. Appl Environ Microbiol 2000; 66:1899–1904.
11. Nieminen SM, Karki R, Auriola S, et al. Isolation and identification of *Aspergillus fumigatus* mycotoxins on growth medium and some building materials. Microbiology 2002; 68:4871–75.
12. Jarvis BB. Chemistry and toxicology of molds isolated from water-damaged buildings. In: DeVries JW, Trucksess MW, Jackson LS (Eds). Mycotoxins and Food Safety. New York: Kluwer Academic/Plenum Publishers, 2002; pp 43–52.
13. Burge, HA. Bioaerosols: prevalence and health effects in the indoor environment. J Allergy Clin Immunol 1990; 86:687–704.
14. Richard JL, Plattner RD, May J, et al. The occurrence of ochratoxin A in dust collected from a problem household. Mycopathologia 1999; 146:99–103.
15. Skaug MA, Eduard W, Stormer FD. Ochratoxin A in airborne dust and fungal conidia. Mycopathologia 2000; 151:93–95.
16. Smoragiewicz W, Cossete B, Boutrard A, et al. Trichothecene mycotoxins in the dust of ventilation systems in office buildings. Int Arch Occup Environ Health 1993; 65:113–17.
17. Tuomi T, Saarinen L, Reijula K. Detection of polar and macrocyclic trichothecene mycotoxins from indoor environments. Analyst 1998; 123:1835–41.
18. Johanning E, Gareis M, Nielsen K, et al. Airborne mycotoxins sampling and screening analysis. Proceedings of the 9th International Conference on Indoor Air Quality and Climate (Indoor Air 2002), Monterey, California, June 30–July 5, 2002. Santa Cruz, CA: Indoor Air 2002 Conference Secretariat.
19. Notermans S, Dufrenne J, Wijands LM, et al. Human serum antibodies to extracellular polysaccharides (EPS) of molds. J Med Vet Mycol 1998; 26:41–48.
20. Gunnbjornsdottir MI, Norback D, Plaschke P, et al. The relationship between indicators of building dampness and respiratory health in young Swedish adults. Respir Med 2003; 97:301–07.
21. Savilahti R, Uitti J, Laippala P, et al. Respiratory morbidity among children following renovation of water-damaged school. Arch Environ Health 2000; 55:405–10.
22. Jaakkola M, Nordman H, Pilpari R, et al. Indoor dampness and molds and development of adult-onset asthma: a population-based incident case-control study. Environ Health Perspect 2002; 110:543–47.
23. Hodgson MJ, Morey P, Leung WY, et al. Building-associated pulmonary disease from exposure to *Stachybotrys chartarum* and *Aspergillus versicolor*. J Occup Environ Med 1998; 40(3):241–49.
24. Croft WA, Jastromski BM, Croft AL, et al. Clinical confirmation of trichothecene mycotoxicosis in patient urine. J Environ Biol 2002; 23:301–20.
25. Kilburn KH. Inhalation of moulds and mycotoxins. Eur J Oncol 2002; 7:197–202.
26. Anyanwu EC, Campbell AW, Vojdani A. Neurophysiological effects of chronic indoor environmental mold exposure on children. Scientific World J 2003; 3:281–90.
27. Ribes JA, Vanover-Sames CL, Baker DJ. Zygomycetes in human disease. Clin Microbiol Rev 2000; 13:236–301.
28. Grossi P, Farina C, Fiocchi R, et al. Prevalence and outcome of invasive fungal infections in 1,963 thoracic organ transplant recipients: a multi-center retrospective study. Italian Study Group of Fungal Infections in Thoracic Organ Transplant Recipients. Transplantation 2000; 70:112–16.
29. Anaissie EJ, Stratton SL, Dignani MC, et al. Pathogenic *Aspergillus* species recovered from a hospital water system: 3-year prospective study. Clin Infect Dis 2002; 34:780–89.
30. Fraser RS. Pulmonary aspergillosis: pathologic and pathogenetic features. Pathol Ann 1993; 28:231–77.
31. Eucker J, Sezer O, Graf B, et al. Mucormycoses. Mycoses 2001; 44:254–60.
32. Taylor MJ, Pnikaue JU, Sherris DA, et al. Detection of fungal organisms in eosinophilic mucin using a fluorescein-labeled chitin-specific binding protein. Otolaryngol Head Neck Surg 2002; 127:377–83.
33. Lander F, Meyer HW, Norm S. Serum IgE specific to moulds, measured by basophil histamine release, is associated with building-related symptoms in damp buildings. Inflamm Res 2001; 50:227–31.
34. Karlsson-Borga A, Jonsson P, Rolfsen W. Specific IgE antibody to 16 widespread mold genera in patients with suspected mold allergy. Ann Allergy 1989; 63:521–26.
35. Zureik M, Neukirch C, Leynaert B, et al. Sensitisation to airborne moulds and severity of asthma: cross sectional study from European Community respiratory health survey. Br Med J 2002; 325:411–14.
36. Sumi Y, Natura H, Takeuchi M, et al. Granulomatous lesions in the lung induced by inhalation of mold spores. Virchows Arch 1994; 424:661–68.
37. Ojanen T. Class specific antibodies in serodiagnosis of farmer's lung disease. Br J Ind Med 1992; 49:332–36.
38. Erkinjuntti-Pekkanen R, Reiman M, Kokkarinen JI, et al. IgG antibodies, chronic bronchitis, and pulmonary function values in farmer's lung patients and matched controls. Allergy 1999; 54:1181–87.
39. Patel AM, Ryu JH, Reed CE. Hypersensitivity pneumonitis: current concepts and future questions. J Allergy Clin Immunol 2001; 108:661–70.
40. Gareis M. Cytotoxicity testing of samples originating from problem buildings. In: Johanning E, Yang CS (Eds). Proceedings of the International Conference on Fungi and Bacteria in Indoor Environments: Health Effects, De-

tection and Remediation. Saratoga Springs, NY, October 1994. Albany, NY: Eastern New York Occupational Health Program, 1995; pp 139–44.
41. Nagata T, Suzuki H, Ishigami N, et al. Development of apoptosis and changes in lymphocyte subsets in thymus, mesenteric lymph nodes and Peyer's patches of mice orally inoculate with T-2 toxin. Exp Toxicol Pathol 2001; 52:3309–15.
42. Jones C, Ciacci-Zanella JR. Zhang V, et al. Analysis of fumonisin B1-induced apoptosis. Environ Health Perspect 2002; 109(suppl 2):315–20.
43. Poapolathep A, Ohtsuka R, Kiatipattanasakul W, et al. Nivalenol-induced apoptosis of thymus, spleen, and Peyer's patches of mice. Exp Toxicol Pathol 2002; 53:441–46.
44. Desai K, Sullards MC, Allegood J, et al. Fumonisins and fumonisin analogs as inhibitors of ceramide synthase and inducers of apoptosis. Biochim Biophys Acta 2002; 1585:188–92.
45. Jakab GJ, Hmieleski RR, Hemenway DR, et al. Respiratory aflatoxicosis: suppression of pulmonary and systemic host defenses in rats and mice. Toxicol Appl Pharmacol 1994; 125:198–205.
46. Berek L, Petri IB, Mesterhazy A, et al. Effects of mycotoxins on human immune functions in vitro. Toxicol In Vitro 2001; 15:25–30.
47. Bondy GS, Petska JJ. Immunomodulation by fungal toxins. J Toxicol Environ Health B Crit Rev 2000; 3:109–43.
48. Pace JG. Effect of T-2 mycotoxin on rat liver mitochondria electron transport system. Toxicon 1983; 21: 675–80.
49. Pace JG. T-2 mycotoxin inhibits mitochondrial protein synthesis. Toxicon 1998; 26:77–85.
50. Hoehler D, Marquardt RR, McIntosh AR, et al. Induction of free radicals in hepatocytes, mitochondria and microsomes of rats by ochratoxin A and its analogs. Biochim Biophys Acta 1997; 1357:225–33.
51. Sajan MP, Satav JG, Battacharya RK. Effect of aflatoxin B1 in vitro on rat liver mitochondrial respiratory functions. Indian J Exp Biol 1997; 35:1187–90.
52. Schwartz GG. Does ochratoxin A cause testicular cancer? Cancer Causes Control 2002; 13:91–100.
53. Dominguez-Malagon H, Gaytan-Graham S. Hepatocellular carcinoma: an update. Ultrastruct Pathol 2001; 25:497–516.
54. Pfohl-Leszkowicz A, Petkova-Bocharova T, Chernozemsky IN, et al. Balkan endemic nephropathy and associated urinary tract tumours: a review on aetiolgical causes and the potential role of mycotoxins. Food Addit Contam 2002; 19:282–302.
55. Pfohl-Leszkowicz A, Grosse Y, Kane A, et al. Differential DNA adduct formation and disappearance in three mouse tissues after treatment with mycotoxin ochratoxin A. Mutat Res 1993; 289:265–73.
56. Petkova-Bochatrova T, Stoichev II, Chernozemsky IN, et al. Formation of DNA adducts in tissue of mouse progeny through transplancental contamination and/or lactation after administration of single does of ochratoxin A to the pregnant mother. Environ Mol Mutagen 1998; 32:155–62.
57. Hsieh LL, Hsieh TT. Detection of aflatoxin B1-DNA adducts in human placenta and cord blood. Cancer Res 1993; 53:1278–80.
58. Niranjan BF, Bhat NK, Avadhani NG. Preferential attack of mitochondrial DNA by aflatoxin B1 during hepatocarcinogenesis. Science 1982; 214(4528):73–75.
59. Monod M, Capoccia S, Lechene B, et al. Secreted proteases from pathogenic fungi. Int J Med Microbiol 2002; 292:405–19.
60. Vesper SJ, Dearborn DG, Elidemir O, et al. Quantification of siderophore and hemolysin from *Stachybotrys chartarum* strains, including a strain isolated from the lung of a child with pulmonary hemorrhage and hemosiderosis. Appl Environ Microbiol 2000; 66:2678–81.
61. Kordula T, Banbula A, Macomson J, et al. Isolation and properties of Stachyrase A, a chymotrypsin-like serine proteinase from *Stachybotrys chartarum*. Infect Immun 2002; 70:419–21
62. Ebina K, Ichinowatari S, Yokota K. Studies on toxin *Aspergillus fumigatus*. Vol 22. Fashion of binding Asp-hemolysin to human erythrocytes and Asp-hemolysin-binding proteins of erythrocyte membranes. Microbiol Immunol 1985; 29:91–101.
63. Kudo Y, Ootani T, Kumangai T, et al. A novel oxidized low-density lipoprotein-binding protein, Asp-hemolysin recognizes lysophosphatydylcholine. Biol Pharm Bull 2002; 25:787–90
64. Ziem G, McTamney J. Profile of patients with chemical injury and sensitivity. Environ Health Perspect 1997; 105:417–36.
65. Boyuma A. Isolation of mononuclear cells and granulocytes from blood. Scand J Clin Lab Invest 1968; 21(suppl 97):77–81.
66. Mossman T. Rapid colorimetric assay for cellular growth and survival application to proliferation and cytotoxic assays. J Immunol Methods 1983; 35:1949–54.
67. Nakamura RM, Tucker ES. Antibodies as reagent. In: Henry JD (Ed). Diagnosis and Management by Laboratory Methods. Philadelphia, PA: W.B. Saunders, 1982; pp 122–77.
68. Edgington TS, Dalessio DJ. The assessment by immunofluorescence methods of human antimyelin antibodies in man. J Immunol 1970; 105:1949–54.
69. Thrasher JD, Heuser G, Broughton A. Immunological abnormalities in humans chronically exposed to chlorpyrifos. Arch Environ Health 2002; 57:181–87.
70. Bombardier CH, Buchwald D. Chronic fatigue, chronic fatigue syndrome, and fibromyalgia. Med Care 1996; 34: 924–30.
71. Buchwald D, Pearlman T, Umali J, et al. Functional status in patients with chronic fatigue syndrome, other fatiguing illness, and healthy individuals. Am J Med 1996; 101:364–70.
72. Bartley GB. The epidemiologic characteristics and clinical course of ophthalmopathy associated with autoimmune thyroid diseases in Olmsted County, Minnesota. Trans Am Ophthalmol Soc 1994; 92:477–588.
73. Valentini G, Black C. Systemic sclerosis. Best Pract Res Clin Rheumatol 2002; 16:807–16.
74. Alamanos Y, Voulgari PV, Siozos C, et al. Epidemiology of systemic lupus erythematosus in northwest Greece 1982–2001. J Rheumatol 2003; 30:731–35.
75. Branham WS, Dial SL, Moland CL, et al. Phytoestrogens and mycoestrogens bind to the rat uterine estrogen receptor. J Nutr 2002; 132:658–64.
76. Kuiper GG, Lemmen JG, Carlsson G, et al. Interaction of estrogenic chemicals and phytoestrogens with estrogen receptor *B*. Endocrinology 1998; 139:4252–63.
77. Mendelson G. Measurement of conscious symptom exaggeration by questionnaire: a clinical study. J Psychosom Res 1987; 31:703–11.
78. Allaz AF, Vannotti M, Desmeules J, et al. Use of the label "litigation neurosis" in patients with somatoform pain disorder. Gen Hosp Psychiatry 1998; 20:91–97.

79. Tait RC, Margolis RG, Krause SJ, et al. Compensation status and symptoms reported by patients with chronic pain. Arch Phys Med Rehabil 1988; 69:1027–29.
80. Lambier AM, Durinx C, Scharpe S, et al. Dipeptidyl-peptidase IV from bench to bedside: an update on structural properties, functions, and clinical aspects of the enzyme DPP IV. Crit Rev Clin Lab Sci 2003; 40:209–94.
81. Vojdani A, Ghoneum M, Brautbar N. Immune alteration associated with exposure to toxic chemicals. Toxicol Ind Health 1992; 8:239–54.
82. Abbas AK, Lichtman AH, Pober JS. Cellular and Molecular Immunology. 2nd ed. Philadelphia, PA: W.B. Saunders, 1994; p 108.
83. Oczenski W, Krenn H, Jilch R, et al. HLA-DR as a marker for increased risk for systemic inflammation and septic implications after cardiac surgery. Intensive Care Med 2003; 19:1253–57.
84. Visman MY, Bocher BS, Peebles RS, et al. Expression of activation markers on alveolar macrophages in allergic asthmatics after endobronchial or whole-lung challenge. Clin Immunol 2002; 104:77–85.
85. Cope RA. Exploring the reciprocal relationship between immunity and inflammation in chronic inflammatory arthritis. Rheumatology (Oxford) 2003; 42:716–31.
86. McGeer PL, Itagaki S, McGeer EG. Expression of the histocompatibility glycoprotein HLA-DR in neurological diseases. Acta Neuropathol (Berl) 1988; 76:550–57.
87. Pipiernak M. Natural CD4+CD25+ regulatory T cells. Their role in the control of superantigen responses. Immunol Rev 2001; 182:180–89.
88. Malek TR. The main function of IL-2 is to promote the development of T regulatory cells. J Leukoc Biol 2003; 74(6):961–65.
89. von Schilling C. Immunotherapy with anti-CD20 compounds. Semin Cancer Biol 2003; 13:211–22.
90. Anolik J, Sanz I, Looney RJ. B cell depletion therapy in systemic lupus erythematosus. Curr Rheumatol Rep 2003; 5:350–56.
91. Dales R, Miller D, White J, et al. Influence of residential fungal contamination on peripheral blood lymphocyte populations in children. Arch Environ Health 1998; 53: 190–95.
92. Nuntharatanapong N, Suramana T, Chaemthavorn S, et al. Increase in tumour necrosis factor-alpha and a change in the lactate dehydrogenase isoenzyme pattern in plasma of workers exposed to aflatoxin-contaminated feeds. Arh Hig Rada Toksikol 2001; 52:191–98.
93. Thrasher JD, Broughton A, Madison R. Immune activation and autoantibodies in humans with long-term exposure to formaldehyde. Arch Environ Health 1990; 45: 217–23.
94. Rose NR, Mackay IR. Autoimmune Diseases. New York: Academic Press, 1985.
95. Dalekos GN, Zachou K, Liaskos C, et al. Autoantibodies and defined target autoantigens in autoimmune hepatitis: an overview. Eur J Intern Med 2002; 13:292–303.
96. Muratori P, Muratori L, Agostinelli D, et al. Smooth muscle antibodies in type 1 autoimmune hepatitis. Autoimmunity 2002; 35:497–500.
97. Kristenson BO, Andersen PL, Wiik A. Autoantibodies and vascular events in essential hypertension: a five-year longitudinal study. J Hypertens 1984; 2:19–24.
98. Andersen I, Andersen P, Graudal H. Smooth-muscle antibodies in rheumatoid arthritis. Acta Pathol Microbiol Scand [C] 1980; 83:131–35.
99. Cimolai N, Cheong AC. Anti-smooth muscle antibody in clinical human and experimental *Mycoplasma pneumoniae* infection. J Appl Microbiol 1997; 82:625–30.
100. Butland RJ, Cole P, Citron KM, et al. Chronic bronchial suppuration and inflammatory bowel disease. Q J Med 1981; 50:63–75.
101. Nakamura RM, Chisari FV, Edgington TS. Laboratory tests for diagnosis of autoimmune diseases. Prog Clin Pathol 1975; 6:177–203.
102. Oehling A, Diegurez I, Crisci CD. Anti-smooth muscle antibodies in bronchial asthma and chronic bronchitis. Allergol Immunopathol (Madr) 1979; 7:433–38.
103. Willison HJ, Nobuhiro Y. Peripheral neuropathies and anti-glycolipid antibodies. Brain 2002; 125:2591–2625.
104. Vojdani A, Vojdani E, Cooper E. Antibodies to myelin basic protein, myelin oligodendrocytes, peptides, a-B-crystallin, lymphocyte activation and cytokine production in patients with multiple sclerosis. J Intern Med 2001; 254:1–12.
105. Connolly AM, Pestronk A. Anti-tubulin autoantibodies in acquired demyelinating polyneuropathies. J Infect Dis 1997; 176(suppl 2):S157–59.
106. Briani C, Berger JS, Latov N. Antibodies to chondroitin sulfate C: a new detection assay and correlations with neurological diseases. J Neuroimmunol 1998; 84:117–21.
107. Dabby, R, Weimer LH, Hays AP, et al. Antisulfatide antibodies in neuropathy. Clinical and electrophysiologic correlates. Neurology 2000; 54:1448–52.
108. Alaedini A, Sander HW, Hays AP, et al. Antiganglioside antibodies in multifocal acquired sensory and motor neuropathy. Arch Neurol 2003; 60:42–46.
109. Rook GA, Zumla A. Gulf War syndrome: Is it due to a systemic shift in cytokine balance towards a TH2 profile? Lancet 1997; 349:1831–33.
110. Rosenbaum ME, Vojdani A, Susser M, et al. Improved immune activation markers in chronic fatigue and immune dysfunction syndrome (CFIDS) patients treated with thymic protein A. J Nutr Environ Med 2001; 11:241–47.
111. Mok CC, Lau CS. Pathogenesis of systemic lupus erythematosus. J Clin Pathol 2003; 56:481–90.
112. Lordan JL, Buccheri F, Richter A, et al. Cooperative effects of Th2 cytokines and allergen on normal and asthmatic bronchial epithelial cells. J Immunol 2002; 169: 407–14.

Psychological, Neuropsychological, and Electrocortical Effects of Mixed Mold Exposure

B. ROBERT CRAGO
Neurobehavioral Health Services
Tucson, Arizona
MICHAEL R. GRAY
Progressive Health Care Group
Benson, Arizona
LONNIE A. NELSON
Department of Psychology
University of Arizona
and
Neurobehavioral Health Services
Tucson, Arizona

MARILYN DAVIS
Neurobehavioral Health Services
Tucson, Arizona
LINDA ARNOLD
ImmunTox, LLC
Benson, Arizona
JACK D. THRASHER
Sam-1 Trust
Alto, New Mexico

ABSTRACT. The authors assessed the psychological, neuropsychological, and electrocortical effects of human exposure to mixed colonies of toxigenic molds. Patients (N = 182) with confirmed mold-exposure history completed clinical interviews, a symptom checklist (SCL-90-R), limited neuropsychological testing, quantitative electroencephalogram (QEEG) with neurometric analysis, and measures of mold exposure. Patients reported high levels of physical, cognitive, and emotional symptoms. Ratings on the SCL-90-R were "moderate" to "severe," with a factor reflecting situational depression accounting for most of the variance. Most of the patients were found to suffer from acute stress, adjustment disorder, or post-traumatic stress. Differential diagnosis confirmed an etiology of a combination of external stressors, along with organic metabolically based dysregulation of emotions and decreased cognitive functioning as a result of toxic or metabolic encephalopathy. Measures of toxic mold exposure predicted QEEG measures and neuropsychological test performance. QEEG results included narrowed frequency bands and increased power in the alpha and theta bands in the frontal areas of the cortex. These findings indicated a hypoactivation of the frontal cortex, possibly due to brainstem involvement and insufficient excitatory input from the reticular activating system. Neuropsychological testing revealed impairments similar to mild traumatic brain injury. In comparison with premorbid estimates of intelligence, findings of impaired functioning on multiple cognitive tasks predominated. A dose-response relationship between measures of mold exposure and abnormal neuropsychological test results and QEEG measures suggested that toxic mold causes significant problems in exposed individuals. Study limitations included lack of a comparison group, patient selection bias, and incomplete data sets that did not allow for comparisons among variables.

<Key words: dose-response relationship, neuropsychological testing, Symptom Checklist-90-R, quantitative electroencephalograph, toxic encephalopathy, toxic mold>

NEUROTOXICITY can cause irreversible nervous system damage related to cell death or permanent alterations of cell structure and receptor sensitivity. Clinical signs are classified as organi/c mental impairments, seizures, movement disorders, involvement of cranial nerves or spinal peripheral nerves, and neuromuscular dysfunction.[1] Neurotoxic exposure and injury are assessed by careful neurological and neuropsychological evaluation, complemented with functional imaging of the brain.

Occupants of mold-infested structures develop multiorgan symptoms that involve the upper and lower respiratory systems, central and peripheral nervous systems, skin, gastrointestinal tract, connective tissue, immune system, and musculoskeletal system.[2-18] Complaints of neurocognitive dysfunction are prevalent among the symptoms reported.[2,3,18–22] A large body of literature exists on the effects of various neurotoxins on neuropsychological functioning, including cognitive impairment.[23] However, only 4 studies have reported measurements of neurobehavioral changes related to mold and mycotoxin exposure.[18,20–22] The changes described include impairments in balance, reaction time, cognition, verbal learning, recall, visual spatial learning, memory, attention/concentration, and set shifting.

Only a few complementary neuroimaging studies have been published in regard to assessment of the effects of mixed mold exposure on the central nervous system. Adolescents with suspected acoustic mycotic neuroma resulting from environmental exposure to toxic molds had abnormal brainstem evoked potentials.[2] In another study, abnormal electroencephalogram (EEG) examinations in 7 of 10 patients exposed to toxic mold were reported. All 10 patients had frontal-temporal theta wave activity, which indicated diffuse changes characteristic of metabolic encephalopathy. Abnormal brainstem auditory evoked potentials were demonstrated in 9 of these patients, and 4 of the 10 patients showed clear abnormalities.[3]

All of the patients in the current study reported wide-ranging symptoms, including headache, dizziness, visual changes, cognitive impairment, and emotional dysregulation. Their illness has been defined as "mixed mold mycotoxicosis."[19]

It is likely that studies of mold-induced neurotoxicity will yield findings similar to other studies of neurotoxicity from other causes. Single photon emission computed tomography (SPECT) has been used to complement and define the effects of toxic exposure on the central nervous system. A review of the literature on the use of SPECT scans following neurotoxic exposure confirmed that abnormalities can exist from months to years after exposure has ceased, and can involve asymmetrical abnormalities with hypoperfusion in the frontal, parietal, and temporal lobes.[24] Moreover, in 33 workers with encephalopathy following toxic exposure, 94% had abnormal SPECT scans. The most frequent areas of abnormality were the temporal lobes (67.7%), frontal lobes (61.3%), basal ganglia (45.2%), thalamus (29.0%), parietal lobes (12.9%), and motor strip (9.7%).[25]

In recognition of the complexity of health problems associated with mixed mold exposure, a multicenter investigation of patients with chronic health complaints from mold exposure was undertaken. We used generally accepted, standardized, detailed health and environmental history questionnaires, environmental monitoring data, physical examination, accepted pulmonary function testing protocols, routine clinical chemistry, standardized measures of specific immune markers (T, B, and natural killer [NK] cells), measures of antibodies to molds, neuropsychological testing, and 19-channel quantitative EEG (QEEG). The results of this project are being reported in a series of papers. The study presented herein was conducted to assess the psychological, neuropsychological, and electrocortical effects of mixed toxic mold exposure.

Materials and Method

Patients. Adult patients (N = 182) with a history of exposure to mixed colonies of molds and their associated mycotoxins (confirmed with environmental and serologic testing) as a result of structural water intrusion in residential, workplace, or school-based settings,[19] were included in this multicenter study of data gathered from chart review. All patients had been referred for evaluation of health problems related to toxic mold exposure. Because of the prominence of neurological symptoms and complaints of cognitive dissonance, patient assessments included the following neuropsychological and neurophysiological evaluations: a structured psychometric symptom checklist, neuropsychological testing, and QEEGs. The patients (age 42.7 ± 16 yr [mean ± standard deviation]) were evaluated from September 1999 through June 2003. The group comprised 126 females (age 39.3 ± 18.1 yr) and 83 males (age 36.6 ± 21.1 yr). All of the patients were evaluated for this study. Test results were compared against a national normative database in all cases. Inasmuch as our study was based on data from chart reviews, the numbers of subjects for each measure varied slightly.

Measures of psychological distress. Each patient was evaluated with a standard clinical interview and a psychometric self-report symptom inventory—the Symptom Checklist 90-R (SCL-90-R).[26] Patients rated each item on a 5-point scale of distress, ranging from *not at all distressed* (0) to *extremely distressed* (5). Their ratings were computer-scored and produced normalized t scores for 9 symptom dimensions (i.e., Somatization, Obsessive-Compulsive, Interpersonal Sensitivity, Depression, Anxiety, Hostility, Phobic Anxiety, Paranoid Ideation, and Psychoticism) and 3 global indices (i.e., Global Severity Index, Positive Symptom Distress Index, and Positive Symptom Total). For comparison, we used an adult nonpsychiatric patient normative database because it best represented the individuals examined in this study.

Neuropsychological testing. We selected neuropsychological tests on the basis of specific patient complaints, clinical efficiency, and time constraints. All tests chosen had appropriate and comparable norms for adults and children. To obtain estimates of premorbid and general intellectual functioning for all patients, we administered the Vocabulary subtest from the Wechsler Adult Intelligence Scale, 3rd ed. (WAIS-III) and/or from the Wechsler Abbreviated Scale of Intelligence (WASI).[27]

Three subtests from the WAIS-III were administered. The Digit Span subtest measures attention and working memory capacity. The Digit Symbol Coding subtest measures visual motor learning and psychomotor speed. The Symbol Search subtest is an indicator of visual scanning and concentration.

Two subtests from the Delis-Kaplan Executive Function System (D-KEFS) were selected to measure executive or higher-level cognitive functions.[28] The D-KEFS Color-Word Test produces baseline measures of color naming and word reading to compare with executive measures of inhibition and inhibition switching. The D-KEFS Trail Making Test provides baseline measures of visual scanning, number sequencing, letter sequencing, and motor speed, as well as an executive measure of number-letter switching.

The Integrated Visual and Auditory Continuous Performance Test (IVA-CPT) is a computerized test designed to evaluate auditory and visual attention over time.[29] The IVA-CPT produces global composite scores consisting of a Response Control Quotient (a positive way to describe the problem of response inhibition) and an Attention Quotient (a positive way to describe problems of inattention, loss of focus, and slow processing speed). Resulting measures are normed with a mean of 100 and a standard deviation of 15.

QEEG. Brain electrical activity was recorded from 19 cortical positions in accordance with the International 10/20 electrode-placement system, using a Lexicor Neurosearch Digital EEG acquisition system (Lexicor Research Center [Boulder, Colorado]). Electrodes were positioned on the scalp using appropriately sized electrocaps. We used impedance measurements for each cortical site to ensure accurate data collection. The raw data were edited for artifacts and then subjected to quantitative and neurometric analyses of amplitude, power, and mean frequency, using the NxLink database.[30–32] (The NxLink database has received a 510(k) clearance from the FDA [July 1998, #K974748], indicating that construction of the database was scrutinized for good manufacturing practices, and signifying the legitimacy of marketing claims made concerning the database.)

The NX-Link database uses the following grouped band frequencies: delta (1.5–3.5 Hz), theta (3.5–7.5 Hz), alpha (7.5–12.5 Hz), and beta (12.5–25.0 Hz). Absolute power magnitude is the amount of electrical activity at each frequency band. Relative power is a proportion of the total absolute power across the different frequency bands. Measures of power reflect different levels of cortical electrical activity and offer insight into cortical regional differences in activation, functional differentiation and integration, and corticothalamic regulation. Decreases in mean frequency are commonly referred to as "slowing," indicating that the average speed of oscillation is decreased relative to the population normal values for that frequency band. The reverse is true of higher-than-expected average frequency, which would be considered "accelerated."

For each QEEG variable, the 5 cortical sites or positions with the most deviant scores were chosen to create a distinct QEEG summary variable. The effects of toxic mold exposure on cortical functioning were related primarily to prefrontal and frontal dysfunctions, with some involvement of other cortical regions. We used these summary variables as predictor variables in forward stepwise regression analyses to determine the amount of variance accounted for by these symptom sets for each neuropsychological testing finding, SCL-90-R factor, and exposure measure.

Data preparation included examination of medication effects. Anxiolytic and narcotic medications were found to have significant effects on some QEEG variables; therefore, our analysis and interpretation excluded such cases.

Measures of exposure and dose-response relationship. Our analyses also took into account the degree of exposure to toxic molds for each individual in the sample. For each patient, we used the following predictor variables: average hours per day present in the building, days of exposure, maximum exposure (calculated as: hr/day × days of exposure), average number of colony forming units (CFUs) per m^3 from all available air samples in a given building, average CFUs × maximum exposure, and whether *Stachybotrys* was present (yes = 1, no = 0). Scores for each of these measures were then used as predictor variables in a forward stepwise regression analysis to determine the amount of variance accounted for by each of these measures of exposure on the SCL-90-R factors, the neuropsychological test measures, and the QEEG results.

Statistical analysis. Repeated measures analyses of variance (ANOVAs) were used to examine the relationships between the QEEG measures, neuropsychological testing results, SCL-90-R factors, and measures of exposure.[33]

Results

Psychological disturbance (SCL-90-R). The descriptive statistics summarizing the group means for 106 of the toxic-mold–exposed population revealed that most of the patients reported a wide range of somatic, affective, and cognitive symptoms, as well as a very high level of general distress. Figure 1 shows that all SCL-90-R scores were significantly elevated. The Global Severity Index had a mean average *t* score of 67—almost 2 standard deviations above the norm. The 4 highest scores on subtests occurred on the Obsessive-Compulsive, Somatization, Depression, and Anxiety subscales.

Examination of the items that made up each of the SCL-90-R scaled scores revealed considerable overlap in symptoms across the categories of cognitive, affective, and somatic symptoms. A factor analysis of all items defined 6 factors. After examination of the indi-

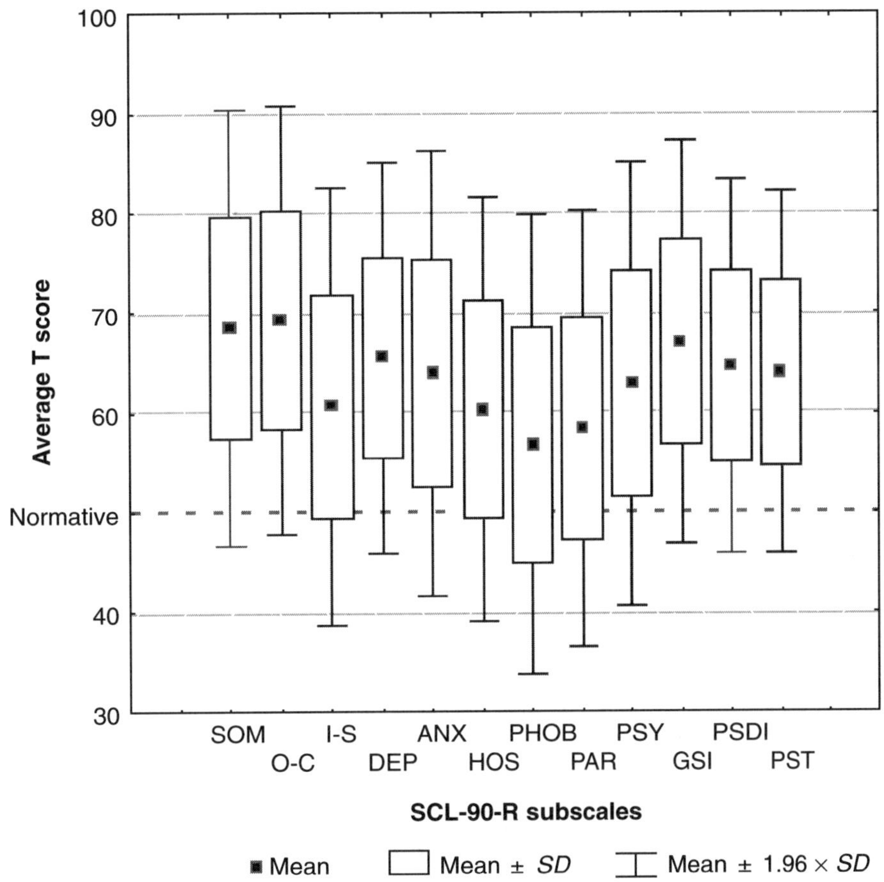

Fig. 1. Average T scores on the Symptom Checklist 90-R (SCL-90-R), demonstrating the significantly high level of psychological distress experienced by the study group as a whole. ***Notes:*** **SOM = Somatic, O-C = Obsessive-Compulsive, I-S = Interpersonal Sensitivity, DEP = Depression, ANX = Anxiety, HOS = Hostility, PHOB = Phobia, PAR = Paranoia, PSY = Psychoticism, GSI = Global Symptom Index, PSDI = Positive Symptom Distress Index, and PST = Positive Symptom Total.**

vidual items composing these factors, we numbered and labeled them as follows. *Factor 1: Melancholic Depression* accounted for 32.7% of the total variance in responses. It contained symptom items commonly associated with or reported in depressed states, such as feelings of guilt, worthlessness, hopelessness, and interpersonal distress. *Factor 2: Somatic Complaints* accounted for 6.2% of the total variance in responses. It is straightforward in name and content. *Factor 3: Cognitive Distortion* accounted for 5.5% of the total variance in responses. Described symptoms are usually associated with more severe mental problems, such as paranoia, hypersensitivity, and nervous irritability. *Factor 4: Affective Loss of Control* accounted for 4.9% of the total variance in responses. It contained items reflecting more severe symptoms of agitated depression. *Factor 5: Anxious Depression* accounted for 3.7% of the total variance in responses and contained symptoms that suggested increased arousal, agitated depression, and somatic complaints, such as sleep disturbance, fatigue, tension, nausea, "feeling blue," and worrying too much. *Factor 6: Cognitive Complaints,* which accounted for 3.4% of the total variance in responses, contained 2 items describing concerns about cognitive processes. It is noteworthy that only Factor 1 accounted for any significant amount of variance.

At the time of our analysis, item-by-item data were available for only 52 of the 106 patients who had taken the SCL-90-R. We used scores for each of the 6 factors as predictor variables in forward stepwise regression analyses to determine the amount of variance that these symptom sets accounted for in each neuropsychological testing finding and QEEG finding.

Neuropsychological testing vs. normative data. Patients' levels of completed education (if they were born after World War II) and a measure of vocabulary (from the WAIS-III subtest scaled score) are widely accepted methods for estimating premorbid levels of intellectual functioning.[34] The average level of education for the 109 adults in this sample was 14.57 yr, indicating a slightly higher than average level of premorbid functioning. This finding agreed with the average WAIS-III vocabulary scores for this sample, which was almost 1 standard deviation above normal.

We found statistically significant differences between the high average range scaled score for the group's estimate of premorbid intelligence (the Vocabulary subtest from the WAIS-III) and the group's scaled scores for performance on the D-KEFS Color-Word subtests of Color Naming, Word Reading, Inhibition, Inhibition-Switching ($p < 0.0000$), and the group's scaled scores on the D-KEFS Trail Making Test subtests of Visual Scanning, Number Sequencing, Number-Letter Switching, and Motor Speed ($p < 0.0005$) (Fig. 2). Statistically significant differences ($p < 0.05$) were also found between the scaled score for the Vocabulary subtest from the WAIS-III and the group's scaled scores for performance on other WAIS-III subtest measures of "fluid" intelligence (e.g., attention, working memory, visual motor learning, speed, and visual scanning).

The IVA-CPT, which tests response inhibition and the ability to maintain attention over time, has a normative population mean of 100 and a standard deviation of 15. Figure 3 summarizes the results of the normative scores on the IVA-CPT. All scores were below expectations, given the group's estimated level of intelligence on the WAIS-III Vocabulary subtest and the average level of education. Attention scores were especially depressed. We also found dissociation between response control and attention across both auditory and visual domains of the IVA-CPT. Attentional functioning in this group was significantly impaired ($p < 0.0126$) compared with response inhibition capacities.

QEEG measures vs. normative data. Without regard to any categorical predictors, the group as a whole exhibited a definite pattern of slowing in the faster beta frequency ($F(18, 2{,}574) = 2.8707$; $p = 0.00005$) and acceleration of the slower delta ($F(18, 2{,}592) = 2.9554$, $p = 0.00003$) and theta ($F(18, 2{,}592) = 3.1680$; $p = 0.00001$) frequencies. The QEEG summary variables for these measures identified frontal cortical positions as the most significantly deviant (mean frequency delta: Fp1 Fp2 F3 F4 F8; mean frequency theta: Fp1 Fp2 F3 F4 C4; mean frequency beta: Fp1 Fp2 F7 F8 T3).

There were significant increases in absolute ($F(18, 2{,}538) = 12.007$; $p = 0.0000$) and relative ($F(18, 2{,}520) = 9.5775$; $p = 0.0000$) power alpha, and in absolute ($F(18, 2{,}538) = 6.5807$; $p = 0.00000$) and relative ($F(18, 2{,}520) = 3.4143$; $p = 0.00000$) power theta. Again, the QEEG summary variables identified the frontal cortical

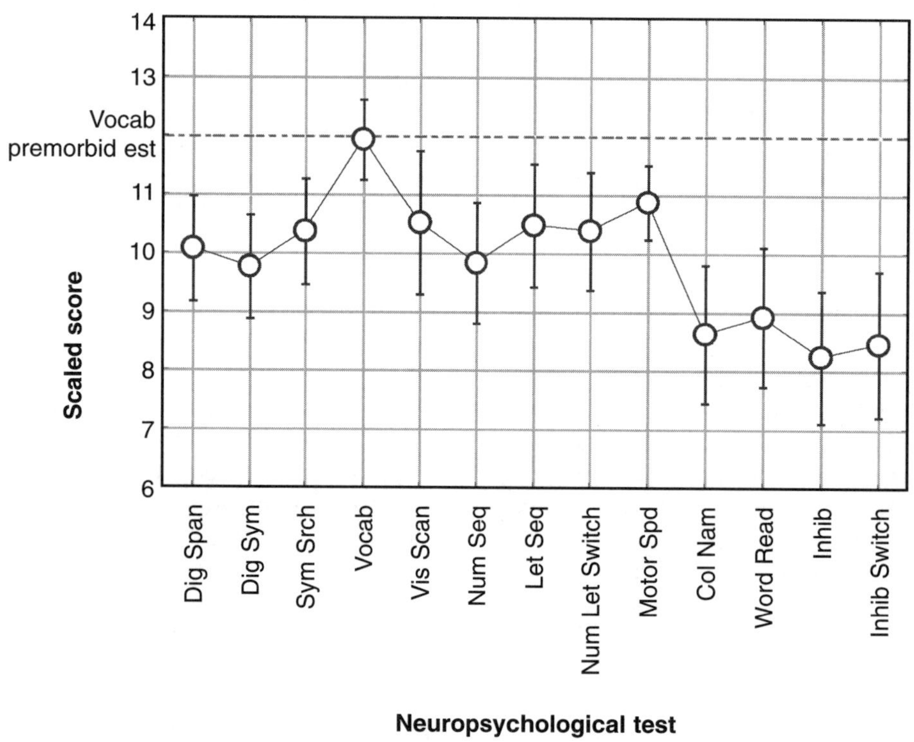

Fig. 2. Results of neuropsychological testing, compared with normative data. Premorbid intelligence, estimated with the Wechsler Adult Intelligence Scale, 3rd ed. (WAIS III) Vocabulary subtest, was compared with results for WIAS III subtests (Digit Span, Digit Symbol Coding, and Symbol Search); Delis-Kaplan Executive Function System (D-KEFS) Trail Making subtests (Visual Scanning, Number Sequencing. Letter Sequencing, Number-Letter Switching, and Motor Speed); and D-KEFS Color Word subtests (Word Reading, Inhibition, and Inhibition Switching).

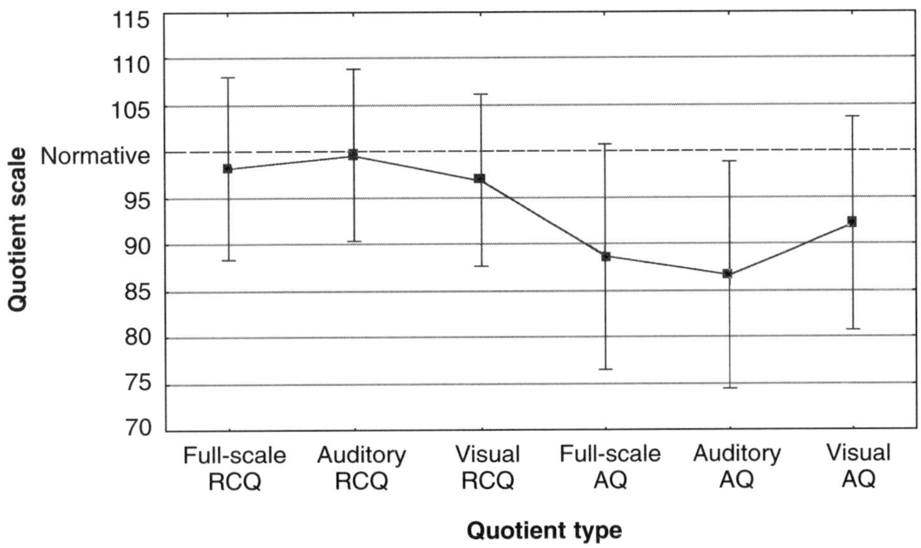

Fig. 3. Normative scores, by analysis of variance, on the Integrated Visual and Auditory Continuous Performance Test of response inhibition and attention. All scores were below expectations given the group's intelligence levels (estimated with the Wechsler Adult Intelligence Scale, 3rd ed. [WAIS III] Vocabulary subtest) and average level of education. Attention Quotient scores were significantly lower than Response Control Quotient scores. *Notes:* $F(5, 410) = 2.9448$, $p = 0.01263$. Vertical bars denote 95% confidence intervals. RCQ = Response Control Quotient, and AQ = Attention Quotient.

areas as being the most deviant (absolute alpha: Fp1 Fp2 F4 F8 Fz; relative alpha: Fp1 Fp2 F4 Fz Cz; absolute theta: F3 F4 F7 F8 Fz; relative theta = Fp1 Fp2 F7 F8 F4).

The absolute power *Z* scores in the delta frequency band across the scalp were significantly decreased relative to the normative values ($F(18, 2{,}538) = 2.4576$; $p = 0.00058$). Specifically, the decreases were most notable in the prefrontal, parietal, and occipital areas (i.e., Fp1, Fp2, Cz, P3, P4, Pz, O1, and O2)

Exposure measures vs. SCL-90-R, QEEG, and cognitive measures. Measures of exposure were highly predictive of neuropsychological test performance, moderately predictive of QEEG measures, and only slightly predictive of measures of subjective stress on the SCL-90-R.

There were predictive, nonsignificant trends for performance on the D-KEFS Trail Making Test and the Color-Naming subtest (in which maximum exposure, average CFUs, and whether *Stachybotrys* was present in the environment yielded an adjusted multiple $R^2 = .240$, $p < 0.070$); the D-KEFS Word Reading subtest (in which maximum exposure and whether *Stachybotrys* was present in the environment yielded an adjusted multiple $R^2 = .220$, $p < 0.053$); and the D-KEFS Color-Word Inhibition (classic Stroop) subtest (in which maximum exposure, average CFUs, and whether *Stachybotrys* was present in the environment yielded an adjusted multiple $R^2 = .235$, $p < 0.073$).

Table 1 shows the significant predictive power ($p < 0.05$) observed for the D-KEFS Trail Making subtests of visual scanning, letter sequencing, number-letter sequencing, and motor speed; the D-KEFS Color-Word Inhibition/Switching subtest; the WAIS-III Digit Symbol Coding and Symbol Search subtests; and the IVA-CPT full-scale attention quotient and the visual and auditory attention quotients.

It should be kept in mind in reviewing these findings that the sample sizes are smaller than those for other analyses reported herein, which can affect the degrees of freedom and the variability accounted for by these scores. Although some of the values reported in Table 1 are quite high (e.g., WAIS-III Symbol Search subtest adjusted multiple $R^2 = .787$), it is still safe to conclude that measures of exposure account for a significant amount of the variance in neuropsychological test performance on the basis of these findings.

Table 2 shows that 6 of the 8 QEEG summary variables were related significantly to measures of mold exposure. Four of the 6 variables were measures of power in the theta and alpha frequencies. Significant predictive power was found for estimates of degree of exposure and for the QEEG variables of mean frequency delta, relative power theta, relative power alpha, absolute power delta, absolute power theta, and absolute power alpha.

Exposure measures predicted scores on only 2 of the SCL-90-R factors. The only statistically significant relationship was found between CFU × maximum exposure and a combination of SCL-90-R Factor 3 (Cognitive Distortion) and Factor 6 (Cognitive Complaints), which together yielded an adjusted multiple R^2 of 0.291, $p <$

Table 1.—Relationships between Measures of Exposure and Performance on Neuropsychological Tests

Test	*n*	Variables in model	Significant β weights*	Adjusted multiple R^2	*p*
D-KEFS Trail Making Test: Visual Scanning	19	Hours of exposure, days of exposure, maximum exposure, average CFU, *Stachybotrys* present	Hours of exposure = 0.836 Days of exposure = 3.32	. 350	0.054
D-KEFS Trail Making Test: Letter Sequencing	19	Hours of exposure, days of exposure, maximum exposure, CFU × maximum exposure	Days of exposure = 4.99 Maximum exposure = –4.6	.326	0.047
D-KEFS Trail Making Test: Number-Letter Switching	19	Hours of exposure, days of exposure, maximum exposure, CFU × maximum exposure, *Stachybotrys* present	Hours of exposure =0 .997 Days of exposure = 5.12 Maximum exposure = -4.6	.399	0.035
D-KEFS Trail Making Test: Motor Speed	19	Hours of exposure, days of exposure, maximum exposure, average CFU, CFU × maximum exposure, *Stachybotrys* present	*Stachybotrys* present = 0.963 Days of exposure = 5.08 Maximum exposure = –4.7 CFU × maximum exposure = 0.626 Hours of exposure = 0.642	.705	0.001
D-KEFS Color-Word Inhibition: Switching	18	*Stachybotrys* present, maximum exposure	*Stachybotrys* present = 0.623 Maximum exposure = 0.546	.298	0.028
WAIS-III: Digit Symbol Coding	36	Hours of exposure, days of exposure	Hours of exposure = –0.47 Days of exposure = 0.325	.259	0.002
WAIS-III: Symbol Search	15	Hours of exposure, maximum exposure, average CFU, CFU × maximum exposure	Hours of exposure = –0.64 Maximum exposure = 0.598 Average CFU = 0.804 CFU × maximum exposure = –0.71	.787	0.0004
IVA-CPT: Full-Scale Attention Quotient	30	Maximum exposure, average CFU, CFU × maximum exposure	Maximum exposure = 0.399 Average CFU = –0.40	.300	0.006
IIVA-CPT: Visual Attention Quotient		Maximum exposure, average CFU, CFU × maximum exposure	Average CFU = –0.41	.276	0.009
IIVA-CPT: Auditory Attention Quotient		Maximum exposure, average CFU	Maximum exposure = 0.469	.278	0.004

Notes: The fluctuating sample sizes reflect the fact that the statistics program selected only cases from the chart review for which all data in the model were present. CFU = colony-forming unit.
*$p < 0.05$.

Table 2.—Predictive Power for Estimates of Degree of Exposure and Quantitative Electroencephalogram (QEEG) Variables

QEEG variable	*n*	Variables in model	Significant β weights*	Adjusted multiple R^2	*p*
Mean frequency delta	62	Average CFU	Average CFU = –0.29	.069	0.022
Relative theta	62	CFU × maximum exposure, *Stachybotrys* present, days of exposure, average CFU	None	.096	0.043
Relative alpha	62	CFU × maximum exposure, *Stachybotrys present*, hours/day of exposure, maximum exposure	CFU × maximum exposure = 0.344	.124	0.020
Absolute delta	62	*Stachybotrys* present	*Stachybotrys* present = 0.270	.057	0.034
Absolute theta	62	*Stachybotrys* present	*Stachybotrys* present = 0.255	.049	0.046
Absolute alpha	62	Average CFU, CFU × maximum exposure	None	.086	0.026

Note: CFU = colony-forming unit.
*$p < 0.05$.

0.0124. That is, 29.1% of the cognitive distortion and cognitive complaints were predicted by the variance observed in CFU × maximum exposure. A nonsignificant trend was found between hours of exposure and Factor 3 (Cognitive Distortion), which yielded an adjusted R^2 of 0.064, $p < 0.096$.

QEEG and neuropsychological test performance. We observed several statistically significant relationships between the QEEG summary variables and performance on neuropsychological tests. The data in Table 3 demonstrate that significant QEEG predictors of test performance were found for measures of number and letter sequencing, number-letter switching, motor speed, response inhibition, and visual attention. The specific tests were the D-KEFS Trail Making Number Sequencing, Letter Sequencing, Number-Letter Switching, and Motor Speed subtests; the D-KEFS Color-Word Inhibition subtest; and the IVA-CPT Visual Attention Quotient. QEEG variables showing the greatest predictive power were those that involved the theta or alpha frequency bands and, most often, the mean frequency theta or relative power theta summary variables.

Minimal QEEG predictors of test performance or nonsignificant trends ($p < 0.10$ through $p < 0.051$) were found for measures of visual scanning, color naming, attention, and response inhibition (WAIS-III Symbol Search subtest, D-KEFS Color-Word Color Naming subtest, IVA-CPT Attention Quotient, IVA-CPT Response Control Quotient, and IVA-CPT Auditory Response Control Quotient).

SCL-90-R and cognitive performance. We could address only partially the mitigating influence of psychological factors on cognitive test performance in this study. However, sufficient cases were available to compare the results of the SCL-90-R with Digit Symbol Coding and the results of the IVA-CPT scales.

The SCL-90-R scales and factors did not predict performance on Digit Symbol Coding (adjusted R^2 = .002). The SCL-90-R scales of depression and anxiety were found to have the opposite effect on the IVA-CPT Attention Quotient, even though these SCL-90-R scales correlate in this data set at R = .81.

We observed a significant relationship for Factor 1 (Melancholic Depression) and Factor 5 (Anxious Depression) with the IVA-CPT full-scale Attention Quotient (adjusted R^2 = .241, $p < 0.04$), Auditory Attention Quotient (adjusted R^2 = .200, $p < 0.06$), and Visual Attention Quotient (adjusted R^2 = .305, $p < 0.02$). In all of these cases, Factor 1 had a significant *negative* beta weight, whereas Factor 5 had a significant *positive* beta weight. Increases in Factor 1 were correlated with decreases in the attention quotient scores, whereas increases in Fac-

Table 3.—Quantitative Electroencephalogram (QEEG) Predictors of Neuropsychological Test Performance

Test	n	QEEG variables in model	Significant β weights*	Adjusted multiple R^2	p
D-KEFS Trail Making Test: Number Sequencing	65	Mean frequency theta, absolute alpha, mean frequency beta	Mean frequency theta = 0.571 Absolute alpha = –0.44	.220	0.0004
D-KEFS Trail Making Test: Letter Sequencing	65	Mean frequency theta, absolute alpha, mean frequency beta	Mean frequency theta = 0.503 Absolute alpha = –0.43	.169	0.002
D-KEFS Trail Making Test: Number-Letter Switching	65	Absolute delta, mean frequency theta, mean frequency beta, relative alpha, absolute theta, absolute alpha, relative theta	Mean frequency theta = 0.502 Relative alpha = –1.1 Absolute theta = –1.9 Absolute alpha = 1.83 Relative theta = 0.886	.193	0.007
D-KEFS Trail Making Test: Motor Speed	65	Absolute delta, relative theta, mean frequency delta	Absolute delta = –0.31 Relative theta = 0.271	.110	0.017
D-KEFS Color-Word Inhibition	65	Mean frequency theta, absolute theta, relative theta, absolute delta, a bsolute alpha, relative alpha	Mean frequency theta = 0.560 Absolute theta = –3.2 Relative theta = 1.46 Absolute delta = 0.471 Absolute alpha = 2.91 Relative alpha = –1.6	.348	0.00002
IVA-CPT Visual Attention Quotient	98	Absolute alpha, relative alpha, absolute theta, relative theta, mean frequency theta, mean frequency delta, absolute delta	Mean frequency theta = 0.419 Absolute theta = –2.0 Relative theta = 0.904 Absolute alpha = 1.62 Relative alpha = –0.82	.137	0.005

*$p < 0.05$.

tor 5 were correlated with increases in the attention score. The sample size available for each analysis (n = 19) made these findings tentative at best.

SCL-90-R and QEEG. We also explored the effect of stress on cortical electrical activity. A significant relationship was found between only 2 of the 12 QEEG summary variables and SCL-90-R factors that accounted for small amounts of the variance on item analysis.

A positive beta weight was observed between relative power theta and Factor 6 (Cognitive Complaints), whereas a negative beta weight was seen between relative power theta and Factor 5 (Anxious Depression); adjusted multiple $R^2 = .107$, $p < 0.025$. That is, patients' reports of concern regarding cognitive abilities (Factor 6) were related positively to increasing relative power theta—often an indication of decreased cognitive arousal or efficiency. This can be interpreted as consistent with the negative relationship observed for Factor 5, in which symptoms of increased arousal, tension, and anxiety can indicate increased cortical arousal and vigilance and, therefore, decreased relative power theta.

Significant correlation (adjusted multiple $R^2 = .112$, $p < 0.052$) was also found between the mean frequency beta summary score—a marker associated with cognitive efficiency (faster = greater efficiency; slower = lesser efficiency)—and SCL-90-R Factor 4 (Affective Loss of Control) and Factor 5 (Anxious Depression). Factor 3 (Cognitive Distortion) and Factor 6 (Cognitive Complaints) showed negative relationships with this cortical indicator.

Our results suggested that, to a limited extent, increased frontal cortical arousal (as indicated by decreased relative power theta and increased mean frequency beta) was associated with increased anxiety and less complaints of cognitive deficits. Furthermore, decreased frontal cortical arousal, as indicated by increased relative power theta and decreased mean frequency beta, was associated with increased cognitive complaints and decreased anxiety.

Discussion

Psychological distress. Patients—including multiple family members—exposed to toxic molds reported moderate to severe levels of psychological distress related to the development of a wide range of physical, cognitive, and emotional symptoms. Problems included the frustration of trying to find knowledgeable and appropriate medical care, interference with social and work life, temporary or permanent abandonment of homes and possessions, financial stress, and anxiety and helplessness as a result of continuing poor health.

Most of these patients, in absence of any significant premorbid psychiatric problems, could be diagnosed as suffering from acute stress, adjustment disorder, or post-traumatic stress. *Only 3.8% of our sample population reported significant premorbid psychiatric or neuropsychiatric problems* (e.g., history of major depression, post-traumatic stress disorder, seizure disorder, closed head injury). Individuals with significant historical problems were eliminated from the data analysis in order to minimize the effect of such problems in this patient population.

The patients in this study showed a significant level of psychological stress, with depression being the only factor to account for a substantial amount of variance in the reported symptoms. There were limited significant relationships between some of the SCL-90-R factors and QEEG findings. Although a strong relationship was found between the average number of CFUs present in the environment and the subjectively reported cognitive difficulties on 2 of the SCL-90-R factors, it must be remembered that these 2 factors accounted for only 8.9% of the total variance on the item analysis. The process of differential diagnosis supports the conclusion that the individuals in this study suffered severe psychological distress resulting from a combination of overwhelming personal stress and poor health, with the mitigating influence of organically based central nervous system deregulation of emotions as a result of toxic or metabolic encephalopathy.

At this point, it is difficult to say what the cost of these deficits might be in terms of productivity or personal relationships, although it would clearly be significant, to a reasonable medical and scientific certainty, if the distress experienced by the patients in this sample is any indicator.

One limitation in our study was that patient reports could not be validated independently. We considered the reports to be credible, however, given that the patients were interviewed individually and completed a questionnaire; there was little motive to deceive in this clinical situation. Another measure of credibility is the fact that patients who were involved in litigation did not report more symptoms than nonlitigants.[19]

Neuropsychological testing. The results of the neuropsychological testing produced positive findings for impaired cognitive functioning on a wide variety of tasks, when compared with premorbid estimates of intelligence. The pattern and severity of results are similar to, and corroborate, the results obtained by other researchers who have conducted preliminary studies of cognitive functioning in toxic-mold–exposed individuals.[18,20–22] This pattern is also similar to that for individuals diagnosed with mild traumatic brain injury.[21]

Prior researchers have reported that symptoms of depression can impair neuropsychological test performance.[21,35,36] Baldo[21] assessed cognitive functioning and depression in a small group of 10 mold-exposed patients and found a significant relationship between cognitive impairment and depression. In the current study, in agreement with previous research, both anxiety and depression were indicative of stress, but only depression had a detrimental effect on attention. One interpretation may be that symptoms of anxiety are associated with higher levels of vigilance, whereas symptoms of melancholic depression are associated with lower arousal and mental sluggishness. However, this preliminary and limited analysis does not offer conclusive evidence regarding the effects of psychological distress on neuropsychological test performance in this sample. The limitation of incomplete data sets for all of the neuropsychological tests prevents a clear interpretation at this time.

We acknowledge the complex relationships among toxic mold exposure, impaired cognition, psychosocial stressors, poor physical health, and emotional factors. Our overall findings, however, lend support to the hypothesis that patients' cognitive deficits are frequently related to underlying organic deficits caused by toxic mold exposure. Most critical for this study was the significant and consistent effect of mold-exposure measures on the results of cognitive testing. Also, abnormalities in cortical electrical activity—primarily in the frontal and prefrontal lobes—were significantly and consistently related to deficits in cognitive functioning and mold-exposure measures. An additional factor that suggests that the observed cognitive deficits were of organic origin is the lack of variability of deficits between tests usually most influenced by psychological factors (e.g., measures of attention or working memory) and tests not usually influenced by psychological factors (e.g., color naming or word reading). Corroboration was also found in a study that examined the influence of personality traits on neuropsychological test performance in toxic encephalopathy cases vs. healthy referent cases. Persson et al.[37] concluded that the neuropsychological performance decrements in toxic encephalopathy cases were not related to elevated mental stress, but were dominated by the effects of organic brain impairment.

It is important to address the fact that most of the patients in this study subjectively reported moderate to severe ratings of cognitive impairment, rather than mild to moderate as measured by the testing. The pattern of deficits commonly seen in mild traumatic brain injury is very similar to that found in mold-exposed individuals.[20,21,27] This phenomenon—clinically referred to as "brain fog"—is also common in individuals who suffer from multiple chemical sensitivity.[38] Patients reported a loss of their sense of self, of their usual ways of doing things, and even of their personality. They were painfully aware of their deficits and were constantly frustrated by their loss of cognitive efficiency and frequent mistakes. This can be understood as a disturbance or dysfunction of the frontal cortical areas, as implicated in the QEEG findings and the relationship of exposure data to test performance in this study. In humans, the sense of self is organized in the frontal brain areas.[39] For these reasons, we recommend that studies or clinical evaluations of cognitive functioning in mold-exposed patients employ functional imaging techniques to assess organic dysfunction.

QEEG. The results of the QEEG data recorded from mold-exposed patients indicate a restriction in the range of functioning (narrowed frequency bands) of the frontal lobes, that is, increased (accelerated) mean frequency of the slower frequencies (delta and theta) and decreased (slowed) higher frequencies (beta). These changes indicate a collapse toward the middle of the frequency spectrum. Such findings, coupled with the increased levels of absolute and relative power theta and alpha in the frontal sites, indicate a hypoactivation of the frontal cortex. The latter may result from brainstem involvement and may indicate insufficient excitatory input from the reticular activating system anatomically seated in the midbrain. Deviant QEEG findings of this magnitude should not have been observed in frontal lobe functioning without some insult to the functioning of the neural systems that depend on integrative coordination from the frontal lobes. These findings are consistent with other functional imaging studies mentioned earlier.[1,2,3,24,25]

Measures of toxic mold exposure were related significantly to QEEG findings, and both measures of exposure and QEEG measures were related significantly to cognitive test performance. Psychological factors appeared to have only a limited relationship to QEEG results, reflecting the arousal level of the frontal lobes. This finding supports the conclusion that exposure to toxic mold results in central nervous system dysfunction, as measured by QEEG.

The use of QEEG and neurometrics in research and clinical practice has been the subject of some controversy, although recent opinions and evidence describe more strengths than weaknesses.[32,40–43] Our decision to use QEEG and neurometrics included the facts that they are noninvasive, relatively inexpensive, free from cultural and ethnic factors, have good reliability and validity, and are highly sensitive for detecting dementias and encephalopathy.[44,45]

The most significant limitation of the NxLink database is its exclusive reliance on banded EEG. Findings restricted to narrow frequencies are obscured with the use of relatively wide bands normed in the database. Plans for future investigations include examination of other indices of QEEG activity, such as coherence and phase lag relationships, and examination of the data using other databases that allow single-frequency analysis.

Summary and Conclusions

Patients exposed to toxic molds reported high levels of physical, cognitive, and emotional symptoms. Ratings on the SCL-90-R were moderate to severe, with a factor reflecting depression accounting for most of the variance. Most patients could be diagnosed as suffering from problems of acute stress, adjustment disorder, or post-traumatic stress. Impaired cognitive functioning was observed on multiple cognitive tasks, compared with premorbid estimates of intelligence, in a pattern of impairment similar to that for mild traumatic brain injury. The QEEG findings indicated a hypoactivation of the frontal cortex, suggesting brainstem involvement and insufficient excitatory input from the reticular activating system. QEEG measures were correlated with neuropsychological test results.

Findings of a dose-response relationship between measures of exposure and the outcome of neuropsychological tests and QEEG measures suggest that evaluation of neuropsychological and neurobehavioral deficits in mold-exposed patients should consider degree of exposure, organic-based central nervous system dysfunction, and psychological variables. Differential diagnosis supported an etiology of organic-based dysregulation of emotions and cognitive functioning as a result of toxic or metabolic encephalopathy, with some degree of mitigation by psychological variables, especially depression.

Additional work is needed to examine the effect of the length of time since last exposure to toxic mold on outcome measures. Patients have reported some reduction in symptoms when they have been removed from continued exposure. How much of a reduction, under what circumstances, and for which patients have not been fully determined.

Limitations of this study included lack of a comparison group that underwent the same testing as the mold patients, and small sample size. The lack of a comparison group was mitigated by the fact that the measures used in this study were all compared with the published normative databases developed for each of the tests. With respect to sample size, the fact that the analyses were performed on data gathered from chart review resulted in inconsistent sample sizes and the variable sample sizes reported in some of the stepwise regression analyses and descriptive statistics. For this reason, the relative contribution to the variance accounted for by any given variable can be compared only with the relative contributions of variables for which a similar number of cases were available.

Future research will include expanded use of neuropsychological testing, QEEG measures, exposure measures, correlations with immune parameters, an increase in sample size, and more complete data analysis. The use of causal modeling and path analysis may improve the interpretability of the results, allowing the multidirectional relationships that exist within this complex topic to be modeled in several different ways.

* * * * * * * * * *

Submitted for publication October 15, 2003; revised; accepted for publication May 14, 2004.

Requests for reprints should be sent to B. Robert Crago, Ph.D., Neurobehavioral Health Services, 5363 E. Pima Street, Suite 100, Tucson, AZ 85712.

E-mail: bcbrain1@msn.com

* * * * * * * * * *

References

1. Van Sweden B, Niedermeyer E. Toxic encephalopathies. In: Niedermeyer E, Lopes DaSilva F (Eds). Electroencephalography: Basic Principles, Clinical Applications, and Related Fields. Baltimore, MD: Lippincott Williams and Wilkins, 1993; pp 643–51.
2. Anyanwu E, Campbell AW, High W. Brainstem auditory evoked response in adolescents with acoustic mycotic neuroma due to environmental exposure to toxic molds. Int J Adolesc Med Health 2002; 14:67–76.
3. Anyanwu EC, Campbell AW, Vojdani A. Neuropsychological effects of chronic indoor environmental toxic mold exposure on children. Scientific World Journal 2003; 3:281–90.
4. Croft WA, Jarvis BB, Yatawara CS. Airborne outbreak of trichothecene toxicosis. Atmos Environ 1986; 20:549–52.
5. Johanning E, Biagini R, Hull DL, et al. Health and immunology study following exposure to toxigenic fungi (*Stachybotrys chartarum*) in a water-damaged office environment. Int Arch Occup Environ Health 1996; 68: 207–18.
6. Gunnbjornsdottir MI, Norback D, Paschke P, et al. The relationship between indicators of building dampness and respiratory health in young Swedish adults. Respir Med 2003; 97:301–07.
7. Savilahti R, Uitti J, Laippala P, et al. Respiratory morbidity among children following renovation of water-damaged school. Arch Environ Health 2000; 55:405–10.
8. Ebboj NE, Hansen MO, Sigaard T, et al. Building-related symptoms and molds: a two-step intervention study. Indoor Air 2002; 12: 272–77.
9. Seuri M, Husman K, Kinnunen H, et al. An outbreak of respiratory diseases among workers at a water-damaged building—a case report. Indoor Air 2000; 10:138–45.
10. Flannigan B, McCabe EM, McGarry F. Allergenic and toxigenic microorganisms in houses. J Appl Bact Sym 1991; 70:61–73.
11. Jaakkola M, Nordman H, Pilpari R, et al. Indoor dampness and molds and development of adult-onset asthma: a population-based incident case-control study. Environ Health Perspect 2002; 110:543–47.
12. Kurup V, Shen HD, Banerjee B. Respiratory fungal allergy. Microbes Infect 2000; 2:1101–10.
13. Zureik M, Neukirch C, Leynaert B, et al. Sensitisation to airborne moulds and severity of asthma: cross sectional study from European Community respiratory health survey. Br Med J 2002; 325:411–14.
14. Hodgson, MJ, Morey P, Leung WY, et al. Building-associated pulmonary disease from exposure to *Stachybotrys chartarum* and *Aspergillus versicolor*. J Occup Environ Med 1998; 40(3):241–49.
15. Patel AM, Ryu JH, Reed CE. Hypersensitivity pneumonitis: current concepts and future questions. J Allergy Clin Immunol 2001; 108:661–70.
16. Fan LL. Hypersensitivity pneumonitis in children. Curr Opin Pediatr 2002; 14:323–26.

17. Croft WA, Jastromski BM, Croft AL, et al. Clinical confirmation of trichothecene mycotoxicosis in patient urine. J Environ Biol 2002; 23:301–20.
18. Kilburn KH. Inhalation of molds and mycotoxins. Eur J Oncol 2002; 7:197–202.
19. Gray MR, Thrasher JD, Crago R, et al. Mixed mold mycotoxicosis: immunological changes in humans following exposure in water-damaged buildings. Arch Environ Health 2003; 58(7):410–20.
20. Gordon WA, Johanning E, Haddad L. Cognitive impairment associated with exposure to toxigenic fungi, health effects. In: Johanning E (Ed). Bioaerosols, Fungi and Mycotoxins: Health Effects, Assessment, Prevention and Control. Albany, NY: Eastern New York Occupational and Environmental Health Center, 1999; pp 94–98.
21. Baldo JV. Neuropsychological performance of patients following mold exposure. Appl Neuropsychol 2002; 9(4): 193–202.
22. Auger P, Pepin P, Miller JD, et al. Health effects, pathology, epidemiology. In: Johanning E (Ed). Bioaerosols, Fungi and Mycotoxins: Health Effects, Assessment, Prevention and Control. Albany, NY: Eastern New York Occupational and Environmental Health Center, 2001; pp 131–38.
23. Hartmann DE. Neuropsychological Toxicity: Identification and Assessment of Human Neurotoxic Syndromes. 2nd ed. New York: Plenum Press, 1995.
24. Heuser G, Mena I. Neuro-SPECT in neurotoxic chemical exposure demonstration of long-term functional abnormalities. Toxicol Ind Health 1998; 14:813–27.
25. Callender TJ, Morrow L, Subramanian K, et al. Three dimensional metabolic imaging in patients with toxic encephalopathy. Environ Res 1993; 60:295–319.
26. Derogatis LR. SCL-90-R: Administration, Scoring and Procedures Manual. 3rd ed. Bloomington, MN: National Computer Systems, Inc., 1994.
27. Wechsler Adult Intelligence Scale. 3rd ed. San Antonio, TX: The Psychological Corporation, 1997.
28. Delis DC, Kaplan E, Kramer JH. Delis-Kaplan Executive Function System. San Antonio, TX: The Psychological Corporation, 2001.
29. Sanford J. The Intermediate Visual and Auditory Continuous Performance Test. Richmond, VA: Braintrain, 1995.
30. John ER. Functional Neuroscience. Vol 2. Neurometrics: Clinical Applications of Quantitative Electrophysiology. Hillsdale, NJ: Lawrence Erlbaum, 1977.
31. NxLink Neurometric Analysis System. Richmond, WA: NxLink Ltd, 2001.
32. Johnstone J, Gunkelman J. Use of databases in QEEG evaluation. J Neurother 2003; 7:31–52.
33. Statistica 6.0 for Windows. Tulsa, OK: Statsoft, Inc., 1984–2002.
34. Hebben N, Milberg W. Essentials of Neuropsychological Assessment. New York: Wiley, 2002.
35. Austin M, Mitchell P, Goodwin GM. Cognitive deficits in depression: possible implications for functional neuropathology. Br J Psychiatry 2001; 178:200–06.
36. Elliot C, Greene RL. Clinical depression and implicit memory. J Abnorm Psychol 1992; 101:572–74.
37. Persson R, Osterberg K, Karlson B, et al. Influence of personality traits on neuropsychological test performance in toxic encephalopathy cases and healthy referent subjects. Neurotoxicology 2000; 5:667–75.
38. Ziem G, McTamney J. Profile of patients with chemical injury and sensitivity. Environ Health Perspect 1997; 105: 417–36.
39. Schore AN. Affect Regulation and the Origin of the Self: The Neurobiology of Emotional Development. Hillsdale, NJ: Lawrence Erlbaum, 1994.
40. Nuwer M. Assessment of digital EEG, quantitative EEG, and EEG brain mapping: report to the American Academy of Neurology and American Clinical Neurophysiology Society. Neurology 1997; 49:277–92.
41. Hoffman DA, Lubar JF, Thatcher RW, et al. Limitations of the American Academy of Neurology and American Clinical Neurophysiology Society paper on QEEG. J Neuropsychiatry Clin Neurosci 1999;11(3):401–07.
42. Thatcher RW, Moore N, John ER, et al. QEEG and traumatic brain injury: rebuttal of the American Academy of Neurology 1997 report by the EEG and Clinical Neuroscience Society. Clin Electroencephalogr 1999; 30:94–98.
43. Thatcher RW, Biver CJ, North DM. Quantitative EEG and the Fry and Daubert Standards of Admissibility. Clin Electroencephalogr 2003; 34:39–53.
44. Hughes JR, John ER. Conventional and quantitative electroencephalography in psychiatry. J Neuropsychiatry Clin Neurosci 1999; 11:190–208.
45. Rosen I. Electroencephalography as a diagnostic tool in dementia. Dement Geriatr Cogn Disord 1997; 81:10–16.

Neural Autoantibodies and Neurophysiologic Abnormalities in Patients Exposed to Molds in Water-Damaged Buildings

ANDREW W. CAMPBELL
Medical Center for Immune and Toxic Disorders
Spring, Texas
JACK D. THRASHER
Sam-1 Trust
Alto, New Mexico
ROBERTA A. MADISON
Department of Health Sciences
California State University
Northridge, California

ARISTO VOJDANI
Immunosciences Lab., Inc.
Beverly Hills, California
MICHAEL R. GRAY
Progressive Healthcare Group
Benson, Arizona
AL JOHNSON
Integrative Neurology
Richardson, Texas

ABSTRACT. Adverse health effects of fungal bioaerosols on occupants of water-damaged homes and other buildings have been reported. Recently, it has been suggested that mold exposure causes neurological injury. The authors investigated neurological antibodies and neurophysiological abnormalities in patients exposed to molds at home who developed symptoms of peripheral neuropathy (i.e., numbness, tingling, tremors, and muscle weakness in the extremities). Serum samples were collected and analyzed with the enzyme-linked immunosorbent assay (ELISA) technique for antibodies to myelin basic protein, myelin-associated glycoprotein, ganglioside GM_1, sulfatide, myelin oligodendrocyte glycoprotein, α-B-crystallin, chondroitin sulfate, tubulin, and neurofilament. Antibodies to molds and mycotoxins were also determined with ELISA, as reported previously. Neurophysiologic evaluations for latency, amplitude, and velocity were performed on 4 motor nerves (median, ulnar, peroneal, and tibial), and for latency and amplitude on 3 sensory nerves (median, ulnar, and sural). Patients with documented, measured exposure to molds had elevated titers of antibodies (immunoglobulin [Ig]A, IgM, and IgG) to neural-specific antigens. Nerve conduction studies revealed 4 patient groupings: (1) mixed sensory–motor polyneuropathy (n = 55, abnormal), (2) motor neuropathy (n = 17, abnormal), (3) sensory neuropathy (n = 27, abnormal), and (4) those with symptoms but no neurophysiological abnormalities (n = 20, normal controls). All groups showed significantly increased autoantibody titers for all isotypes (IgA, IgM, and IgG) of antibodies to neural antigens when compared with 500 healthy controls. Groups 1 through 3 also exhibited abnormal neurophysiologic findings. The authors concluded that exposure to molds in water-damaged buildings increased the risk for development of neural autoantibodies, peripheral neuropathy, and neurophysiologic abnormalities in exposed individuals.

<Key words: mold exposure, mycotoxins, neural antibodies, neuropathy, neurophysiology>

WATER INTRUSION into houses and office buildings leads to the growth of molds and bacteria, which are known to produce toxic byproducts that include endotoxins (lipopolysaccharides), β-D-glucans, and mycotoxins (e.g., trichothecenes, ochratoxins, and aflatoxins, tremorgens), as well as volatile organic compounds.[1–8] These compounds have been found in water-damaged buildings and homes, and in artificially infested building materials.[7–11] Indoor air can be contaminated with mold spores and hyphae fragments.[12] In addition, mycotoxins have been identified in ventilation duct particulate matter or dust, and in the air of buildings in which occupants and pets have experienced adverse health effects related to mold exposure.[13–23]

Molds and mycotoxins affect the respiratory tract, kidneys, liver, and skin, as well as the immune and nervous systems.[24–45] Neurotoxic mycotoxins include ergot alkaloids, trichothecenes, citreovirdin, patulin, fumonisins, and tremorgens.[46–54] Tremorgens affect the brainstem[46] and stellate ganglion, and the basket and Purkinje cells of the cerebellum.[47] Mycotoxins affect neuroreceptor sites (e.g., gamma-aminobutyric acid [GABA] receptor site[48] and inositol 1,4,5-trisphophate receptor[49]), inhibit acetylcholinesterase,[50] release excitatory neurotransmitters (e.g., glutamate, aspartate, GABA, and serotonin),[51,52] and block biosynthesis of complex sphingolipids through inhibition of ceramide synthase.[53,54] They are also mitochondrial toxins and apoptotic agents.[43,44,53]

The symptoms and health problems associated with mold-infested, water-damaged buildings involve multiple organs, including the upper and lower respiratory tracts, gastrointestinal tract, circulatory system, and the central nervous system (CNS) and peripheral nervous system (PNS).[2,5,24–37] Recent studies have shown that mold exposure has adverse effects on the nervous system. Some mycotoxins have been shown to be tremorgenic and are suspected as causative agents in woodtrimmer's disease[20] and tremorgenic encephalopathy;[21] mycotoxins present in household environments have been found to affect dogs.[22,23]

Two patterns of neurobehavioral impairment attributable to mold exposure have been described. Kilburn[26] reported on 10 individuals who had impaired balance, reaction time, color discrimination, visual fields, cognition, verbal recall, and trail making. A different group of 10 subjects exhibited impairments in all but measures of color discrimination and visual fields. Abnormalities in electroencephalograph (EEG) theta and delta activity, visual evoked potentials, and brainstem evoked potentials have been reported in children exposed to molds.[27] The EEG changes in the children were specific to the frontotemporal area of the brain, suggesting a metabolic encephalopathy. Six individuals had abnormal nerve conduction. In addition, abnormal brainstem auditory evoked potentials have been described in 4 children with suspected mycotic neuromas who were exposed to mixed molds, including *Stachybotrys chartarum* and *Aspergillus species*.[57] Moreover, both neurobehavioral[31,57] and correlated quantitative EEG[31] changes indicative of right frontal lobe involvement have been reported in patients with chronic exposure to mold in water-damaged buildings. Mold exposure has also been implicated in optic neuritis[57] and multifocal choroiditis.[58] Finally, demyelination of the CNS has been reported following exposure to ibotenic acid,[59] abuse of "magic mushrooms" (Psilocybe),[60] and gliotoxin.[61] Because stachylysin has been found in human serum following exposure to *S. chartarum*,[62] and mycotoxins are present in indoor air and bioaerosols,[13–19] it is imperative that health complaints of occupants exposed to molds in water-damaged buildings be taken seriously and be investigated with appropriate diagnostic testing.

This communication describes 119 mold-exposed patients who had multiorgan symptoms and peripheral neuropathy. Complaints included severe fatigue, decreased muscle strength, sleep disturbances, numbness and tingling of extremities (with and without tremors of the fingers and hands), and severe headache. Patients had abnormal neurological examinations. Ninety-nine of these individuals had abnormal nerve conduction velocities (NCVs) in association with autoantibodies against 9 neural antigens, whereas 20 had normal test results. We present data on motor neuropathy, sensory neuropathy, and mixed sensory–motor polyneuropathy, as well as increased antibodies to neural antigens.

Materials and Method

Patients. The study population consisted of 119 patients (79 females and 40 males; mean age ± standard deviation [*SD*] = 41.3 ± 12.9 yr). The patients had health complaints and proven environmental exposure to molds in their homes and/or workplaces. Mold exposure was documented by Aerotech Laboratories (Phoenix, Arizona). All patients were interviewed one-on-one by the principal author (AWC) regarding exposure history, as well as health problems and symptoms for each organ system (e.g., CNS, PNS, respiratory, skin, musculoskeletal). Mold-specific serum antibody tests for *S. chartarum, Penicillium, Aspergillus, Cladosporium, Alternaria*, and *Chaetomium*, performed on each patient by Immunosciences Lab., Inc. (Beverly Hills, California), verified exposure to molds. Some of these data have been reported previously.[63–65]

We studied patients who had symptoms of peripheral neuropathy (e.g., tingling, tremors, loss of sensation in extremities). Blood was drawn for serology testing for neural antigens. NCV tests were performed at or near the time of initial presentation as follows: all 119 patients were tested at 10.8 ± 41 days; patients with abnormal NCVs (n = 99) were tested at 11.5 ± 44 days; and patients with normal NCVs (n = 20, controls) were tested at 7.5 ± 18 days.

Blood samples. Peripheral venous blood was collected and shipped at ambient temperature to Immunosciences Lab., Inc. (Beverly Hills, California). Autoantibodies (immunoglobulin IgG, IgM, and IgA) against 9 neural antigens were assessed for each patient.

Neural antigens. Myelin basic protein (MBP), myelin-associated glycoprotein (MAG), ganglioside (GM_1), chondroitin sulfate ($CONSO_4$), α-B-crystallin (crystallin), and tubulin were purchased from Sigma-Aldrich (St. Louis, Missouri). Neurofilament antigen (NAF) was purchased from Boehringer Mannheim Roche (Indianapolis, Indiana). MBP peptides 87-206 and myelin oligodendroctye glycoprotein (MOG) peptides 21-4-, 61-80 were synthesized by Research Genetics (Huntsville, Alabama).

Controls for neural antigens. The controls for determination of the mean ± *SD* and 95% confidence intervals (CIs) for the neural antigens consisted of 500 healthy adult blood donor volunteers. The controls were of similar age and sex distribution as the 119 patients.

ELISA testing. We used enzyme-linked immunosorbent assay (ELISA) to test for antibodies against 9 different neural-specific antigens, as reported previously.[66–68] Briefly, antigens were dissolved in methanol at a concentration of 1.0 mg/ml and then diluted 1:100 in 0.1 M carbonate-bicarbonate buffer (pH 9.5). Then, 50 µl of the mixture was added to each well of a polystyrene flat-bottom ELISA plate. Plates were incubated overnight at 4 °C and then washed 3 times with 20 mM Tris-buffered saline (TBS) containing 0.05% Tween 20 (pH 7.4). The nonspecific binding of immunoglobulins was prevented by adding a mixture of 1.5% bovine serum albumin (BSA) and 1.5% gelatin in TBS and then incubating for 2 hr at room temperature, followed by incubation overnight at 4 °C. Plates were washed as described above, and serum samples diluted 1:100 in 1% BSA-TBS were added to duplicate wells and incubated for 2 hr at room temperature. Sera from patients with multiple sclerosis (MS), polyneuropathies, and other neurological disorders with known high titers of IgG, IgM, and IgA against different neurological antigens were used to rule out nonspecific antibody activities of inter-assay and intra-assay variability. Plates were washed, and peroxidase-conjugated goat antihuman IgG, IgM, or IgA antiserum (KPI [Gaithersburg, Maryland]), diluted 1:400 in 1% BSA-TBS, was added to each well. The plates were incubated for an additional 2 hr at room temperature. After washing 5 times with TBS-Tween buffer, the enzyme reaction was started by the addition of 100 µl of *o*-phenylenediamine in citrate-phosphate buffer containing hydrogen peroxide diluted to 1:10,000 (pH 5.0). After 45 min, the reaction was stopped with 50 µl of 2N sulfuric acid. The optical density was read at 492 nm with a microtiter reader (Dynex Laboratories [Chantilly, Virginia]). Several control wells containing all reagents except human serum were used to detect nonspecific binding.

We calculated coefficients of intra-assay variation by running 5 samples 8 times in 1 assay. Coefficients of inter-assay variation were determined by measuring the same samples in 6 consecutive assays. This replicate testing established the validity of the ELISAs, determined the appropriate dilutions with minimal background, and detected IgG, IgM, and IgA against different antigens. Sera from 500 asymptomatic blood donors in southern California were used to calculate expected ranges at 95% CI.

Neurophysiological tests. Bilateral peripheral nerve studies involving nerve conduction and central response (F wave) were performed on the 119 patients in accordance with accepted techniques of the American Society of Electroneurodiagnostic Technologists (Kansas City, Missouri) and the American Neurological Association (Minneapolis, Minnesota).[69–71] The testing was performed under the direct supervision of, and interpreted by, a board-certified neurologist. Onset latency (ms), amplitude (µV), and velocity (m/sec) were recorded for 4 motor nerves (median, ulnar, peroneal, and tibial). The peak latency (ms) and amplitude (µV) were recorded for 3 sensory nerves (median, ulnar, and sural). F wave and H reflex were recorded for the median, ulnar, peroneal, and tibial nerves. The studies were conducted with a TECA Synergy Multimedia Electromyograph with multisync color SVGA monitor and Delux stimulator probe (TECASyngery, Synergy Version 8.2 [Oxford Instruments {Surry, U.K.}]). The motor axons of peripheral nerves that innervate somatic muscle were evaluated by recording the response following electrical stimulation.

Statistical analysis. We performed critical 2-tailed *t* tests on neural autoantibodies. Odds ratios (ORs) were calculated for the data to determine the percentage of individuals with antibody titers that exceeded the maximum expected laboratory range (95% CI) for each neural autoantibody. For this calculation, data for patients with abnormal and normal NCVs were combined.

Results

Neural autoantibodies. The mean ± *SD* of autoantibodies for each isotype (IgA, IgM, and IgG) against each neural antigen for patients with abnormal NCVs (n = 99), normal NCVs (n = 20), and asymptomatic blood donor controls (n = 500) are given in Table 1. Rather than repeating the data for each antineural antigen isotype, the salient features will be outlined briefly. In general, the highest isotype titers detected were MBP, MAG, tubulin, and NAF. These were followed by the other 5 neural antigens: GM_1, sulfatide, MOG, crystallin, and $CONSO_4$.

We performed critical 2-tailed *t* tests for each isotype titer against neural antigens, comparing abnormal vs. normal patients, abnormal patients vs. controls, and normal patients vs. controls for each isotype (statistical data not shown). With respect to IgG titers, the only significant difference between abnormal and normal patients was NAF ($p < 0.01$). IgG titers for all isotypes for abnormal and normal patients differed significantly from controls ($p < 0.001$). IgM titers for neural antigens were significantly different between abnormal and normal patients for glutamate receptor ($p < 0.01$), tubulin ($p < 0.01$), NAF ($p < 0.01$), and $CONSO_4$ ($p < 0.05$). IgM titers against each neural antigen for abnormal patients vs. controls ($p < 0.001$) and for normal patients vs. controls ($p < 0.001$) were significantly different. The only significant difference between patients with abnormal vs. normal IgA titers was NAF ($p < 0.05$). Comparison of abnormal and normal patients vs. controls revealed IgA titers for each neural antigen which were significantly different ($p < 0.01$), except for normal crystallin titers ($p < 0.05$).

The percentages of individuals with autoantibodies for each isotype that exceeded the laboratory expected range at 95% CI against the neural antigens are presented in Table 2. IgG titers for abnormal patients exceeded the 95% CI for sulfatide (17.2%), MOG (10.1%), crystallin (10.1%), glutamate receptor (11.1%), tubulin (57.6%), $CONSO_4$ (27.3%), and NAF (6.1%); MBP (4%), MAG (4%), and GM_1 (0%) did not exceed their expected ranges. Similar observations for normal patients were made for sulfatide (20%), MOG (20%), crystallin (20%), tubulin (30%), and $CONSO_4$ (20%), except for MBP (0%), GM_1 (0%), glutamate receptor (0%), and NAF (0%). IgM titers for abnormal patients exceeded those of controls for all neural antigens (range = 17.2%–42.4%) except GM_1 (0%). In normal patients, IgM titers did not exceed control values for GM_1 (0%), sulfatide (5%), MOG (5%), glutamate receptor (0%), and tubulin (5%), whereas MBP (20%), MAG (30%), crystallin (25%), $CONSO_4$ (20%), and NAF (20%) exceeded control values. IgA autoantibodies in abnormal patients for each neural antigen exceeded control values for MBP (20.3%), MAG (23.2%), crystallin (8.1%), glutamate receptor (7.1%), tubulin (8.1%), $CONSO_4$ (10.5%), and NAF (10%). With respect to normal patients, only MAG (15%), tubulin (10%), and NAF (10%) exceeded values for controls.

The percentage of patients who had ORs that exceeded the 95% CI are given in Table 3. The ORs for IgG were not significant for MBP (0.66) and MAG (1.05), whereas those for all other neural autoantibodies were significant as follows: sulfatide (3.36), crystallin (6.53), glutamate receptor (10.08), tubulin (55.1),

Table 1.—Autoantibody Titers against 9 Neural Antigens in Patients with Abnormal Nerve Conduction Velocities (NCVs) (n = 119) and Those with Normal NCVs (n = 20), vs. Asymptomatic Controls (N = 500), for Each Isotype

	IgG						IgM						IgA					
	Abnormal		Normal		Controls		Abnormal		Normal		Controls		Abnormal		Normal		Controls	
Neural autoantibody	$\bar{x}$	*SD*	$\bar{x}$	*SD*	$\bar{x}$	*SD*	$\bar{x}$	*SD*	$\bar{x}$	*SD*	$\bar{x}$	*SD*	$\bar{x}$	*SD*	$\bar{x}$	*SD*	$\bar{x}$	*SD*
MBP	63.3	21.6	54.7	18.3	27.0	12.2	50.8	24.3	47.0	9.2	25.1	13.2	18.5	6.0	16.8	4.7	7.2	3.7
MAG	63.6	18.8	69.0	18.2	26.0	7.3	51.0	16.1	48.1	12.4	24.5	8.1	19.4	7.6	17.5	7.4	8.3	2.4
GM_1	15.8	6.7	16.4	7.0	11.1	2.8	16.4	6.3	15.6	4.5	10.4	3.3	12.3	6.2	12.6	4.4	11.5	3.6
Sulfatide	16.7	6.2	17.5	6.7	12.2	3.4	17.2	6.3	15.1	4.3	11.3	3.7	13.3	9.1	12.4	4.0	9.4	3.1
MOG	15.9	5.9	17.5	6.6	8.9	5.5	18.2	6.3	16.8	4.7	8.4	5.5	12.0	3.5	11.4	2.3	7.9	4.6
Crystallin	16.1	5.8	16.5	6.7	11.9	2.5	18.7	6.9	19.0	8.2	12.1	2.9	13.6	7.5	13.2	4.7	11.1	4.3
Glutamate	10.2	3.4	9.2	1.5	7.0	2.8	10.6	3.7	9.3	1.2	7.6	2.0	9.5	3.7	8.5	2.1	8.1	2.1
Tubulin	64.1	16.8	58.7	11.9	23.7	9.2	41.9	18.5	33.0	7.8	18.1	7.9	17.9	16.1	14.7	3.7	9.8	2.6
$CONSO_4$	11.5	4.7	10.2	2.1	7.7	2.2	11.8	4.9	10.4	2.0	5.6	2.1	11.0	7.7	9.2	1.4	7.0	2.7
NAF	61.6	23.9	53.3	10.7	26.4	11.3	53.1	20.7	44.1	12.0	24.7	10.3	18.2	8.2	16.4	5.4	8.7	3.4

Notes: Ig = immunoglobulin, $\bar{x}$ = mean, *SD* = standard deviation, MBP = myelin basic protein, MAG = myelin-associated glycoprotein, GM_1 = ganglioside, MOG = myelin oligodendroctye glycoprotein, $CONSO_4$ = chondroitin sulfate, and NAF = neurofilament antigen.

Table 2.—Percentages of Individuals with Autoantibody Titers that Exceeded Expected Ranges (95% Confidence Intervals), for Patients with Abnormal Nerve Conduction Velocities (NCVs) and Those with Normal NCVs, vs. Asymptomatic Controls, for Each Isotype

	IgG			IgM			IgA		
Neural autoantibody	Abnormal (%)	Normal (%)	Controls (%)	Abnormal (%)	Normal (%)	Controls (%)	Abnormal (%)	Normal (%)	Controls (%)
MBP	4.0	0	5	34.3	20	2	20.2	5	1
MAG	4.0	5	4	42.4	30	3	23.2	15	2
GM_1	0.0	0	2	0.0	0	4	4.0	5	5
Sulfatide	17.2	20	6	17.2	5	5	3.0	5	2
MOG	10.1	20	5	25.3	5	4	1,0	0	3
Crystallin	10.1	20	2	27.3	25	4	8.1	5	3
Glutamate	11.1	0	1	15.2	0	0	7.1	0	2
Tubulin	57.6	30	2	39.4	5	2	8.1	10	1
$CONSO_4$	27.3	20	2	33.3	20	1	10.5	5	2
NAF	6.1	0	3	41.4	20	2	18.2	10	1

Notes: Ig = immunoglobulin, MBP = myelin basic protein, MAG = myelin-associated glycoprotein, GM_1 = ganglioside, MOG = myelin oligodendroctye glycoprotein, $CONSO_4$ = chondroitin sulfate, and NAF = neurofilament antigen.

$CONSO_4$ (17.26), and NAF (5.15). The OR for GM_1 could not be calculated because of the 0 value for the patients. The ORs for IgM neural autoantibodies were significant for all antigens, and ranged from 3.39 to 44.6. The ORs for GM_1 and glutamate receptor were not calculated because of 0 values for the controls. With respect to IgA autoantibodies, the ORs for sulfatide (1.7) and MOG (0.82) were not significant, whereas those for the other neural antigens were significant as follows: MBP (21.2), MAG (13.7), crystallin (2.65), glutamate receptor (3.03), tubulin (9.08), $CONSO_4$ (4.99), and NAF (2). GM_1 could not be calculated because of 0 values; the 95% CI for MOG was not calculated because of the value of 1 in controls.

The percentages of individuals with autoantibody titers that exceeded the maximum 95% CI for expected laboratory ranges, along with ORs, are presented in Table 4. We calculated these data for abnormal, normal, and control patients as follows: If an individual had only 1 isotype against a neural antigen (i.e., IgG), that person was given the same score as an individual with 2 or more isotypes (i.e., IgG + IgM + IgA). The percentages of individuals with autoantibodies against each neural antigen were highest among the abnormal

Table 3.—Odds Ratios (ORs) and 95% Confidence Intervals (CIs) for Autoantibodies for Each Isotype Presented in Table 1

	IgG		IgM		IgA	
Neural autoantibody	OR	95% CI	OR	95% CI	OR	95% CI
MBP	0.66	1.92, 0.22	22.98	47.9, 11.0	21.2	57.4, 7.8
MAG	1.05	3.86, 0.39	21.86	40.9, 11.59	13.7	6.2, 28.8
GM_1	—*	—*		—*		
Sulfatide	3.36	6.1, 1.84	3.39	5.6, 2.05	1.7	0.52, 5.5
MOG	2.53	4.95, 1.25	6.7	12.2, 3.5	0.82	—†
Crystallin	6.53	15.2, 2.83	8.83	16.4, 4.9	2.65	1.13, 6.17
Glutamate	10.08	29.4, 3.3	—*		3.06	8.17, 1.14
Tubulin	55.1	112.4, 27.1	24.8	50.9, 11.8	9.08	38.02, 26.8
$CONSO_4$	17.26	36.6, 8.2	44.6	116.7, 16.4	4.99	2.05, 11.9
NAF	5.15	13.06, 1.88	29.8	62.2, 14.9	20.0	7.26, 54.6

Notes: Ig = immunoglobulin, MBP = myelin basic protein, MAG = myelin-associated glycoprotein, GM_1 = ganglioside, MOG = myelin oligodendroctye glycoprotein, $CONSO_4$ = chondroitin sulfate, and NAF = neurofilament antigen. For these calculations, patients with abnormal nerve conduction velocities (NCVs) were grouped with those with normal NCVs and compared vs. controls.
*Not calculated because of zero values (refer to Table 2).
†Not calculated because of the value of 1 in a cell (refer to Table 2).

Table 4.—Percentages of Individuals with 1 or More Isotypes for Each Neural Antigen that Exceeded the Maximum 95% Confidence Interval (CI) Titer

Neural autoantibody	Abnormal (*n* = 119) (%)	Normal (*n* = 20) (%)	Controls (*n* = 500) (%)	OR	95% CI
MBP	47.5	20	8	8.63	11.5, 6.4
MAG	54.5	45	9	11.4	16.4, 7.8
GM_1	4.0	5	11	0.91	1.7, 0.48
Sulfatide	30.3	25	13	2.8	4.4, 1.7
MOG	32.3	25	12	3.3	4.85, 2.2
Crystallin	35.4	35	9	5.5	8.0, 2.7
Glutamate	26.3	0	3	9.0	17.6, 4.6
Tubulin	69.7	35	5	33.58	56.8, 19.1
$CONSO_4$	46.5	30	5	14.75	25.3, 8.58
NAF	50.5	25	5	16.3	16.3, 12.4

Notes: MBP = myelin basic protein, MAG = myelin-associated glycoprotein, GM_1 = ganglioside, MOG = myelin oligodendroctye glycoprotein, $CONSO_4$ = chondroitin sulfate, and NAF = neurofilament antigen. The percentages were determined as follows: If a patient had 1 isotype (e.g., immunoglobulin [Ig]G, IgM, or IgA), that patient was given the same score as a patient with 2 or more isotypes. Thus, the total percentages for patients with abnormal nerve conduction velocities (NCVs), normal NCVs, and controls were less than the totals for each isotype as presented in Table 1. For simplicity, the data for patients with abnormal and normal NCVs were combined to obtain the odds ratios (ORs).

patients (range = 26.3%–69.0%) when compared with controls (range = 3%–13%), with the exception of GM_1 (4%). Similarly, the normal patients had an increased percentage of individuals with higher titers (range = 20%–45%) when compared with the controls, with the exception of GM_1 (5%) and glutamate receptor (0%). The ORs were significant (95% CI) for all neural antigens, except for GM_1 (0.91). The ORs were highest for tubulin (33.58) and decreased, in descending order, for NAF (16.3), $CONSO_4$ (14.75), MAG (11.4), glutamate receptor (9.0), MBP (8.63), crystallin (5.5), MOG (3.3), and sulfatide (3.3).

NCV testing. No changes or abnormalities were observed for F and H waves in the abnormal or normal patients (data not shown).

The data obtained from the NCV studies for motor nerves and sensory nerves are summarized in Tables 5 and 6. Patients with abnormal findings comprised 3 groups, as follows: (1) mixed sensory–motor polyneuropathy ($n = 55$), (2) motor neuropathy ($n = 17$), and (3) sensory neuropathy ($n = 27$). There were 20 patients with no abnormalities (controls). We compared the data obtained for the 20 controls with data for the 3 groups of abnormal patients (mixed, motor, and sensory neuropathies), for statistical purposes.

Results for the mixed polyneuropathy group differed significantly from controls. Latencies for the median (4.2 ± 1.2 ms, $p < 0.001$), ulnar (3.13 ± 1.1 ms, $p < 0.05$), peroneal (5.1 ± 1.4 ms, $p < 0.001$), and tibial motor nerves (5.5 ± 2.9 ms, $p < 0.001$) were signifi-

Table 5.—Neuropathies Experienced by Patients with Abnormal Nerve Conduction Velocity (NCV) Measurements ($n = 119$), vs. Controls with Normal Values ($n = 20$), by NCV Parameter Tested

	Neuropathy with abnormal NCV							
	Mixed ($n = 55$)		Motor ($n = 17$)		Sensory ($n = 27$)		Controls ($n = 20$)	
NCV parameter	$\bar{x}$	*SD*	$\bar{x}$	*SD*	$\bar{x}$	*SD*	$\bar{x}$	*SD*
Median latency (ms)	4.2	1.2*	3.6	1.1	3.4	0.36	3.3	0.4
Median amplitude (μV)	9.5	4.3	7.6	3.3†	10.8	2.6	10	3.3
Median velocity (m/sec)	56.5	7.8‡	55.9	5.3‡	58.1	7.4	61.5	7.1
Ulnar latency (ms)	3.13	1.1†	3.7	2.05‡	2.7	0.38	2.56	0.53
Ulnar amplitude (μV)	9.9	3.7	9.9	2	9.9	1.8	10	3.13
Ulnar velocity (m/sec)	60.9	9.9§	60.4	7.4‡	65.1	6.6	66.2	5.4
Peroneal latency (ms)	5.1	1.4*	5.5	1.2*	4.7	0.52*	3.9	0.86
	($n = 54$)				($n = 26$)			
Peroneal amplitude (μV)	5.6	3.2	4.2	1.9§	6.05	2.2	6.6	3
	($n = 54$)				($n = 26$)			
Peroneal velocity (m/sec)	52.5	12.7	47.5	13.3†	52.2	5.8	58.2	13.8
	($n = 54$)				($n = 26$)			
Tibial latency (ms)	5.5	2.9*	6.3	3§	4.2	0.55	4	1
	($n = 53$)							
Tibial amplitude (μV)	9.8	6	8.1	4.8†	13.6	4.8	12.7	6.8
	($n = 53$)							
Tibial velocity (m/sec)	46.6	11.3†	47.3	5.5†	46.6	4.4§	52.3	8.3
	($n = 53$)							
Median latency (ms)	4.2	1.3*	3	0.16	4.2	1.5*	3.08	0.17
	($n = 54$)		($n = 16$)					
Median amplitude (μV)	27.1	15.9	40.3	15.2	32.4	20.5	35.1	15.6
	($n = 54$)		($n = 16$)					
Ulnar latency (ms)	3.7	1.41*	2.7	0.2	4.6	2.3*	2.54	0.25
	($n = 54$)		($n = 16$)					
Ulnar amplitude (μV)	27	18.5†	31.9	14.9	28.1	18.6	35	15.3
	($n = 54$)		($n = 16$)					
Sural latency (ms)	4.9	3.2‡	3.8	0.28	4.4	1.4†	3.8	0.32
	($n = 44$)		($n = 11$)		($n = 24$)		($n = 17$)	
Sural amplitude (m/sec)	11	8.4†	14.2	5.9	11.2	10.5†	21.7	19.2
	($n = 44$)		($n = 11$)		($n = 24$)		($n = 17$)	

Notes: With respect to the 20 normal patients, no abnormal values were observed for each measurement except for the tibial motor nerve measurements, in which the velocity was slightly reduced (41 ms), with the cutoff for normal being > 41 ms. For cases in which the number of patients was not the same as that shown in the column heading, the actual number (*n*) is given within the table.

*$p < 0.001$.
†$p < 0.05$.
‡$p < 0.02$.
§$p < 0.01$.

Table 6.—Percentages of Individuals with Various Numbers (0–7) of Nerves Showing Abnormal Nerve Conduction Measurements (Latency Onset, Amplitude, or Velocity), for Mixed (Motor and Sensory), Motor, and Sensory Neuropathies

	Neuropathy		
No. of nerves with abnormal measurements	Mixed (n = 55) (%)	Motor (n = 17) (%)	Sensory (n = 27) (%)
0	0.0	0.0	11.1
1	1.8	41.2	33.3
2	38.2	47.1	33.7
3	23.6	5.9	18.5
4	21.8	5.9	—
5	10.9	—	—
6	1.8	—	—
7	1.8	—	—

Notes: Among the 20 controls, 1 individual had an abnormal measurement for the tibial motor nerve (velocity = 41 ms), resulting in an abnormal rate of 0.4% for all motor nerve measurements. No abnormalities were seen among the controls for the other 4 motor nerves or the 3 sensory nerves.

cantly increased vs. controls. Amplitudes for all motor nerves were not significantly different from controls. Velocities for the median (56.5 ± 7.8 m/sec, $p < 0.02$), ulnar (60.9 ± 9.9 m/sec, $p < 0.01$), and tibial (46.6 ± 11.3 m/sec, $p < 0.05$) motor nerves were significantly decreased vs. controls. Latencies for the median (4.2 ± 1.3 ms, $p < 0.001$), ulnar (3.7 ± 1.41 ms, $p < 0.02$), and sural (4.9 ± 3.2 ms, $p < 0.02$) sensory nerves were significantly increased vs. controls. Amplitudes for the ulnar (27 ± 18.5 μV, $p < 0.05$) and sural (11 ± 8.4 μV, $p < 0.5$) sensory nerves were significantly decreased compared with controls.

In patients with only motor nerve neuropathy, latencies for ulnar (3.7 ± 2.05 ms, $p < 0.02$), peroneal (5.5 ± 1.2 ms, $p < 0.001$), and tibial (6.2 ± 3 ms, $p < 0.01$) nerves were significantly increased; amplitudes for the median (7.6 ± 3.3 μV, $p < 0.05$), peroneal (4.2 ± 1.9 μV, $p < 0.01$), and tibial (8.1 ± 4.8 μV, $p < 0.5$) nerves were significantly decreased; and velocities for the median (55.9 ± 5.3 m/sec, $p < 0.01$), ulnar (60.4 ± 7.4 m/sec, $p < 0.02$), peroneal (47.5 ± 13.3 m/sec, $p < 0.05$), and tibial (47.3 ± 5.5 m/sec, $p < 0.05$) nerves were significantly slower than the controls. Latencies and amplitudes for the sensory nerves (median, ulnar, and tibial) were not significantly different from control values.

In patients with only sensory neuropathy, latencies for the median (4.2 ± 1.5 ms, $p < 0.001$), ulnar (4.6 ± 2.3 ms, $p < 0.001$), and sural (4.4 ± 1.4 ms, $p < 0.05$) nerves were significantly increased vs. controls; amplitudes of the sural nerve (11.2 ± 10.5 μV, $p < 0.05$) were significantly decreased; and all neurophysiological measurements tended to differ from the control values. None of the measurements for motor nerves in this group—except for peroneal latency ($p < 0.001$) and amplitude ($p < 0.01$)—were different from the controls.

Table 6 summarizes the data for the percentages of patients with various numbers of nerves that demonstrated abnormal conduction. In those patients with mixed neuropathy, all nerves had abnormal measurements with a distribution as follows: 1 involved nerve (5.5%), 2 involved nerves (38.2%), 3 involved nerves (23.6%), 4 involved nerves (21.8%), and 5 or more involved nerves (14.5%). Of those patients who exhibited motor neuropathy, 41.2% had only 1 involved nerve, whereas 58.9% had 2 or more involved nerves. Patients with sensory neuropathy had the following distribution: 11.1% had nerves with no abnormal findings, 33.3% had only 1 nerve with abnormal measurements, and 52.2% had 2 or more nerves with abnormal measurements.

Discussion

All patients in this study had documented exposure to molds in their homes and/or workplaces. They also had significantly elevated antibodies to molds and to mycotoxins, which confirmed exposure.[63–65] In addition, multiple organ symptoms were present, as reported previously.[63] In this particular group of patients, additional health complaints consisted of symptoms of peripheral neuropathy (i.e., tingling, numbness, tremors, and muscle weakness in the extremities). Thus, we evaluated these patients for the presence of antibodies to 9 neural antigens, as well as for evidence of abnormalities in peripheral nerve conduction. All patients had significant increases in autoantibodies against neural antigens (Tables 1–4). Abnormalities in latencies, amplitudes, and velocities of selected peripheral nerves (Tables 5 and 6), and peripheral neuropathy, were observed in 99 patients, whereas 20 symptomatic patients had normal NCV measurements.

Examination of the patients' antibody titers revealed that IgG antibody titers to the neural antigens between patients with abnormal vs. normal NCVs were not significantly different, with the exception of NAF ($p < 0.01$). However, when compared with healthy controls, the difference between IgG titers for abnormal vs. normal patients was highly significant ($p < 0.001$). In contrast, IgM titers in abnormal patients were consistently elevated when compared with normal patients, with significant differences for $CONSO_4$ ($p < 0.05$), glutamate receptor ($p < 0.01$), tubulin ($p < 0.01$), and NAF ($p < 0.01$). Autoantibodies in both abnormal and normal patients showed significant elevations compared with controls ($p < 0.001$). With respect to IgA antibodies, NAF titers were significantly elevated in abnormal vs. normal patients. The titers for both abnormal and normal patients were significantly elevated compared with controls, with the exception of GM_1 and glutamate receptor. Thus, we concluded that exposure to molds, and symptoms of peripheral neuropathy, are associated with autoantibodies to 9 different neural antigens. These data support and extend the observations of Gray et al.,[30] who demonstrated increased antibodies to myelin and NAF in mold-exposed individuals.

Autoantibodies to neural antigens have been reported for several neurological conditions.[66,72] Pestronk et al.[66] confirmed that elevated titers of MAG antibodies in patients are relatively specific for sensory and motor polyneuropathy syndromes with demyelination. NCVs were used to confirm the demyelinating changes in these patients. Of the patients studied, 92% with IgM antibodies to MAG had physiologic evidence of demyelination.[66] MS patients have shown antibodies to myelin, MOG, MBP, α-B-crystallin, and complement-mediated demyelination.[68,73] Anti-ganglioside, anti-glycolipid, anti-sulfatide, anti-MAG, anti-tubulin, and anti-$CONSO_4$ antibodies have been demonstrated in motor, sensory, and polyneuropathies with demyelination.[72–86] IgM isotypes to sulfatide,[74,75,77] ganglioside,[78] and MAG[83] are correlated with electrophysiological peripheral nerve abnormalities. Moreover, antigangliosides and galactocerebroside antibodies are associated with infections from *Campylobacter jejuni* and *Mycoplasma pneumoniae* in patients with Guillain-Barre syndrome.[81] In addition, IgM, anti-MAG, anti-glycolipids, and anti-NAF antibodies are present in individuals with chronic demyelinating conditions of the nervous system.[67,73,83,86] Thus, we suggest that the presence of autoantibodies to neural antigens in our patients is the result of exposure to toxic metabolites[13–23,60,61] of molds, or may result from an infectious process. The presence of abnormal T and B cell function of the immune system[30] in nasal,[41] pulmonary,[39,40] and neurologic[87,88] infections by molds supports this conclusion. Finally, we have observed increased T cell activation; C3 and C4 complements; and IgA, IgM, and IgG immune complexes in 33 patients who had chronic exposure to molds in a water-damaged building, which suggests an inflammatory process (manuscript forthcoming).

The ORs in Table 3 are also revealing. IgM isotypes against the neural antigens had ORs consistently greater (range = 3.39–44.6) than those for IgG isotypes (range = 0.66–17.26), with the exception of antitubulin (55.1 vs. 24.8). The ORs indicate an increased risk of developing antineural antibodies, but also may suggest that IgM isotypes are more consistent with symptomatic active or progressive neuropathy than are IgG isotypes, and may represent an ongoing acute or subacute process. As mentioned earlier, IgM antibodies to various neural antigens have been associated with neurophysiological and pathological changes characteristic of various neuropathies. Finally, the ORs in Table 4 for the percentages of individuals with 1 or more isotypes against the neural antigens show a relative increased risk (range = 2.8–33.58) of developing autoantibodies. The only exception was GM_1 autoantibodies (OR = 0.91). Thus, we concluded that individuals exposed to molds in a water-damaged building have an increased risk of developing antineural antibodies. Additional work is needed to determine at what point these processes become irreversible.

Our neurophysiological data revealed 3 different types of peripheral neuropathies: mixed sensory–motor polyneuropathy (55 abnormal patients), motor neuropathy (17 abnormal patients), and sensory neuropathy (27 abnormal patients), as well as patients who exhibited symptoms but had no abnormal electrophysiological findings (20 normal controls) (Table 5). The differences between the 20 normal patients and the 99 abnormal patients are likely attributable to the significant increase in IgG and IgM autoantibodies to NAF, tubulin, glutamate receptor, and $CONSO_4$ observed in the abnormal patients. The role that IgA antibodies play is unclear at this time. Additional observations are needed to clarify the role of each isotype (IgA, IgM, and IgG) and to determine which neural autoantibodies contribute to the observed neuropathies.

The increased latencies for motor and sensory nerves observed in the 55 patients with mixed neuropathy suggest a demyelinating process.[83] The increased latencies were accompanied by a decrease in velocities for the median, ulnar, peroneal, and tibial nerves. All three sensory nerves (median, ulnar, and sural) exhibited increased latencies and decreased amplitudes. Thus, the polyneuropathy observed in these patients appears to be a demyelinating process with decreased number and size of fibers (decreased amplitudes) and chronic involvement of the nerve (decreased velocities).[72,83] Those with motor neuropathies (17 patients) had decreases in latencies (ulnar, peroneal, and tibial nerves), decreased amplitudes (median, peroneal, and tibial nerves), and decreased velocities (median, ulnar, peroneal, and tibial nerves). This appears to be a diffuse neuropathy and may involve some demyelination.[89] Finally, those with sensory neuropathies (27 patients) had increased latencies for all 3 nerves, whereas the sural nerve had a decreased amplitude. The increased latencies and decreased amplitude of the these nerves suggest that demyelination is occurring.[90]

The severity of the neuropathies experienced by the patients in our study is implicit as a result of the involvement of several nerves (Table 6). With respect to the mixed-neuropathy patients, only 1.8% had abnormalities in only 1 nerve, whereas 38.2% had at least 2 nerves involved. The remaining patients (59.5%) had 3 or more nerves with abnormal neurophysiological recordings. Impairments in the patients with motor neuropathy were slightly less dramatic, with 41.2% having a single nerve involvement, and the remainder having 2 or more nerves involved. Finally, in those patients with sensory neuropathy, 33.3% had 1 nerve and 52.2% had

2 or more nerves involved. Thus, we concluded that the neuropathies in these patients were severe and in many cases involved several nerves.

In summary, 119 individuals exposed to mold colonies in water-damaged buildings were found to have autoantibodies directed against 9 different neural antigens. Neurophysiological recordings for latencies, amplitudes, and velocities on 4 motor nerves and 3 sensory nerves revealed peripheral neuropathies in 99 patients (83%). Three abnormal conditions were found: mixed sensory–motor polyneuropathy, motor neuropathy, and sensory neuropathy. We recommend that mold-exposed individuals with symptoms of neuropathy be evaluated for antibodies against neural antigens and for neurophysiological abnormalities. Additional work is needed to correlate and clarify the extent of the peripheral nerve pathology and demyelination, as well as the role of neural autoantibodies in this process.

* * * * * * * * * *

The authors thank Nina Immers for her kind technical support during the data gathering and tabulation for this study.

Submitted for publication April 5, 2004; revised; accepted for publication May 14, 2004.

Requests for reprints should be sent to Andrew M. Campbell, M.D., Medical Center for Immune and Toxic Disorders, 25010 Oakhurst, #200, Spring, TX 77386.

E-mail: md@immunotoxicology.com

* * * * * * * * * *

References

1. Gravesen S, Nielsen PA, Iverson R, et al. Microfungal contamination of damp buildings—examples of constructions and risk materials. Environ Health Perspect 1999; 107 (Suppl 3):505–08.
2. Croft WA, Jarvis BB, Yatawara CS. Airborne outbreak of trichothecene toxicosis. Atmos Environ 1986; 20:549–52.
3. Peltola J, Andersson MA, Haahtela T, et al. Toxic-metabolite–producing bacteria and fungus in an indoor environment. Appl Environ Microbiol 2001; 67:3269–74.
4. Shelton BF, Kirkland KH, Flanders WE, et al. Profiles of airborne fungi in buildings and outdoor environments in the United States. Appl Environ Microbiol 2002; 68: 1743–53.
5. Johanning E, Biagini R, Hull D-L, et al. Health and immunology study following exposure to toxigenic fungi (*Stachybotrys chartarum*) in a water-damaged office environment. Int Arch Occup Environ Health 1996; 68:207–18.
6. Andersson MA, Nikulin M, Kooljalg U, et al. Bacteria, molds, and toxins in water-damaged building materials. Appl Environ Microbiol 1997; 63:387–93.
7. Nielsen KF, Gravesen S, Neilsen PA. Production of mycotoxins on artificially and naturally infested building materials. Mycopathologia 1999; 145:43–56.
8. Claeson AS, Levin H, Blomquist G, et al. Volatile metabolites from microorganisms grown on humid building materials and synthetic media. J Environ Monit 2002; 4:667–72.
9. Tuomi T, Reijut K, Johnsson T, et al. Mycotoxins in crude building materials from water-damage buildings. Appl Environ Microbiol 2000; 66:1899–1904.
10. Nieminen SM, Karki R, Auriola S, et al. Isolation and identification of *Aspergillus fumigatus* mycotoxins on growth medium and some building materials. Microbiology 2002; 68:4871–75.
11. Jarvis BB. Chemistry and toxicology of molds isolated from water-damaged buildings. In: DeVries JW, Trucksess MW, Jackson LS (Eds), Mycotoxins and Food Safety. New York: Kluwer Academic/Plenum Publishers, 2002; pp 43–52.
12. Burge HA. Bioaerosols: prevalence and health effects in the indoor environment. J Allergy Clin Immunol 1990; 86: 687–704.
13. Richard JL, Plattner RD, May J, et al. The occurrence of ochratoxin A in dust collected from a problem household. Mycopathologia 1999; 146:99–103.
14. Skaug MA, Eduard W, Stormer FD. Ochratoxin A in airborne dust and fungal conidia. Mycopathologia 2000; 151:93–95.
15. Smoragiewicz W, Cossete B, Boutrard A, et al. Trichothecene mycotoxins in the dust of ventilation systems in office buildings. Int Arch Occup Environ Health 1993; 65:113–17.
16. Tuomi T, Saarinene L, Reijula K. Detection of polar and macrocyclic trichothecene mycotoxins from indoor environments. Analyst 1998; 123:1835–41.
17. Johanning E, Gareis M, Nielsen K, et al. Airborne mycotoxins sampling and screening analysis. Proceedings of the 9th International Conference on Indoor Air Quality and Climate (Indoor Air 2002), Monterey, California, June 30–July 5, 2002. Santa Cruz, California: Indoor Air 2002 Conference Secretariat.
18. Fischer G, Muller T, Ostrowksi R, et al. Mycotoxins as exposure parameters in bioaerosols of composting sites. Schriftenr Ver Wasser Boden Lufthyg 1999; 104:149–62.
19. Fischer G, Muller T, Ostrowksi R, et al. Mycotoxins of *Aspergillus fumigatus* in pure culture and in native bioaerosols from compost facilities. Chemosphere 1999; 38:1745–55.
20. Land CJ, Hult K, Fuchs R, et al. Tremorgenic mycotoxins from *Aspergillus fumigatus* as possible occupational health problems in sawmills. Appl Environ Microbiol 1987; 53:787–90.
21. Gordon KE, Masotti RE, Waddel WR. Tremorgenic encephalopathy: a role of mycotoxins in the production of CNS disease in humans? Can J Neurol Sci 1993; 20: 237–39.
22. Boysen SR, Rozanski EA, Chan DL, et al. Tremorgenic mycotoxicosis in four dogs from a single household. J Am Vet Med Assoc 2002; 221(10):1441–44, 1420.
23. Young KL, Villar D, Carson TL, et al. Tremorgenic mycotoxin intoxication with penitrem A and roquefortine in two dogs. J Am Vet Med Assoc 2003; 222:52–53, 35.
24. Hodgson MJ, Morey P, Leung W-Y, et al. Building-associated pulmonary disease from exposure to *Stachybotrys chartarum* and *Aspergillus versicolor*. J Occup Environ Med 1998; 40:241–49.
25. Croft WA, Jastromski BM, Croft AL, et al. Clinical confirmation of trichothecene mycotoxicosis in patient urine. J Environ Biol 2002; 23:301–20.
26. Kilburn KH. Inhalation of moulds and mycotoxins. Eur J Oncol 2002; 7:197–202.
27. Anyanwu EC, Campbell AW, Vojdani A. Neurophysiological effects of chronic indoor environmental mold exposure on children. Scientific World Journal 2003; 3: 281–90.

28. Mahmoudi M, Gershwin ME. Sick building syndrome. III. *Stachybotrys chartarum.* J Asthma 2000; 37:1191–98.
29. Walinder R, Wieslander, G, Norback D, et al. Nasal lavage biomarkers: effects of water damage and microbial growth in an office building. Arch Environ Health 2001; 56:30–36.
30. Gray MR, Thrasher JD, Crago R, et al. Mixed mold mycotoxicosis: immunological changes in humans following exposure in water-damaged buildings. Arch Environ Health 2003; 58(7):410–20.
31. Crago BR, Gray M, Nelson LA, et al. Psychological, neuropsychological, and electrocortical effects of mixed mold exposure. Arch Environ Health 2003; 58(8):452–63.
32. Johanning E, Landsbergis P, Gareis M, et al. Clinical experience and results of a sentinel health investigation related to indoor fungal exposure. Environ Health Perspect 1999; 107(Suppl 3):189–94.
33. Jaakkola M, Nordman H, Pilpari R, et al. Indoor dampness and molds and development of adult-onset asthma: a population-based incident case-control study. Environ Health Perspect 2002; 110:543–47.
34. Gent JF, Ren P, Belanger K, et al. Levels of household mold associated with respiratory symptoms in the first year of life in a cohort at risk for asthma. Environ Health Perspect 2002; 110:A781–86.
35. Hayes AW. Mycotoxins: a review of biological effects and their role in human diseases. Clin Toxicol 1980; 17: 45–83.
36. Wannemacher RW Jr, Weiner SL. Trichothecene mycotoxins. In: Sidell FR, Takafuji ET, Franz DR (Eds). Medical Aspects of Chemical and Biological Warfare. Falls Chruch, VA: Office of the Surgeon General, 1997; pp 655–76.
37. Pfohl-Leszkowicz A, Petkova-Bocharova T, Chernozemsky IN, et al. Balkan endemic nephropathy and associated urinary tract tumours: a review on aetiological causes and the potential role of mycotoxins. Food Addit Contam 2002; 19:282–302.
38. Bondy GS, Pestka JJ. Immunomodulation by fungal toxins. J Toxicol Environ Health B Crit Rev 2000; 3:109–43.
39. Fraser RS. Pulmonary aspergilliosis: pathologic and pathogenic features. Pathol Annu 1993; 28(Pt 1):231–77.
40. Patel AM, Ryu JH, Reed CE. Hypersensitivity pneumonitis: current concepts and future questions. J Allergy Clin Immunol 2001; 108:661–70.
41. Thrasher RD, Kingdom TT. Fungal infections of the head and neck: an update. Otolaryngol Clin North Am 2003; 36:577–94.
42. Dominguez-Malagon H, Gaytan-Graham S. Hepatocellular carcinoma: an update. Ultrastruct Pathol 2001; 25: 497–516.
43. Sajan MP, Satav JG, Bhattacharya RK. Effect of aflatoxin B in vitro on rat liver mitochondrial respiratory functions. Indian J Exp Biol 1997; 35:1187–90.
44. Hoehler, D, Marquardt RR, McIntosh AR, et al. Induction of free radicals in hepatocytes, mitochondria and microsomes of rats by ochratoxin A and its analogs. Biochim Biophys Acta 1997; 1357:225–33.
45. Hussein HS, Brasel JM. Toxicity, metabolism, and impact of mycotoxins on humans and animals. Toxicology 2001; 167:101–34.
46. Nishiyama M, Kuga T. Pharmacological effects of tremorgenic mycotoxin fumitremorgin A. Jpn J Pharmacol 1986; 40:481–89.
47. Cavanagh JB, Holton JL, Nolan CC, et al. The effects of tremorgenic mycotoxin penitrem A on the rat cerebellum. Vet Pathol 1998; 35:53–63.
48. Selala MI, Daelemans F, Schepens PJ. Fungal tremorgens: the mechanism of action of single nitrogen containing toxins—a hypothesis. Drug Chem Toxicol 1989; 12: 237–57.
49. Longland CL, Dyer JL, Michelangell F. The mycotoxin paxilline inhibits the cerebellar inositol 1,4,5-trisphosphate receptor. Eur J Pharmacol 2000; 408:219–25.
50. Chen JW, Luo YL, Hwang MJ, et al. Territrem B, a tremorgenic mycotoxin that inhibits acetylcholinesterase with a noncovalent yet irreversible binding mechanism. J Biol Chem 1999; 274:34916–23.
51. Peterson DW, Bradford HF, Mantle PG. Actions of tremorgenic mycotoxin on amino acid transmitter release in vivo. Biochem Pharmacol 1982; 31:2807–10.
52. Norris PJ, Smith CC, De Belleroche J, et al. Actions of tremorgenic fungal toxins on neurotransmitter release. J Neurochem 1980; 34:33–42.
53. Desai K, Sullards MC, Allegood J, et al. Fumonisins and fumonisin analogs as inhibitors of ceramide synthase and inducers of apoptosis. Biochim Biophys Acta 2002; 1585:188–92.
54. Merrill AH, Sullards MC, Wang E, et al. Sphingolipid metabolism: roles in signal transduction and disruption by fumonisins. Environ Health Perspect 2001; 109(Suppl 2): 83–89.
55. Anyanwu E, Campbell A, High W. Brainstem auditory evoked response in adolescents with acoustic mycotic neuromas due to environmental exposure to toxic molds. Int J Adolesc Med Health 2002; 24:67–76.
56. Baldo JV, Ahmand L, Ruff R. Neuropsychological performance of patients following mold exposure. Appl Neuropsychol 2002; 9:193–202.
57. Campbell AW, Anyanwu EC, Vojdani A. Combination of high-dose immunoglobulins and itraconozole in treating chronic mycotic deymelinating optic neuritis. Scientific World Journal. 2003; 3:64–66.
58. Rudich R. Santilli J, Rockwell WJ. Indoor mold spore exposure: a possible factor in the etiology of multifocal choroiditis. Am J Opthalmol 2003; 135:402–04.
59. Coffey PJ, Perry VH, Allen Y, et al. Ibotenic acid induced demyelination in the central nervous system: a consequence of a local inflammatory response. Neurosci Lett 1988; 84:178–84.
60. Spengos K, Schwartz A, Hennerici M. Multifocal demyelination after magic mushroom abuse. J Neurol 2000; 224–25.
61. Shields SA, Gilson JM, Balkemore WF, et al. Remyelination occurs as extensively but more slowly in old rats compared to young rats following gliotoxin-induced CNS demyelination. Glia 1999; 28:77–83.
62. Van Emon JM, Reed AW, Yike I, et al. ELISA measurement of stachylysin™ in serum to quantify human exposures to the indoor mold *Stachybotrys chartarum.* J Occup Environ Med 2003; 45:582–91.
63. Vojdani A, Campbell A, Kashanian A, et al. Antibodies against molds and mycotoxins after exposure to toxigenic fungi in a water-damaged building. Arch Environ Health 2003; 58:324–36.
64. Vojdani A, Thrasher, JD, Madison RA, et al. Antibodies to molds and satratoxin in individuals exposed in water-damaged buildings. Arch Environ Health 2003; 58(7): 421–32.
65. Vojdani A, Kashanian A, Vojdani E, et al. Saliva secretory IgA antibodies against molds and mycotoxins in patients exposed to toxigenic fungi. Immunopharmacol Immunotoxicol 2003; 25:595–614.
66. Pestronk A, Griffin J, Feldman EL, et al. Polyneuropathy

syndromes associated with serum antibodies to sulfatide and myelin-associated glycoprotein. Neurology 1991; 41: 357–62.
67. Vojdani A, Campbell A, Anyanwu E, et al. Antibodies to neural-specific antigens in children with autism: possible cross-reaction with encephalitogenic proteins from milk, *Chlamydia pneumoniae* and *Streptoccocus* A group. J Neuroimmunol 2002; 129:168–77.
68. Vojdani A, Vojdani E, Cooper E. Antibodies to myelin basic protein, myelin oligodendrocytes peptides, α-B-crystallin, lymphocyte activation and cytokine production in patients with multiple sclerosis. J Intern Med 2003; 254:1–12.
69. Dumitru D, Zwarts MJ, Amato AA (Eds). Electrodiagnostic Medicine. Philadelphia: Hanley & Belfus, 1995.
70. Oh SJ. Electrodiagnosis in Diseases of Nerve and Muscle: Prinicples and Practice. 2nd ed. Philadelphia: FA Davis, 1989.
71. DeLisa JA, Mckenzie K, Baran EM. Manual of Nerve Conduction Velocity and Clinical Neurophysiology. 3rd ed. New York: Raven Press, 1994.
72. Steck AJ, Murray N, Dellagi K, et al. Peripheral neuropathy associated with monoclonal IgM autoantibody. Ann Neurol 1987; 22:764–67.
73. Storch MK, Piddlesden S, Haltia M, et al. Multiple sclerosis: in situ evidence for antibody- and complement-mediated demyelination. Ann Neurol 1999; 43:465–71.
74. Nemni R, Fazio R, Quattrini, A, et al. Antibodies to sulfatide and to chondroitin sulfate C in patients with chronic sensory neuropathy. J Neuroimmunol 1993; 43:79–86.
75. Carpo M, Meucci N, Allaria S, et al. Antisulfatide IgM antibodies in peripheral neuropathy. J Neurol Sci 2000; 176: 144–50.
76. Briani C, Berger JS, Latov N. Antibodies to chondroitin sulfate C: a new detection assay and correlations with neurological diseases. J Neuroimmunol 1998; 84: 117–21.
77. Dabby R, Weimer LH, Hays AP, et al. Antisulfatide antibodies in neuropathy. Neurology 2000; 54:1448–52.
78. Alaedini A, Sander HW, Hays AP, et al. Antiganglioside antibodies in multifocal acquired sensory and motor neuropathy. Arch Neurol 2003; 60:42–46.
79. Connolly AM, Pestronk A. Anti-tubulin autoantibodies in acquired deymelinating polyneuropathies. J Infect Dis 1997; 176(Suppl 2):S157–59.
80. Lopate G, Parks BJ, Goldstein J, et al. Polyneuropathies associated with high titre antisulphatide antibodies: characteristics of patients with and without serum monoclonal proteins. J Neurol Neurosurg Psychiatry 1997; 62:581–85.
81. Hao A, Saidi T, Kuroki S, et al. Antibodies to gangliosides and galactocerebroside in patients with Guillian-Barre syndrome with preceding *Campylobacter jejuni* and other identified infections. J Neuroimmunol 1998; 81:116–20.
82. Rosenbluth J, Moon D. Dysmyelination induced in vitro by IgM antisulfatide and antigalactoside monoclonal antibodies. J Neurosci Res 2003; 71:104–09.
83. Busby M, Donaghy M. Chronic dysimmune neuropathy. A subclassification based upon the clinical feature of 102 patients. J Neurol 2003; 250:714–24.
84. McConnel R, Delgado-Tellez E, Cuadra R, et al. Organophosphate neuropathy due to methamidophos: biochemical and neurophysiological markers. Arch Toxicol 1999; 73:296–300.
85. Salih M, Nixon NB, Dawes PT, et al. Prevalence of antibodies to neurofilament polypeptides in patients with rheumatoid arthritis complicated by peripheral neuropathy. Clin Exp Rheumatol 1008; 16:689–94.
86. Willson H, Yuki N. Peripheral neuropathies and anti-glycolipid antibodies. Brain 2002; 125:2591–2625.
87. Walsh TJ, Hier DB, Caplan LR. Aspergillosis of the central nervous system: clinicopathological analysis of 27 patients. Ann Neurol 1985; 18:575–82.
88. Walsh TJ, Hier DB, Caplan LR. Fungal infections of the central nervous system: comparative analysis of risk factors and clinical signs in 57 patients. Neurology 1985; 35:1654–57.
89. Berger T, Rubner P, Schautzer F, et al. Antimyelin antibodies as a predictor of clinically definite multiple sclerosis after a first demyelinating event. N Eng J Med 2003 349:139–45.
90. Reindl M, Linington C, Brehm U, et al. Antibodies against the myelin oligodendrocyte glycoprotein and the myelin basic protein in multiple sclerosis and other neurological diseases: a comparative study. Brain 1999; 122:2047–56.

Antibodies to Molds and Satratoxin in Individuals Exposed in Water-Damaged Buildings

ARISTO VOJDANI
Immunosciences Lab, Inc.
Beverly Hills, California
JACK D. THRASHER
Sam-1 Trust
Alto, New Mexico
ROBERTA A. MADISON
Department of Health Sciences
California State University, Northridge
MICHAEL R. GRAY
Progressive Health Care Group
Benson, Arizona
GUNNAR HEUSER
NeuroMed and NeuroTox Associates
Agoura Hills, California
ANDREW W. CAMPBELL
Medical Center for Immune and Toxic Disorders
Spring, Texas

ABSTRACT. Immunoglobulin (Ig)A, IgM, and IgG antibodies against *Penicillium notatum, Aspergillus niger, Stachybotrys chartarum,* and satratoxin H were determined in the blood of 500 healthy blood donor controls, 500 random patients, and 500 patients with known exposure to molds. The patients were referred to the immunological testing laboratory for health reasons other than mold exposure, or for measurement of mold antibody levels. Levels of IgA, IgM, and IgG antibodies against molds were significantly greater in the patients ($p < 0.001$ for all measurements) than in the controls. However, in mold-exposed patients, levels of these antibodies against satratoxin differed significantly for IgG only ($p < 0.001$), but not for IgM or IgA. These differences in the levels of mold antibodies among the 3 groups were confirmed by calculation of *z* score and by Scheffé's significant difference tests. A general linear model was applied in the majority of cases, and 3 different subsets were formed, meaning that the healthy control groups were different from the random patients and from the mold-exposed patients. These findings indicated that mold exposure was more common in patients who were referred for immunological evaluation than it was in healthy blood donors. The detection of antibodies to molds and satratoxin H likely resulted from antigenic stimulation of the immune system and the reaction of serum with specially prepared mold antigens. These antigens, which had high protein content, were developed in this laboratory and used in the enzyme-linked immunosorbent assay (ELISA) procedure. The authors concluded that the antibodies studied are specific to mold antigens and mycotoxins, and therefore could be useful in epidemiological and other studies of humans exposed to molds and mycotoxins.

<Key words: antibody, *Aspergillus*, ELISA, molds, mycotoxins, *Penicillium*, *Stachybotrys*>

WATER INTRUSION into houses and office buildings leads to the amplification of molds and bacteria.[1–6] These organisms are known to produce toxic byproducts that include endotoxins (lipopolysaccharides), β-D-glucans, and mycotoxins (e.g., trichothecenes, ochratoxins, and aflatoxins), as well as solvents. These compounds have been demonstrated in water-damaged buildings and from artificially contaminated building materials.[7–11] Molds can be aerosolized, resulting in indoor air contaminated with spores and hyphae fragments.[11] In addition, mycotoxins have been identified in ventilation duct dust and in the air in buildings in which occupants have experienced adverse health effects related to mold exposure.[12–18]

Exposure to airborne molds can result in immunologic-mediated illnesses, as well as pathological and inflammatory processes. Immunologic health effects include upper and lower respiratory allergy, asthma, and hypersensitivity pneumonitis.[19–25] Immunoglobulin (Ig)E antibodies to a variety of molds are demonstrable in individuals with suspected allergic sensitivity to molds.[26–29] By using *Stachybotrys* extracts in enzyme-linked immunosorbent assay (ELISA) and radioallergosorbent (RAST) tests, researchers have measured mold-specific IgG and IgE in the blood of individuals exposed to toxigenic fungi.[5] These investigators found no elevated levels of IgG or IgM antibodies to *Stachybotrys chartarum* that had a statistically significant association with health outcome.[5] In a mouse study, an increase in IgA production and IgA nephropathy were reported after injection of trichothecene or vomitoxin.[19] Similar findings were reported when IgE, IgG, and IgA antibodies against *S. chartarum* were identified in patients with asthmatic or mycotoxicosis symptoms.[30] Mycelium extract was used in the ELISA assay, and the *Stachybotrys*-specific IgG and IgA were detected in the group of exposed subjects, but not in control groups. IgA levels were significantly higher ($p < 0.01$) and IgG levels less high ($p < 0.05$) in the patient group than in the control groups. IgE levels did not differ between groups. It has been suggested that exposure to *Stachybotrys* does not cause IgE-mediated allergy in humans and that the IgA response better reflects exposure to the fungus than does the IgG response.[30]

Other investigators have reported that serum IgA level was a more specific factor than IgG level for indicating Farmer's Lung, a disease associated with fungal exposure. They reached the same conclusion concerning antibodies against *Aspergillus fumigatus*.[31,32] When specific IgA was detected in patients' sera, the IgA concentrations in bronchoalveolar lavage were high. This suggests that the concentration of serum IgA may depend on the magnitude of respiratory exposure to fungi.[33] In contrast, IgG antibodies to 8 different molds in subjects with Farmer's Lung disease, compared with healthy farmer controls, were significantly elevated above their healthy counterparts. The Farmer's Lung group had positive titers against several species of mold, whereas the control farmers usually had a positive titer against 1 or 2 microbes. A positive association between IgG antibody levels and chronic bronchitis and diffusing capacity was observed in the Farmer's Lung patients, but not in the control farmers.[34]

Pathological and inflammatory conditions are often caused by molds. Molds and yeast invade tissues in both immunocompromised and normal individuals, establishing an infectious state that is difficult to manage clinically.[35–50] With respect to human infants, *Stachybotrys chartarum* has been associated with pulmonary hemosidersosis.[51–54] The organism *S. chartarum*, isolated from the lungs of an infant, produced a hemolysin (stachylysin), siderophore, and stachyrase A. Apparently, different strains of *S. chartarum* produce different quantities and types of various highly toxic materials, including trichothecenes.[55–61] In an earthworm model, stachylysin increased the permeability of blood vessels, causing leakage through the blood vessel endothelium and walls.[59] Additionally, pathology may result from interference with lung surfactant synthesis by *S. chartarum* spores and isosatratoxin-F in juvenile mice.[62] Ultrastructural changes in Type II alveolar cells (condensed mitochondria, increased cytoplasmic rarefaction, and distended lamellar bodies with irregular shapes) have been observed following exposure to *S. chartarum*.[62–65] Thus, hemolysins, siderophores, and proteases appear to have an important role in the pathogenesis of mold infections.[55–69]

Numerous species of molds, including some found indoors in contaminated buildings, produce mycotoxins.[11,69,70] Controlled laboratory exposure of animals to mycotoxins has pointed to several potential toxic effects on humans (e.g. cytotoxicity, immunosuppression, mitochondrial toxicity, carcinogenicity, nephrotoxicity, and deoxyribonucleic acid [DNA] adducts).[71–87] The adverse health effects from exposure to a mixture of molds and their metabolites are different from those related to a single mycotoxin under controlled conditions. Thus, humans express symptoms that include the central and peripheral nervous systems; neurocognitive deficiencies; problems with skin, upper and lower respiratory tracts, gastrointestinal tract, and skeletal-muscular system; chronic fatigue; and flu-like illnesses.[5,17,23,88–94]

Inasmuch as the respiratory tract is the major route of human exposure to fungi and mycotoxins in water-damaged buildings, we conducted this study to measure specific IgG, IgM, and IgA antibodies to 3 mold species commonly found in such environments. An ELISA procedure to detect antibodies to satratoxin H was also undertaken. Simultaneous detection of antibodies against toxigenic molds and their mycotoxins is considered a biomarker of exposure to molds and their metabolites, and may explain clinical disease processes resulting from mold exposure in the workplace and other environments.[93–95]

Materials and Method

Study subjects. The study population consisted of 500 patients exposed to molds in water-damaged buildings, from 3 different states—143 from California (64 males, 79 females); 159 from Arizona (72 males, 87 females), and 198 from Texas (83 males, 115 females)—who were referred to Immunosciences Lab, Inc. (Beverly Hills, California) by 3 of the authors (MRG, GH, AWC). Patients' ages ranged from 22 to 76 yr. Environmental engineering firms tested approximately 60 damaged buildings by swab samples, tape transfer, and viable microbial activities and reported *Aspergillus*, *Penicillium*, and *Stachybotrys* at a level of > 2,000 colony-forming units per m^3. Patients had lived in these buildings for periods ranging from several weeks to more than 2 yr and had reported unpleasant musty odors to their referring physicians.

The patients had the following symptoms in common: neurological and behavioral symptoms (memory loss, blurred vision, migraine, loss of balance, cognitive deficit), fatigue, nausea, rhinitis, sinusitis, rashes, and allergies. Blood samples were drawn from the patients and sent to Immunosciences Lab (Beverly Hills, California) to test for antibodies against the 3 most common molds detected in their respective buildings. Antibodies to satratoxin H were also determined. For comparison, we used 500 blood samples obtained from healthy, asymptomatic blood donors from California of similar age and sex. An additional 500 individuals, whose blood had been sent to the laboratory for health reasons other than mold exposure, were selected randomly from our patient population. Blood samples from the mold-exposed patients, randomly selected patients, and controls were tested with ELISA for levels of IgG, IgM, and IgA against *Aspergillus niger, Pencillium notatum,* and *Stachybotrys chartarum,* and for satratoxin H antigens.

Preparation of fungal antigen. The protocol we followed for preparation of optimal fungal antigen extracts was based on procedures reported previously.[96–99] Molds (*Stachybotrys chartarum, Penicillium notatum,* and *Aspergillus niger*) were obtained from the American Type Culture Collection (Rockville, Maryland). The molds were first cultured in 2% malt extract agar for 8 days at 25 °C, after which spore suspensions were prepared in 0.1 *M* phosphate-buffered saline (PBS) pH 7.4 containing 0.05% Tween 20 (Sigma Aldrich [St. Louis, Missouri]). One milliliter of each spore suspension was inoculated into 100 ml of 2% malt extract broth (for *Stachybotrys*) or cellulose broth (for the other molds) in glass bottles; the cultures were incubated for 10 days at 25 °C. The mycelium was separated from the broth by centrifugation at 2,000 *g* for 20 min, dried in a vacuum dryer, and stored at –70 °C. Dried mycelium containing spores was suspended at 50 mg/ml in 0.1 *M* PBS pH 7.5 containing 0.02% phenylmethylsulfonyl flouride (PSF) and 0.02% sodium azide. Mycelium suspension was sonicated for 5 sec at an output of 70% using a Virsonic 50 cell disruptor (Virtis Co. [Gardiner, New York]). The sonication step was repeated 10 times for maximum cell disruption, and thereafter the suspension was kept on a shaker for 24 hr at 4 °C. After centrifugation at 4,000 *g*, the supernatant was dialyzed at molecular cut-off 2kD against PBS at 4 °C for 24 hr, lyophilized, and stored at –70 °C. For quality control and reproduction of antigenic preparation of these mold extracts, 20 mg of each was dissolved in 1 ml of 0.01 *M* PBS, the protein content was determined,[28] and the sample's components were analyzed by 15% sodium dodecyl sulfate (SDS) gel electrophoresis. The protein content of our preparations was compared with allergenic extracts for the same molds obtained from Antigen Laboratories, Inc. (Liberty, Missouri). Many more protein bands were present in our preparations than in the commercial antigens; however, a few bands were present in the allergenic extracts but not in our preparations. Consequently, we mixed our antigenic preparations with the Antigen Laboratories extracts, in a ratio of 1:1, and used the combined extracts in our ELISA assays.

Preparation of satratoxin H. Satratoxin H was prepared in accordance with the method of Johanning et al.,[5] with modification in our laboratory. One hundred mg of dried mycelium was extracted with 2 ml of 20% methanol in chloroform at 40 °C, with repeated sonication for 30 min. The extract was passed through a silica gel Column Whatman LPS-1 (Whatman [Clifton, New Jersey]) and washed with 10 ml of 8% methanol in dichloromethane. The eluent was evaporated under a stream of dry nitrogen, and the remaining oily material was dissolved in 1 ml of ethanol and analyzed by reversed-phase high-performance liquid chromatography with a Model 5600 CoulArray Detector with solvent delivery pump Model 580 and an analytical cell that makes use of the 2 porous graphite electrodes (ESA, Inc. [Chelmsford, Massachusetts]). The column was C-18 rainin, 5 μm, 4.6 × 250 mm, with a 15-min gradient of 60–75% methanol in water, flow of 1 ml/min, and monitoring at 260 nm. Two peaks—one at 10.6 and the other at 12.2 min, corresponding in retention time to satratoxin H and other trichothecenes—were obtained. The total satratoxin H and trichothecene obtained in 100 mg of sample were estimated to be about 1.5 μg and 1.7 μg, respectively. Satratoxin H was also obtained from the Department of Microbiology, Texas Tech University, Health Sciences Center (Lubbock, Texas).

Binding of satratoxin to human serum albumin (HSA). Satratoxin was coupled to the carrier protein-HSA with 1-cyclohexyl-3-(2-morpholinoethyl) carbodiimide-metho-4-toluolsufonate (CCMT) and succinic anhydride. One hundred μg of satratoxin in 100 μl pyridine was reacted with 2 mg of succinic anhydride in a 45 °C water bath with off-and-on vortexing. The reaction mixture was then evaporated to dryness. The satratoxin-hemisuccinate was dissolved in 200 μl of dimethyl formamide and added dropwise to a 1 ml solution of 0.1

M carbonate buffer pH 9.5 containing 10 mg of HSA and 2 mg of CCMT. The mixture was kept on the stirrer for 4 hr, followed by the addition of 1 mg CCMT and adjustment of the pH to 7.5 using 1 *M* hydrochloric acid, followed by an additional incubation of 4 hr. Finally, after the addition of another 1 mg of CCMT, the pH was adjusted to 5.5 and the mixture was kept on the stirrer for 4 hr at room temperature. Uncoupled residues of the reagents and derivatives were removed by dialysis at a cutoff of 2,000 against 0.1 *M* PBS pH 7.2 for 48 hr. After centrifugation at 10,000 *g*, binding of satratoxin to HSA was examined by SDS gel electrophoresis. A shift in the location of the HSA band after the addition of mycotoxin was used as evidence for the binding capacity of satratoxin to the carrier protein.

ELISA for detection of IgG, IgM, and IgA against fungal antigens and mycotoxins. The levels of IgG, IgM, and IgA antibodies against antigens of molds in human sera were analyzed by indirect ELISA. Microtiter plates were coated with 0.1 ml of either HSA (in duplicate to serve as controls) or mold extract and satratoxin at a protein concentration of 10 μg/ml. After incubation, washing, and blocking with 2% BSA blocking solution, 0.1 ml of human serum, at optimal dilution of 1:2 for IgE and 1:100 in serum diluent buffer (2% BSA in 0.1 ml PBS plus 0.01% Tween 20) for IgG, IgM, and IgA, were added into the quadruplicate wells of the plates. Plates were incubated at 37 °C for 2 hr and washed 3 times with PBS Tween 20. Then, 0.1 ml of affinity-purified goat anti-human IgG(γ), IgM(μ), or IgA(α) or light chain specific, conjugated with alkaline phosphatase at dilutions of 1:500, were added and incubated at 37 °C for 1 hr. Color development was measured after repeated washing and the addition of 0.1 ml of paranitrophenylphosphate substrate, incubation for 30 min, and stop solution. The intensity of color was measured spectro-photometrically at 405 nm. For each specimen, the ELISA background readings of wells coated with HSA were automatically subtracted from the ELISA readings of the wells coated with mold antigens. The background readings of wells coated with nonspecific antigen (HSA) reacted with the serum and all other reagents and was less than 12% of the absorbencies of the wells coated with mold antigens. Serially diluted sera from rabbits immunized with molds antigens, and from patients following immunotherapy, with assigned values of 400, 3,200, and 12,800 ELISA units against different molds, were used to construct a standard curve. To determine the ELISA values for unknowns, we plotted the mean absorbances obtained from duplicate wells from each calibrator against the antibody concentration, with the absorbance on the vertical axis and concentration on the horizontal axis.

Coefficients of intra-assay and inter-assay variations and optimal dilution of serum for mold and mycotoxin antibodies. Coefficients of intra-assay variations were calculated by running 5 samples 8 times in 1 assay. Coefficients of inter-assay variations were determined by measuring the same samples in 6 consecutive assays. This replicate testing established the validity of the ELISA assays, determined the appropriate dilution with minimal background, and detected serum IgG, IgM, and IgA against different antigens. Two sera from healthy controls and 2 from patients exposed to molds were used to construct control curves.

The optimal dilutions of sera were determined by diluting sera from 20 different controls, and patients sera, 1:25–1:400 in serum diluent buffer and adding them to duplicate wells coated with either HSA or mold antigens. Dilutions between 1:50 and 1:200 resulted in a good linearity; therefore, a serum dilution of 1:100 was used for all IgG, IgM, and IgA antibody assays.

Antibody-specificity testing by absorption of sera. Specificity of the ELISA assay for molds and satratoxin antibodies was confirmed by mold antigens competition. For this, 3 different sera with high levels of IgG, IgM, and IgA antibodies (optical density in ELISA > 0.8) against *Stachybotrys* were used in different test tubes. One ml of each serum was pre-incubated with 1 mg of either HSA, *Stachybotrys, Aspergillus, Penicillium,* or *Alternaria* antigens. After mixing, the tubes were kept for 1 hr in a 37 °C water bath, followed by 1 hr of incubation at 4 °C and centrifugation at 3,000 *g* for 10 min. The supernatant was used for measurement of IgG, IgM, and IgA antibody levels against *Stachybotrys* antigens. ELISA values of unabsorbed serum were compared with those for serum absorbed with HSA or with fungal antigens.

Statistical analysis. The main objective of our data analysis was to examine the differences between the levels of IgG, IgM, and IgA antibodies against different molds and a mycotoxin among 3 groups: (1) controls, (2) random patients, and (3) mold-exposed patients. The IgG, IgM, and IgA were considered as 3 dependent variables with 1 factor variable that divided our samples into the 3 groups mentioned above.

The General Linear Model (GLM) for Windows, version 11.5 (SPSS, Inc. [Chicago, Illinois]), with advanced option, was used in this study. The *z* scores and *p* values were obtained using Hotelling's Trace. In addition, for the post hoc tests, Scheffé's significant difference test was performed on 500 samples in each group. The GLM multivariate procedure provided analysis of variance for multiple dependent variables by 1 or more factor variables.

Results

Fungal contamination of buildings. We were in continuous personal contact with our patients' referring physicians.[93–95,100] Each of the suspect buildings was tested by a different independent environmental firm for the presence of mold and mold spores. All buildings were contaminated with a variety of mold genera. Inasmuch as *Aspergillus* sp., *Pencillium* sp., and *Stachybotrys chartarum* were the most frequently identified molds, we selected *Aspergillus niger*, *Penicillium notatum*, and *Stachybotrys chartarum* as representative organisms. Satratoxin H was incorporated in our study as being representative of trichothecene because this mycotoxin is known to be produced by *S. chartarum*.

Detection of IgG, IgM, and IgA antibodies against molds and satratoxin H. Sera from controls, mold-exposed patients, and randomly selected patients were analyzed for the presence of IgG, IgM, and IgA antibodies to the 3 molds and satratoxin H. The mean titers ± standard deviations (*SD*s), as determined by ELISA for each isotype, along with *z* scores and *p* values, are given in Tables 1–4. The serum dilution used was 1:100, which was determined previously to be optimum.[100] IgG, IgM, and IgA titers against *P. notatum*, *A. niger*, and *S. chartarum* are presented in Tables 1–3, respectively. The *z* scores for all 3 isotypes against *Penicillium*, *Aspergillus*, and *Stachybotrys* were higher than 3.3. IgG, IgM, and IgA antibodies against satratoxin H are given in Table 4. For satratoxin IgG, the differences between the controls and the molds-exposed group were statistically significant ($p < 0.001$; $z = 11.3$).

To examine statistical differences among the 3 groups for the levels of molds and mycotoxins, the post hoc tests (Scheffé's significant difference tests) were performed and were classified under 3 different subsets, as presented in Table 5. Means for groups in homogenous subsets are based on Type III sum of squares (sample size = 1,500, with 500 subsets in each group). Means that are reported in the same subset are statistically similar. For example, means for IgA (satratoxin) for controls, random patients, and mold-exposed patients are 627, 784, and 759, respectively, which are statistically alike. Similarly, the means for *Aspergillus* and *Stachybotrys* IgA in controls and in random patients are classified under the same subset, whereas the means for IgA in mold-exposed groups are statistically different for both controls and random patients. For all other determinations, 3 different subsets are formed, meaning that IgG and IgM against *Penicillium, Aspergillus, Stachybotrys,* and satratoxin are statistically different in controls, random patients, and mold-exposed patients. For IgA against *Penicillium*—although formation of 3 subsets indicates statistical differences among the 3 groups—the IgA values for *Aspergillus* and *Stachybotrys* were reported in 2 subsets, meaning that the mold-exposed group was statistically different from both controls and random patients. And, finally, satratoxin IgA means are classified under the same subset, indicating that no dif-

Table 1.—Antibody Levels (in ELISA Units) in Response to *Penicillium notatum* in Controls, Randomly Selected Patients, and Mold-Exposed Patients, with *z* Scores and *p* Values

	Controls vs. mold-exposed						Random vs. mold-exposed					
	Controls ($n = 500$)		Exposed ($n = 500$)				Controls ($n = 500$)		Exposed ($n = 500$)			
Antibody	$\bar{x}$	*SD*	$\bar{x}$	*SD*	*z*	*p*	$\bar{x}$	*SD*	$\bar{x}$	*SD*	*z*	*p*
IgG	620	535	2,159	2,458	13.7	< 0.001	1,383	1,839	2,159	2,458	5.6	< 0.001
IgM	692	551	1,692	2,442	8.9	< 0.001	1,241	1,530	1,692	2,442	3.5	< 0.001
IgA	640	572	1,256	2,163	6.1	< 0.001	853	1,070	1,256	2,163	3.7	< 0.001

Notes: $\bar{x}$ = mean, *SD* = standard deviation, and Ig = immunoglobulin.

Table 2.—Antibody Levels (in ELISA Units) in Response to *Aspergillus niger* in Controls, Randomly Selected Patients, and Mold-Exposed Patients, with *z* Scores and *p* Values

	Controls vs. mold-exposed						Random vs. mold-exposed					
	Controls ($n = 500$)		Exposed ($n = 500$)				Controls ($n = 500$)		Exposed ($n = 500$)			
Antibody	$\bar{x}$	*SD*	$\bar{x}$	*SD*	*z*	*p*	$\bar{x}$	*SD*	$\bar{x}$	*SD*	*z*	*p*
IgG	618	426	1,795	2,316	11.1	< 0.001	1,349	1,417	1,795	2,316	3.7	< 0.001
IgM	782	420	1,725	2,449	8.5	< 0.001	1,177	1,302	1,725	2,449	4.4	< 0.001
IgA	732	595	1,346	2,456	5.4	< 0.001	849	938	1,346	2,456	4.2	< 0.001

Notes: $\bar{x}$ = mean, *SD* = standard deviation, and Ig = immunoglobulin.

Table 3.—Antibody Levels (in ELISA Units) in Response to *Stachybotrys chartarum* in Controls, Randomly Selected Patients, and Mold-Exposed Patients, with *z* Scores and *p* Values

	Controls vs. mold-exposed						Random vs. mold-exposed					
	Controls (n = 500)		Exposed (n = 500)				Controls (n = 500)		Exposed (n = 500)			
Antibody	$\bar{x}$	*SD*	$\bar{x}$	*SD*	*z*	*p*	$\bar{x}$	*SD*	$\bar{x}$	*SD*	*z*	*p*
IgG	803	530	2,304	2,432	13.5	< 0.001	973	1,234	2,304	2,432	10.9	< 0.001
IgM	629	602	1,940	2,478	11.5	< 0.001	1,115	1,212	1,940	2,478	6.7	< 0.001
IgA	665	665	1,511	2,660	6.9	< 0.001	760	1,086	1,511	2,660	5.8	< 0.001

Notes: $\bar{x}$ = mean, *SD* = standard deviation, and Ig = immunoglobulin.

Table 4.—Antibody Levels (in ELISA Units) in Response to Satratoxin in Controls, Randomly Selected Patients, and Mold-Exposed Patients, with *z* Scores and *p* Values

	Controls vs. mold-exposed						Random vs. mold-exposed					
	Controls (n = 500)		Exposed (n = 500)				Controls (n = 500)		Exposed (n = 500)			
Antibody	$\bar{x}$	*SD*	$\bar{x}$	*SD*	*z*	*p*	$\bar{x}$	*SD*	$\bar{x}$	*SD*	*z*	*p*
IgG	767	641	1,523	1,352	11.3	< 0.001	1,054	1,147	1,523	1,352	5.90	< 0.001
IgM	611	648	1,320	1,590	9.2	< 0.001	1,160	1,170	1,320	1,590	1.80	< 0.060
IgA	715	588	705	868	2.1	< 0.440	747	819	705	868	0.78	< 0.430

Notes: $\bar{x}$ = mean, *SD* = standard deviation, and Ig = immunoglobulin.

ferences were detected among the 3 groups. The statistical differences among the 3 groups for the levels of IgG, IgM, and IgA against molds and mycotoxins were further confirmed by the calculation of exact *z* scores and *p* values, which are provided in Tables 1–4. We considered *p* values < 0.05 and *z* scores > 3.3 to be statistically significant.

Specificity and intra-assay and inter-assay precision. Specificity, and intra-assay and inter-assay precision, for each of the molds, and for several other genera and satratoxin H, have been determined previously.[100] In brief, the coefficients of intra-assay variation calculated for 8 replicates of ELISA assays were 5.7–10.2% for IgG, 5.8–9.2% for IgM, and 5.6–11.3% for IgA. Inter-assay precision was calculated for the same 5 samples assayed in 6 different runs. The inter-assay variations were 7.8–12.7%, 9.5–15.5%, and 10.6–15.3% for IgG, IgM, and IgA, respectively.

Absorption of *Stachybotrys* antibodies with HSA and mold antigens. Similar to our earlier study,[100] 3 different sera with high levels of IgG, IgM, and IgA against *Stachybotrys* were absorbed with nonspecific and specific antigens. Data summarized in Table 6 show that nonspecific proteins, such as HSA, did not change IgG, IgM, and IgA antibody levels against *Stachybotrys*, whereas *Stachybotrys* antigens absorbed the IgG antibody titer levels from 42.7–58%, IgM antibody levels from 21.4–38.5%, and IgA levels from 26.8–34.6%. Similar to IgG, but to a lesser degree, IgM and IgA antibodies were absorbed with *Stachybotrys* antigens. This significant absorption and inhibition of IgG, IgM, and IgA antibodies by fungal antigens is excellent evidence for the specificity of fungal antibodies. Furthermore, other molds antigens, such as *Alternaria*, were not capable of absorbing levels of *Stachybotrys* antibodies, whereas *Penicillium* and *Aspergillus* absorbed only IgG antibody against *Stachybotrys* from 14–20%. This absorption of anti-*Stachybotrys* antibody by *Penicillium* and *Aspergillus* indicates minor antigenic cross-reactivity between these molds.

Discussion

Adverse health effects from exposure to molds in water-damaged buildings can result in allergic reactions, asthma, hypersensitivity pneumonitis, pulmonary infections, and mucous membrane irritation and toxicity. However, despite this variety of adverse health effects, significant emphasis has been placed mainly on Type I allergy and asthma,[20,21,101–103] and not on the other immunopathologic mechanisms involved in the pathogenesis of Types II–IV allergy.[104]

Hundreds of molds, with thousands of antigens, can contaminate indoor air. In addition, some of these

Table 5.—Multiple Comparisons and Means for Groups in Homogenous Subsets Determined with Scheffé's Post Hoc Tests

Mold/mycotoxin and group	IgG subset			IgM subset			IgA subset		
	1	2	3	1	2	3	1	2	3
Penicillium									
Controls	689			748			679		
Random patients		1,547			1,386			944	
Mold-exposed			2,386			1,985			1,841
Aspergillus									
Controls	578			679			688		
Random patients		1,398			1,214		842		
Mold-exposed			2,035			2,147			1,756
Stachybotrys									
Controls	744			617			668		
Random patients		1,083			1,184		815		
Mold-eposed			2,398			2,086			2,153
Satratoxin									
Controls	725			636			627		
Random patients		1,109			1,163		784		
Mold-exposed			1,460		1,405		759		

Notes: Statistical analyses for examination of differences between the levels of immunoglobulin (Ig)G, IgM, and IgA antibodies against molds and a mycotoxin among 3 different groups using the General Linear Model for Windows, version 11.5 (SPSS, Inc. [Chicago, Illinois]). Means for classification of groups in different subsets were based on the sum of squares of sample size of 500 subjects in each of 3 groups. Means reported in the same subset are statistically similar; if they are classified under a different subset it means that controls differed from random patients, as well as from mold-exposed patients.

Table 6.—Optical Densities of Sera with High Levels of Immunoglobulin (Ig)G, IgM, and IgA Antibodies against *Stachybotrys,* before and after Absorption with Nonspecific and Specific Antigens

Absorption status	Sample 1			Sample 2			Sample 3		
	IgG	IgM	IgA	IgG	IgM	IgA	IgG	IgM	IgA
Before absorption									
Optical density	1.54	1.92	1.36	1.10	2.15	1.83	2.30	0.96	0.87
After absorption with:									
Human serum albumin									
Optical density	1.49	1.83	1.27	0.98	2.26	1.95	2.16	0.93	0.91
% inhibition	NS	NS	NS	NS	NS	NS	NS	NS	NS
Stachybotrys chartarum antigens									
Optical density	0.69	1.18	0.89	0.63	1.69	1.34	0.96	0.61	0.57
% inhibition	55.20	38.50	34.60	42.70	21.40	26.80	58.20	36.40	34.50
Aspergillus niger antigens									
Optical density	1.31	1.73	1.27	0.95	1.98	1.76	1.89	0.88	0.78
% inhibition	15.00	10.00	6.60	13.60	8.00	4.00	18.00	8.40	10.30
Penicillium notatum antigens									
Optical density	1.29	1.80	1.26	0.91	1.89	1.81	1.83	0.92	0.79
% inhibition	16.20	6.30	7.30	17.20	12.00	10.90	20.40	4.20	9.20
Alternaria alternata antigens									
Optical density	1.51	1.83	1.29	1.16	2.10	1.72	2.25	0.94	0.80
% inhibition	NS	NS	NS	NS	NS	NS	NS	NS	NS

Note: NS = Nonsignificant.

molds produce potentially toxic metabolites (e.g., mycotoxins and solvents)[3–17,69,70] and can invade tissues (causing aspergilliosis, cryptococcosis, coccidioidomycosis, or pulmonary hemosiderosis)[35–50] by producing proteases, hemolysins, and siderophores.[51–59,61,66–68]

Moreover, the potential cytotoxic action of molds and their metabolites includes DNA adducts,[89,82,85,86] adverse effects on pulmonary surfactant synthesis in rodents,[62–65] mitochondrial toxicity,[77,78,83] apoptosis,[73–76] abnormalities of the human immune system,[93] neurocognitive deficits with changes in electroencephalogram,[89,90,94] peripheral neuropathy,[90] autoimmunity,[93,95] and carcinogenesis.[79,81,87] Because it is becoming increasingly apparent that both atopic and nonatopic individuals experience adverse health reactions to mold exposure unrelated to IgE-mediated sensitivity,[89,90,93–95] biomarkers and clinical tests involving the immune system, the respiratory tree, and the central nervous system must be developed and implemented.[89,93–95,100]

Our data clearly show that exposure to molds in water-damaged buildings leads to the production of IgA, IgM, and IgG antibodies to antigens of *Aspergillus niger, Penicillium notatum, Stachybotrys chartarum,* and satratoxin H in subpopulations of patients, and corroborates previous observations on these and other molds by some of the authors.[100] Previous reports on IgG antibodies against *S. chartarum* and *Aspergillus* have produced equivocal results. In one study,[5] 4 of 48 individuals possibly exposed to *S. chartarum* had IgG antibodies to the organism; other researchers found statistically nonsignificant differences between exposed and control groups.[23,105] On the other hand, a recent cross-sectional comparison of water-damaged or mold-contaminated homes in Finland found fungal-specific IgG concentrations in the sera of patients living in houses both with and without mold, with other cases showing a tendency to exhibit higher antibody levels to most fungi than seen in the control groups.[106]

The differences between our study and others in regard to antibodies to molds might be accounted for by differences in the selection of controls and in the preparation of mold antigens. First, in addition to the controls and patients presented herein, we have tested for antibodies to molds in 500 randomly selected individuals with diagnoses other than mold exposure. The titers to molds IgG, IgM, and IgA, and satratoxin H IgA were intermediate to the controls and patients in this study, but significantly different than both the controls and patients presented herein (Tables 1–5). Some of the controls—who represented a healthy population—probably were exposed to molds. Such exposure likely would result in less statistical significance when the control group is compared with the patients. Therefore, if supposedly true controls (individuals not exposed to molds) could be used in this study, the mean ± *SD* would have been less (estimated at 10–20% lower) than reported in Tables 1–5. We anticipate that this would result in higher *z* scores and lower *p* values, or a greater statistical difference between controls and exposed groups. Moreover, in the selection of controls, one assumes they have a likelihood of exposure. Thus, it is apparent from these observations that selection of controls is critical and that it is probable that clinicians miss molds as a possible cause of the illness for which medical attention is being sought.

The second explanation for the differences between the observations presented herein and those of other investigators likely lies in the preparation of antigens. We prepared our fungal antigens similar to the methods described by others,[5,30] but with some modification.[100] Similar culture and suspension techniques were used, but were modified with repeated sonications during 24 hr of extractions, which resulted in higher yield of fungal antigens in suspension. Comparison of commercially prepared fungal antigens with our preparations showed differences in protein concentrations before and after sonications. Protein concentrations of commercial antigens ranged from 0.4 to 2.1 mg/ml, whereas the protein concentrations used in this study ranged from 0.8 to 3.5 mg/ml. After 10 sonications, our protein concentrations ranged from 5.8 to 16.5 mg/ml. In addition, SDS gel electrophoresis on an equal amount of protein (1 mg) revealed (depending on mold species) 9–22 bands in commercial antigens, 7–16 bands in our preparations before sonications, and 21–36 protein bands after repeated sonications. Therefore, for maximum efficacy, we mixed our mold antigens with commercially available allergenic mold extracts and used 2 µg/well of these mixtures on ELISA plates. With this mixture of mold antigens, we found that human sera diluted 1:100 for IgG, IgM, and IgA in serum diluent resulted in optimum ELISA optical densities. The coefficients of intra-assay and inter-assay variations were less than 16% for all isotypes. IgG antibodies to molds was the lowest (5.7–10.2%) and IgA the highest (5.6–11.3%).[100] Nonspecific proteins (HSA) did not change the levels of these antibodies; however, specific mold antigens extracted from *Stachybotrys* absorbed the IgG antibodies 43–58%, the IgM antibodies 21–38%, and the IgA antibodies 27–34%. Interestingly, when similar absorptions of sera were performed with other mold antigens, absorption of IgG, IgM, and IgA antibodies to *Stachybotrys* with *Alternaria* was insignificant, or less than 10%; however, reaction of the same sera with *Aspergillus* and *Penicillium* was capable of absorbing only the IgG antibody against *Stachybotrys* significantly (14–20%). This minor inhibition of IgG antibody against *Stachybotrys* with *Aspergillus niger* and *Penicillium notatum* antigens is an indication of antigenic cross-reactivity between these molds, which warrants further investigation. The significant inhibition of IgG, IgM, and IgA antibodies against *Stachybotrys* by specific molds and antigens (Table 6) provides further evidence for the specificity of these antibodies. Inhibition of these antibodies with mold antigens, along with the simultaneous presence of IgG, IgM, and IgA antibodies against different molds and satratoxin (Tables 1–4) leads us to conclude that these specific antibodies could be used in subpopulation studies and in epidemiologic investigations of mold and mycotoxin exposure. Furthermore, clinicians may want to consider possible mold exposure in patients who present with multiorgan system abnormalities.

Finally, the detection of IgG antibodies to satratoxin H reveals that the mycotoxin—or the spores and hyphae containing the mycotoxin—can behave as an antigen. This likely occurs by the combination of satratoxin H with carrier mold proteins, and their presentation to the cells involved in the immune system, resulting in subsequent antibody production. Similar observations have been reported for aflatoxin B1, patulin, and ochratoxin A used as a hapten.[107–112]

* * * * * * * * * *

Submitted for publication September 20, 2003; revised; accepted for publication October 30, 2003.

Requests for reprints should be sent to Aristo Vojdani, Ph.D., M.T., 8693 Wilshire Blvd, Suite 200, Beverly Hills, CA 90211

E-mail: immunsci@ix.netcom.com

* * * * * * * * * *

References

1. Gravesen S, Nielsen PA, Iversen R, et al. Microfungal contamination of damp buildings—examples of constructions and risk materials. Environ Health Perspect 1999; 107(suppl 3):505–08.
2. Croft WA, Jarvis BB, Yatawara CS. Airborne outbreak of trichothecene toxicosis. Atmos Environ 1986; 20: 549–52.
3. Peltola J, Andersson MA, Haahtela T, et al. Toxic-metabolite–producing bacteria and fungus in an indoor environment. Appl Environ Microbiol 2001; 67:3269–74.
4. Shelton BF, Kirkland KH, Flanders WE, et al. Profiles of airborne fungi in buildings and outdoor environments in the United States. Appl Environ Microbiol 2002; 68: 1743–53.
5. Johanning E, Biagini R, Hull DL, et al. Health and immunology study following exposure to toxigenic fungi (*Stachybotrys chartarum*) in a water-damaged office environment. Int Arch Occup Environ Health 1996; 68: 207–18.
6. Andersson MA, Nikulin M, Kooljalg U, et al. Bacteria, molds, and toxins in water-damaged building materials. Appl Environ Microbiol 1997; 63:387–93.
7. Nielsen KF, Gravesen S, Nielsen PA. Production of mycotoxins on artificially and naturally infested building materials. Mycopathologia 1999; 145:43–56.
8. Claeson AS, Levin H, Blomquist G, et al. Volatile metabolites from microorganisms grown on humid building materials and synthetic media. J Environ Monit 2002; 4:667–72.
9. Tuomi T, Reijut K, Johnsson T, et al. Mycotoxins in crude building materials from water-damaged buildings. Appl Environ Microbiol 2000; 66:1899–1904.
10. Nieminen SM, Karki R, Auriola S, et al. Isolation and identification of *Aspergillus fumigatus* mycotoxins on growth medium and some building materials. Microbiology 2002; 68:4871–75.
11. Jarvis BB. Chemistry and toxicology of molds isolated from water-damaged buildings. In: DeVries JW, Trucksess MW, Jackson LS (Eds). Mycotoxins and Food Safety. Kluwer Academic/Plenum Publishers, 2002; pp 43–52.
12. Burge HA. Bioaerosols: prevalence and health effects in the indoor environment. J Allergy Clin Immunol 1990; 86:687–704.
13. Richard JL, Plattner RD, May J, et al. The occurrence of ochratoxin A in dust collected from a problem household. Mycopathologia 1999; 146:99–103.
14. Skaug MA, Eduard W, Stormer FD. Ochratoxin A in airborne dust and fungal conidia. Mycopathologia 2000; 151:93–95.
15. Smoragiewicz W, Cossete B, Boutrard A, et al. Trichothecene mycotoxins in the dust of ventilation systems in office buildings. Int Arch Occup Environ Health 1993; 65:113–17.
16. Tuomi T, Saarinene L, Reijula K. Detection of polar and macrocyclic trichothecene mycotoxins from indoor environments. Analyst 1998; 123:1835–41.
17. Jarvis BB, Yatawara CS. Airborne outbreak of trichothecene toxicosis. Atmos Environ 1986; 29:549–52.
18. Johanning E, Gareis M, Nielsen K, et al. Airborne mycotoxins sampling and screening analysis. Proceedings of the 9th International Conference on Indoor Air Quality and Climate (Indoor Air 2002), Monterey, California, June 30–July 5, 2002. Santa Cruz, CA: Indoor Air 2002 Conference Secretariat.
19. Flannigan B, McCabe EM, McGarry F. Allergenic and toxigenic microorganisms in houses. J Appl Bact Sym 1991; 70(suppl):61–73.
20. Jaakkola M, Nordman H, Pilpari R, et al. Indoor dampness and molds and development of adult-onset asthma: a population-based incident case-control study. Environ Health Perspect 2002; 110:543–47.
21. Kurup V, Shen HD, Banerjee B. Respiratory fungal allergy. Microbes Infect 2000; 2:1101–10.
22. Zureik M, Neukirch C, Leynaert B, et al. Sensitization to airborne moulds and severity of asthma: cross sectional study from European Community respiratory health survey. Br Med J 2002; 325(7361):411–14.
23. Hodgson MJ, Morey P, Leung WY, et al. Building-associated pulmonary disease from exposure to *Stachybotrys chartarum* and *Aspergillus versicolor*. J Occup Environ Med 1998; 40(3):241–49.
24. Patel AM, Ryu JH, Reed CE. Hypersensitivity pneumonitis: current concepts and future questions. J Allergy Clin Immunol 2001; 108:661–70.
25. Fan LL. Hypersensitivity pneumonitis in children. Curr Opin Pediatr 2002; 14:323–26.
26. Lander F, Meyer HW, Norm S. Serum IgE specific to moulds, measured by basophil histamine release, is associated with building-related symptoms in damp buildings. Inflamm Res 2001; 50:227–31.
27. Potter PC, Juritz J, Little F, et al. Clustering of fungal allergen-specific IgE antibody responses in allergic subjects. Ann Allergy 1991; 66:149–53.
28. Ezeamuzie CI, Al-Ali S, Kahn M, et al. IgE-mediated sensitization to mould allergens among patients with allergic respiratory diseases in a desert environment. Int Arch Allergy Immunol 2000; 121:300–07.
29. Karlsson-Borga A, Jonsson P, Rolfsen W. Specific IgE antibody to 16 widespread mold genera in patients with suspected mold allergy. Ann Allergy 1989; 63:521–26.
30. Raunio P, Pasanen AL, Husman T, et al. Exposure to *Stachybotrys chartarum* induces immunoglobulin A antibody response in man. In: Johanning E (Ed). Bioaerosols, Fungi and Mycotoxins: Health Effects, Assessment, Prevention and Control. Albany, NY: Eastern New York Occupational and Environmental Health Center, 1999; pp 174–78.
31. Knutsen AP, Mueller KR, Hutcheson PS, et al. Serum

anti-*Aspergillus fumigatus* antibody immunoblot and ELISA in cystic fibrosis with allergic bronchopulmonary aspergilliosis. J Clin Immunol 1994; 93:926–31.
32. Ojanen T, Tukiainen H, Mantyjarvi RA. Class-specific antibodies during follow-up of patients with farmer's lung. Eur Respir J 1990; 3:257–60.
33. Apter AJ, Greenberger PA, Liotta JL, et al. Fluctuation of serum IgA and its subclasses in allergic bronchopulmonary aspergilliosis. J Allergy Clin Immunol 1989; 84: 367–72.
34. Erkinjuntti-Pekkanen R, Reiman M, Kokkarinen JI, et al. IgG antibodies, chronic bronchitis, and pulmonary function values in farmer's lung patients and matched controls. Allergy 1999; 54:1181–87.
35. Marr KA. *Candida* species: emergence of drug-resistant pathogens. Program and Abstracts of the 38th Annual Meeting of the Infectious Diseases Society of America (IDSA). September 2000, New Orleans, Louisiana. Alexandria, VA: IDSA; Abstract S65.
36. Marr KA, Seidel K, White TC, et al. Candidemia in allogeneic blood and marrow transplant recipients: evolution of risk factors after the adoption of prophylactic fluconazole. J Infect Dis 2000; 181:309–16.
37. Wrobel CJ, Chappell ET, Taylor W. Clinical presentation, radiological findings, and treatment results of coccidioidomycosis involving the spine: report on 23 cases. J Neurosurg 2001; 95(suppl 1):33–39.
38. Grossi P, Farina C, Fiocchi R, et al. Prevalence and outcome of invasive fungal infections in 1,963 thoracic organ transplant recipients: a multicenter retrospective study. Italian Study Group of Fungal Infections in Thoracic Organ Transplant Recipients. Transplantation 2000; 70:112–16.
39. Anaissie EJ, Stratton SL, Dignani MC, et al. Pathogenic *Aspergillus* species recovered from a hospital water system: 3-year prospective study. Clin Infect Dis 2002; 34: 780–89.
40. Fraser RS. Pulmonary aspergilliosis: pathologic and pathogenetic features. Pathol Annu 1993; 28:231–77.
41. Ferguson BJ. Mucormycosis of the nose and paranasal sinuses. Otolaryngol Clin North Am 2000; 33:349–65.
42. Eucker J, Sezer O, Graf B, et al. Mucormycoses. Mycoses 2001; 44:253–60.
43. Sumi Y, Nagura H, Takeuchi M, et al. Granulomatous lesions in the lung induced by inhalation of mold spores. Virchows Arch 1994; 424:661–68.
44. Kamei K. Animal models of zygomycosis—*Absidia, Rhizopus, Rhizomucor,* and *Cunninghamella*. Mycopathologia 2001; 152:5–13.
45. Woods JP, Heinecke EL, Luecke JW, et al. Pathogenesis of *Histoplasma capsulatum*. Semin Respir Infect 2001; 16:91–101.
46. Bhargava D, Bhusnurmath B, Sundaran KR, et al. Tonsillar actinomycosis: a clinicopathology study. Acta Trop 2001; 80:163–68.
47. Feldman BS, Snuder LS. Primary pulmonary coccidioidomycosis. Semin Respir Infect 2001; 16:231–37.
48. Liratsopulos G, Ellis M, Nerringer R, et al. Invasive infection due to *Penicillium* species other than *P. marneffei*. J Infect 2002; 45:184.
49. Dosa E, Coczi I, Mojzes EG, et al. Identification and incidence of fungal strains in chronic rhinosinusitis patients. Acta Microbiol Immunol Hung 2002; 49:337–46.
50. Taylor MJ, Pnikaue JU, Sherris DA, et al. Detection of fungal organisms in eosinophilic mucin using a fluorescein-labeled chitin-specific binding protein. Otolaryngol Head Neck Surg 2002; 127:377–83.
51. Etzel RA, Montana E, Sorenson WB, et al. Pulmonary hemorrhage in infants associated with exposure to *Stachybotrys atra* and other fungi. Arch Pediatr Adolesc Med 1998; 152:757–62.
52. Dearborn DG, Yike I, Sorenson WG, et al. An overview of the investigations into pulmonary hemorrhage among infants in Cleveland, Ohio. Environ Health Perspect 1999; 107(suppl):495–99.
53. Montana E, Etzel RA, Allan T, et al. Environmental risk factors associated with pediatric idiopathic pulmonary hemorrhage and hemosiderosis in a Cleveland community. Pediatrics 1997; 99:E5.
54. Montana E, Etzel RA, Dearborn DG, et al. Acute pulmonary hemorrhage in infancy associated with *Stachybotrys atra*, Cleveland, Ohio. Am J Epidemiol 1995; 141:S83.
55. Elidemer O, Colasurdo GN, Rossman SN, et al. Isolation of *Stachybotrys* from the lung of a child with pulmonary hemosiderosis. Pediatrics 1999; 104:964–66.
56. Vesper SJ, Dearborn DG, Elidemer O, et al. Quantification of siderophore and hemolysin from *Stachybotrys chartarum* strains, including a strain isolated from the lung of a child with pulmonary hemorrhage and hemosiderosis. Appl Environ Microbiol 2000; 66:2678–81.
57. Vesper SJ, Magnusson H, Dearborn DG, et al. Initial characterization of the hemolysin stachylysin from *Stachybotrys chartarum*. Infect Immun 2001; 69:912–16.
58. Vesper SJ, Vesper MJ. Stachylysin may be a cause of hemorrhaging in humans exposed to *Stachybotrys chartarum*. Infect Immun 2002; 70:2065–69.
59. Vesper S, Dearborn DG, Yike I, et al. Evaluation of *Stachybotrys chartarum* in the house of an infant with pulmonary hemorrhage: quantitative assessment before, during, and after remediation. J Urban Health 2000; 77: 68–85.
60. Jarvis BB, Sorenson WG, Hintikka EL, et al. Study of toxin production by isolates of *Stachybotrys chartarum* and *Memnoniella echinata* isolated during a study of pulmonary hemosiderosis in infants. Appl Environ Microbiol 1998; 64:3620–25.
61. Kordula T, Banbula A, Macomson J, et al. Isolation and properties of Stachyrase A, a chymotrypsin-like serine proteinase from *Stachybotrys chartarum*. Infect Immun 2002; 70:419–21.
62. Rand TG, Mahoney M, White K, et al. Microanatomical changes in alveolar type II cells in juvenile mice intrathecally exposed to *Stachybotrys chartarum* spores and toxin. Toxicol Sci 2002; 65:239–45.
63. Mason CD, Rand TG, Oulton M, et al. Effects of *Stachybotrys chartarum* (*atra*) conidia and isolated toxin on lung surfactant production and homeostasis. Nat Toxins 1998; 6:27–33.
64. McCrae KC, Rand T, Shaw RA, et al. Analysis of pulmonary surfactants by Fourier-transform infrared spectroscopy following exposure to *Stachybotrys chartarum* (*atra*) spores. Chem Phys Lipids 2001; 110:1–10.
65. Mason CD, Rand TG, Oulton M, et al. Effects of *Stachybotrys chartarum* on surfactant convertase activity in juvenile mice. Toxicol Appl Pharmacol 2001; 172:21–28.
66. Kudo Y, Ootani T, Kumangai T, et al. A novel oxidized low-density lipoprotein-binding protein, Asp-hemolysin recognizes lysophosphatidylcholine. Biol Pharm Bull 2002; 25:787–90.
67. Ebina K, Ichinowatari S, Yokota K. Studies on toxin *Aspergillus fumigatus*. XXII. Fashion of binding Asp-hemolysin to human erythrocytes and Asp-hemolysin-binding proteins of erythrocyte membranes. Microbiol

Immunol 1985; 29:91–101.
68. Monod M, Capoccia S, Lechene B, et al. Secreted proteases from pathogenic fungi. Int J Med Microbiol 2002; 292:405–19.
69. Jarvis BB, Salemme J, Morais A. *Stachybotrys* toxins. Nat Toxins 1995; 3:10–16.
70. Jarvis BB, Zhou Y, Wang S. Toxigenic molds in water-damaged buildings: dechlorogriseofulvins from *Memnoniella echinata*. J Nat Prod 1996; 59:553–54.
71. Gareis M. Cytotoxicity testing of samples originating from problem buildings. In: Johanning E, Yang CS (Eds). Proceedings of the International Conference: Fungi and Bacteria in Indoor Air Environments: Health Effects, Detection and Remediation, Saratoga Springs, New York, October 1994. Albany, NY: Eastern New York Occupational and Environmental Health Center, 1995; pp 139–44.
72. Jakab GJ, Hmieleski RR, Hemenway DR, et al. Respiratory aflatoxicosis: suppression of pulmonary and systemic host defenses in rats and mice. Toxicol Appl Pharmacol 1994; 125:198–205.
73. Nagata T, Suzuki H, Ishigami N, et al. Development of apoptosis and changes in lymphocyte subsets in thymus, mesenteric lymph nodes and Peyer's patches of mice orally inoculated with T-2 toxin. Exp Toxicol Pathol 2001; 52:309–15.
74. Jones C, Ciacci-Zanella JR, Zhang V, et al. Analysis of fumonisin B1-induced apoptosis. Environ Health Perspect 2001; 109(suppl 2):315–20.
75. Poapolathep A, Ohtsuka R, Kiatipattanasakul W, et al. Nivalenol-induced apoptosis of thymus, spleen, and Peyer's patches of mice. Exp Toxicol Pathol 2002; 53: 441–46.
76. Desai K, Sullards MC, Allegood J, et al. Fumonisins and fumonisin analogs as inhibitors of ceramide synthase and inducers of apoptosis. Biochim Biophys Acta 2002; 1585:188–92.
77. Pace JG. Effect of T-2 mycotoxin on rat liver mitochondria electron transport system. Toxicon 1983; 21: 675–80.
78. Pace JG. T-2 mycotoxin inhibits mitochondrial protein synthesis. Toxicon 1988; 26:77–85.
79. Pfohl-Leszkowicz A, Petkova-Bocharova T, Chernozemsky IN, et al. Balkan endemic nephropathy and associated urinary tract tumors: a review on aetiolgical causes and the potential role of mycotoxins. Food Addit Contam 2002; 19:282–302.
80. Pfohl-Leszkowicz A, Grosse Y, Kane A, et al. Differential DNA adduct formation and disappearance in three mouse tissues after treatment with mycotoxin ochratoxin A. Mutat Res 1993; 289:265–73.
81. Schwartz GG. Does ochratoxin A cause testicular cancer? Cancer Causes Control 2002; 13:91–100.
82. Petkova-Bochatrova T, Stoichev II, Chernozemsky IN, et al. Formation of DNA adducts in tissue of mouse progeny through transplacental contamination and/or lactation after administration of single does of ochratoxin A to the pregnant mother. Environ Mol Mutagen 1998; 32: 155–62.
83. Hochler D, Marquardt RR, McIntosh AR, et al. Induction of free radicals in hepatocytes, mitochondria and microsomes of rats by ochratoxin A and its analogs. Biochim Biophys Acta 1997; 1357:225–33.
84. Sajan MP, Satav JG, Battacharya RK. Effect of aflatoxin B1 in vitro on rat liver mitochondrial respiratory functions. Indian J Exp Biol 1997; 35:1187–90.
85. Hisch LL, Hisch TT. Detection of aflatoxin B1-DNA adducts in human placenta and cord blood. Cancer Res 1993; 53:1278–80.
86. Miranjan BF, Bhat NK, Avadhani NG. Preferential attack of mitochondrial DNA by aflatoxin B1 during hepatocarcinogenesis. Science 1982; 214(4528):73–75.
87. Dominguez-Malagon H, Gaytan-Graham S. Hepatocellular carcinoma: an update. Ultrastruct Pathol 2001; 25: 497–516.
88. Croft WA, Jastromski BM, Croft AL, et al. Clinical confirmation of trichothecene mycotoxicosis in patient urine. J Environ Biol 2002; 23:301–20.
89. Kilburn KH. Inhalation of moulds and mycotoxins. Eur J Oncol 2002; 7:197–202.
90. Anyanwu EC, Campbell AW, Vojdani A. Neurophysiological effects of chronic indoor environmental mold exposure on children. Scientific World Journal 2003; 3: 281–90.
91. Mahmoudi M, Gershwin ME. Sick building syndrome. III. *Stachybotrys chartarum*. J Asthma 2000; 37:191–98.
92. Walinder R, Wieslandr G, Norback D, et al. Nasal lavage biomarkers: effects of water damage and microbial growth in an office building. Arch Environ Health 2001; 56:30–36.
93. Gray MR, Thrasher JD, Crago R, et al. Mixed mold mycotoxicosis: immunological changes in humans following exposure in water-damaged buildings. Arch Environ Health 2003; 58(7):410–20.
94. Crago R, Gray MR, Nelson LA, et al. On the neuropsychological and electrocortical impact of mixed mold exposure. Arch Environ Health 2003; 58(8). Forthcoming.
95. Campbell AW, Thrasher JD, Madison RA, et al. Autoantibodies and neurophysiologic abnormalities in patients exposed to molds in water-damaged buildings. Arch Environ Health 2003; 58(8). Forthcoming.
96. Achatz G, Oberkofler H, Lechenauer E, et al. Molecular cloning of major and minor allergens of *Alternaria alternata* and *Cladosporium herbarum*. Mol Immunol 1995; 32:213–27.
97. Paris S, Fitting E, Ramirez E, et al. Comparison of different extraction methods of *Alternaria* allergens. J Allergy Clin Immunol 1990; 85:941–48.
98. Portnoy J, Pacheco F, Ballam Y, et al. The effect of time and extraction buffers on residual protein and allergen content of extracts derived from four strains of *Alternaria*. J Allergy Clin Immunol 1993; 91:930–38.
99. Raunio P, Karkkainen M, Virtanen T, et al. Preliminary description of antigenic components of *Stachybotrys chartarum*. Environ Res 2001; 85:246–55.
100. Vojdani A, Campbell A, Kashanian A, et al. Antibodies against molds and mycotoxins following exposure to toxigenic fungi in a water-damaged building. Arch Environ Health 2003. 58(6):324–36.
101. Brostoff J. Immunological mechanisms. In: Brostoff J, Challacombe SJ (Eds). Food Allergy and Intolerance. Eastborne, England: W.B. Saunders, 1976; pp 433–55.
102. Saxon A, Diaz-Sanchez D, Zhang K. The allergic response in host defense. In: Rich RR, Fleisher TA, Schwartz BD, et al. (Eds). Clinical Immunology. St. Louis, MO: Mosby, 1995; pp 847–69.
103. Gent JF, Ren P, Belanger K, et al. Levels of household mold associated with respiratory symptoms in the first year of life in a cohort at risk for asthma. Environ Health Perspect 2002; 110:A781–86.
104. Gell PGH, Coombs RRA. Clinical Aspects of Immunology. Oxford, U.K.: Blackwell, 1963.
105. Johanning E, Landsbergis P, Gareis M, et al. Clinical ex-

perience and results of a sentinel health investigation related to indoor fungal exposure. Environ Health Perspect 1999; 107(suppl 3):189–94.
106. Hyvarinen A, Reiman M, Meklin T, et al. Fungal exposure and IgG-levels with and without mold problems. In: Johanning E (Ed). Bioaerosols, Fungi and Mycotoxins: Health Effects, Assessment, Prevention and Control. Albany, NY: Eastern New York Occupational and Environmental Health Center, 1999; 99:166–68.
107. Autrup H, Sermet T, Wakhisi J. Evidence for human antibodies that recognize aflatoxin epitope in groups with high and low exposure to aflatoxins. Arch Environ Health 1990; 45:31–34.
108. Wild CP, Jiang YX, Sabbioni G, et al. Evaluation of methods for quantitation of aflatoxin-albumin adducts and their application to human exposure assessment. Cancer Res 1990; 50:245–51.
109. Gathumbi JK, Usleber E, Martlbauer E. Production of ultra-sensitive antibodies against aflatoxin B1. Lett Appl Microbiol 2001; 32:349–51.
110. McElroy JL, Weiss CM. The production of polyclonal antibodies against the mycotoxin derivative patulin hemiglutarate. Can J Microbiol 1993; 39:861–63.
111. Solti L, Salamon F, Barna-Vetro I, et al. Ochratoxin A content of human sera determined by a sensitive ELISA. J Anal Toxicol 1997; 21:44–48.
112. Breitholtz EA, Hult K. Enzyme-linked immunosorbent assay for analysis of ochratoxin A in human plasma samples using antibodies raised against ochratoxin-protein conjugate. IARC Sci Publ 1991; 115:89–92.

Chronic Sinusitis: Defective T-Cells Responding to Superantigens, Treated by Reduction of Fungi in the Nose and Air

DONALD P. DENNIS
Atlanta Center for ENT and Facial Plastic Surgery
Atlanta, Georgia
and
Department of Otolaryngology—Head and Neck Surgery
Northside Hospital
Atlanta, Georgia
and
Department of Otolaryngology—Head and Neck Surgery
Piedmont Hospital
Atlanta, Georgia

ABSTRACT. In this study, the author used endoscopic sinus photography to study the effects of reduction of fungi in the nose, and in environmental air, on the sinus mucosa of 639 patients diagnosed with chronic rhinosinusitis. Sinus mucosal photographs were taken before and after reduction of fungal load in the nose and air, to determine if there was an optimum environmental air fungal load associated with sinus mucosal recovery to normal appearance. Systemic symptoms associated with fungal exposure, which resolved when fungus was removed from the patient and the environmental air and reappeared with recurrent environmental fungal exposure, are also discussed and are termed *systemic fungal symptoms.* Interventions consisted of nasal fungal load reduction with normal saline nasal irrigations and antimicrobial nasal sprays, and environmental air fungal load reduction with high-efficiency particulate air (HEPA) filtration in combination with ionizers or evaporation of a solution of botanical extract. Main outcome measures were obtained with environmental air 1-hr gravity-plate fungal colony counts, laser air particle counts, and endoscopic sinus photography. Blood levels of immunoglobulins IgG and IgE for 7 common molds were also determined. After intervention, 94% of patients who used antimicrobial nasal sprays and who reduced their environmental fungal air count to 0–4 colonies per 1-hr agar gravity-plate exposure (*n* = 365) exhibited normal sinus mucosa by endoscopic exam. Environmental air fungal counts that exceeded 4 colonies resulted in sinus mucosal abnormalities ranging from edema, to pus and/or nasal polyps at higher counts. Neutralization of allergy, and/or surgery, were used as appropriate following implementation of environmental measures. On the basis of these observations, as well as detailed clinical experience and a review of the current literature, the author hypothesizes that the pathogenesis of chronic rhinosinusitis, allergic fungal sinusitis, and systemic fungal symptoms is a genetic defect at the variable beta chain helper T-cell receptor (TCR Vβ) site which requires the presence of an antigen (fungus). Chronic sinusitis patients who have recurring exposure to environmental air that contains fungal concentrations in excess of 4 colonies per 1-hr agar plate exposure appear to have an increased risk of persistent chronic sinusitis and/or systemic symptoms, regardless of the medical treatment provided.

<Key words: allergic fungal sinusitis, mold, rhinosinusitis, sinusitis, superantigen, T-cell receptor>

CHRONIC RHINOSINUSITIS (CRS) affects approximately 37 million Americans, or 1 in 6 (16.3%). It is more common than arthritis (12.47%), orthopedic impairment (12.14%), or hypertension (11.44%). CRS costs patients and insurance companies over $2.4 billion per year for medication, hospitalization, and surgery. In excess of 200,000 sinus surgeries are performed every year in the U.S. Many patients remain refractory to surgery and antibiotic therapy, despite the more than 46.9 million prescription and nonprescription medications ordered annually.[1] Antibiotics are not effective in treating CRS because they target bacterial super-growth and not the underlying fungal problem. In the most recent decade, the rate of CRS occurrence has been increasing steadily,[1] but the pathogenesis of chronic sinusitis has not yet been determined. The standard school of thought is that fungus allergy is involved in fewer than 10% of cases,[1] likely because fungi are visible endoscopically in the nose in less than 10% of cases studied. However, fungi are present microscopically in *93% of cases* examined by culture.[2] By definition, all CRS patients also have secondary bacterial infections. It is likely that the immune reaction to microscopic fungi causes mucosal pitting and mucous stasis, both of which lead to the development of secondary bacterial infections.[2]

In 1999, researchers at the Mayo Clinic demonstrated a causal connection between fungi and CRS. They found that 93% of all CRS cases also met the diagnostic criteria for allergic fungal sinusitis (AFS). It was postulated that an immune reaction to fungi in the mucosa is likely responsible for AFS *and* most CRS.[2] This fact was confirmed by Braun et al.[3] in 2003. In addition, immunoglobulin (Ig)E-mediated hypersensitivity was not present in the majority of cases studied, regardless of whether nasal polyps were present.

Because 93% of CRS cases were found to meet the diagnostic criteria for AFS, it is likely that AFS and CRS represent varying degrees of allergic response to the same fungal antigens, with AFS representing more extensive production of polyps and allergic mucin. Therefore, AFS and CRS will be used synonymously in this article. The Mayo Clinic's diagnostic criteria for AFS are (a) CRS; (b) presence of allergic mucin (clusters of eosinophils and their byproducts, such as Charcot-Leyden crystals and major basic protein); and (c) presence of fungal organisms in the mucin, confirmed by histology or culture, or both.[2] In the present study, radioallergosorbent (RAST) delayed mold reaction testing revealed elevated fungus-specific IgG levels to 2 or more of 7 common fungi in all patients.

In CRS patients, the nasal mucous contains eosinophils, Charcot-Leyden crystals, IgG fungal antibodies, no helper T-lymphocytes, and no antigen-processing cells (APCs).[2] Peripheral blood and nasal mucosa of CRS patients contain fungal-specific elevated IgG and fungal antigens. These antigens activate helper T-lymphocytes in the blood, producing cytokines (interleukin [IL]-5 and IL-13) that recruit eosinophils; lymphocytes from normal controls do not recruit eosinophils in response to fungal antigens.[4] This fact is key to our hypothesis because it means that helper T-cells in CRS patients are bypassing APCs, and are therefore likely to have defective receptor sites.

IL-13 causes immunoglobulin isotype switching (e.g., to IgE and IgG_1) and IgG production by immature B lymphocytes. IL-5 promotes eosinophil activation and activates B cells for terminal differentiation into Ig-secreting cells.[4] Thus, both eosinophils and IgG are ready to migrate through the mucous membrane and attack the mold in the nasal mucous—a classic Type II hypersensitivity reaction (Fig. 1).

Waxman et al.[5] proposed a Gell-Coombs Type III hypersensitivity reaction[6] in *Aspergillus* sinusitis and chronic bronchopulmonary disease. In AFS, the sinus mucosa contains small lymphocytes, plasma cells, and eosinophils—a condition similar to that seen in asthmatic bronchial mucosa.[7]

Systemic immune-complex–mediated disease (Type III hypersensitivity reaction) affects tissues of the kidneys, joints, skin, heart, and serosal surfaces. The reason for this specific organ/tissue predilection is unknown. More work is needed to determine if a Type III hypersensitivity reaction to fungus is responsible for the arthritis and other systemic fungal symptoms seen clinically.

The only real difference between Type II and Type III reactions, then, is in the location of the antigen. Type II reactions are against antigens located on cell surfaces or on extracellular matrix components, whereas Type III reactions occur against soluble or circulating antigens. Systemic lupus erythematosis and most types of glomerulonephritis are the result of Type III immune injury.[8] Therefore, in all types of allergic hypersensitivity reactions, when the antigen (fungus) is removed the reaction stops. This fact forms the basis for our CRS treatment goal: to remove fungal antigen from both the nose and the environmental air, in order to stop the fungal immune reaction and allow the sinus mucosa to recover.

Researchers have hypothesized that the immunopathology affecting more than 16% of the population is the result of a genetic defect in 1 or more of 9 genes (genes 2, 3, 5.1, 6.7, 8, 12, 13.1, 13.2, 17) on the outside of the beta chain variable region of the T-cell receptor site (TCR Vβ genes)[9–11] (Fig. 2). Normally, the antigen (fungus) must first be processed by an APC into smaller fragments, and these fragments must be transported to the human leukocyte antigen (HLA) class II molecule on the surface of the APC, in order for the TCR site to recognize the antigen linked with the HLA II molecule and bind to it,[6] causing the T-cell to release IL-5 and IL-13. Because the helper T-lymphocytes of CRS patients produce IL-5 and IL-13 when presented with fungal antigens, without APCs,[4] the fungus is likely bypassing the APC HLA II molecule (the antigen-specific area) and binding directly to the outside of the TCR Vβ site via defective gene(s) on the TCR Vβ chain. This genetic defect allows a microbe or toxin superantigen to bind to HLA class II histocompatibility molecules on APCs and TCR sites simultaneously, bypassing APC-induced activation of T-cells.[9,10,12,13] The result is pro-

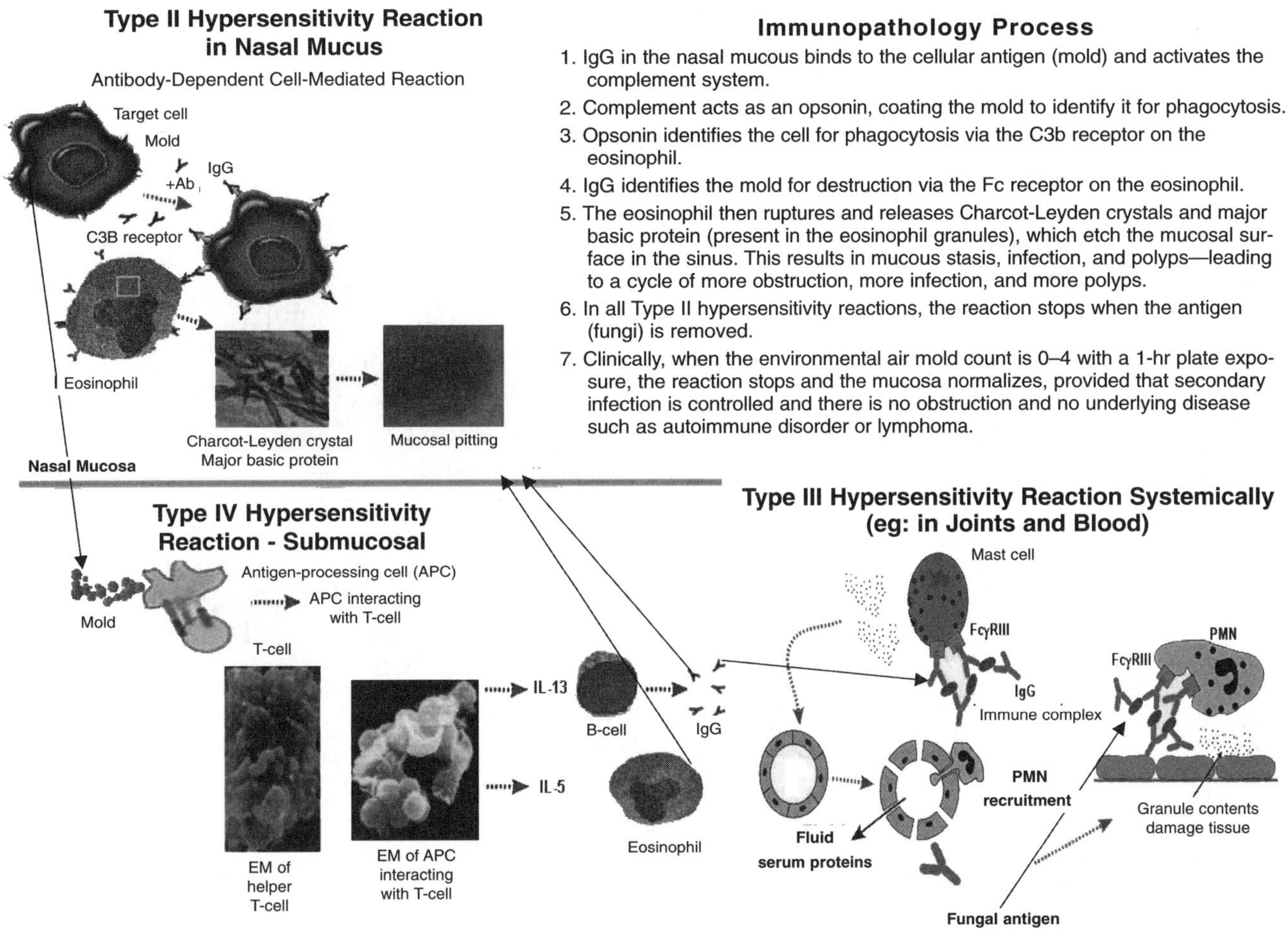

Fig. 1. Hypothesized cellular immunopathology mechanism. Top illustration shows a Gell-Coombs Type II hypersensitivity[6] antibody-dependent cell-mediated reaction. Fungi from nasal mucous migrate into the nasal mucosa, causing a Type IV reaction within the mucosa. Immunoglobulins (Ig)G to fungi, and eosinophils, migrate from the mucosa to the nasal mucous, perpetuating the Type II reaction. In patients with a variable β chain helper T-cell receptor site defect, more fungal exposure results in production of more fungal-specific IgG, which drives a Type III hypersensitivity reaction to cause additional systemic fungal symptoms. *Notes*: PMN = polymorphonuclear cell, and EM = electron micrograph.

found activation of up to 30% of the body's total T-cells, in contrast with the normal T-cell response of < 0.01%.[14] The subsequent superantigen-induced T-cell activation can cause systemic toxicity and is the likely mechanism for fungal-induced sinusitis and systemic disease.[9]

Superantigens are thought to be associated with systemic disease (e.g., human retrovirus in breast cancer,[9,15] Type I diabetes, and multiple sclerosis[16]). Superantigens have also been linked causally to psoriasis[17–19] and rheumatoid arthritis.[20,21] Some fungi are thought to function as superantigens[4,9] Our clinical observations have supported such an association because both the CRS and systemic fungal symptoms (e.g., arthritis, memory loss, 6th nerve palsy, dizziness, hearing loss, seizures, fatigue, vision problems, severe headaches, gastroesophageal reflux disease, gastrointestinal disturbances, asthma, IgG subclass deficiency, and fibromyalgia) have resolved with the use of antimicrobial nose sprays and environmental fungal removal protocol, and return when the protocols are discontinued or when fungal levels in air rise. This retrospective observational study and literature review were conducted to formulate an hypothesis for the cause of CRS and systemic symptoms associated with fungal exposure.

Materials and Method

Study subjects. Between 1989 and 2003, 639 patients age 8–76 yr (average age = 46 yr; 381 females and 258 males) diagnosed with CRS who had failed traditional antibiotic and/or surgical therapy were studied in our private practice otolaryngology clinic. Findings required for inclusion in the study were (a) a diagnosis of CRS on the basis of patient history, (b) abnormal endoscopic sinus exam, and (c) abnormal sinus computed

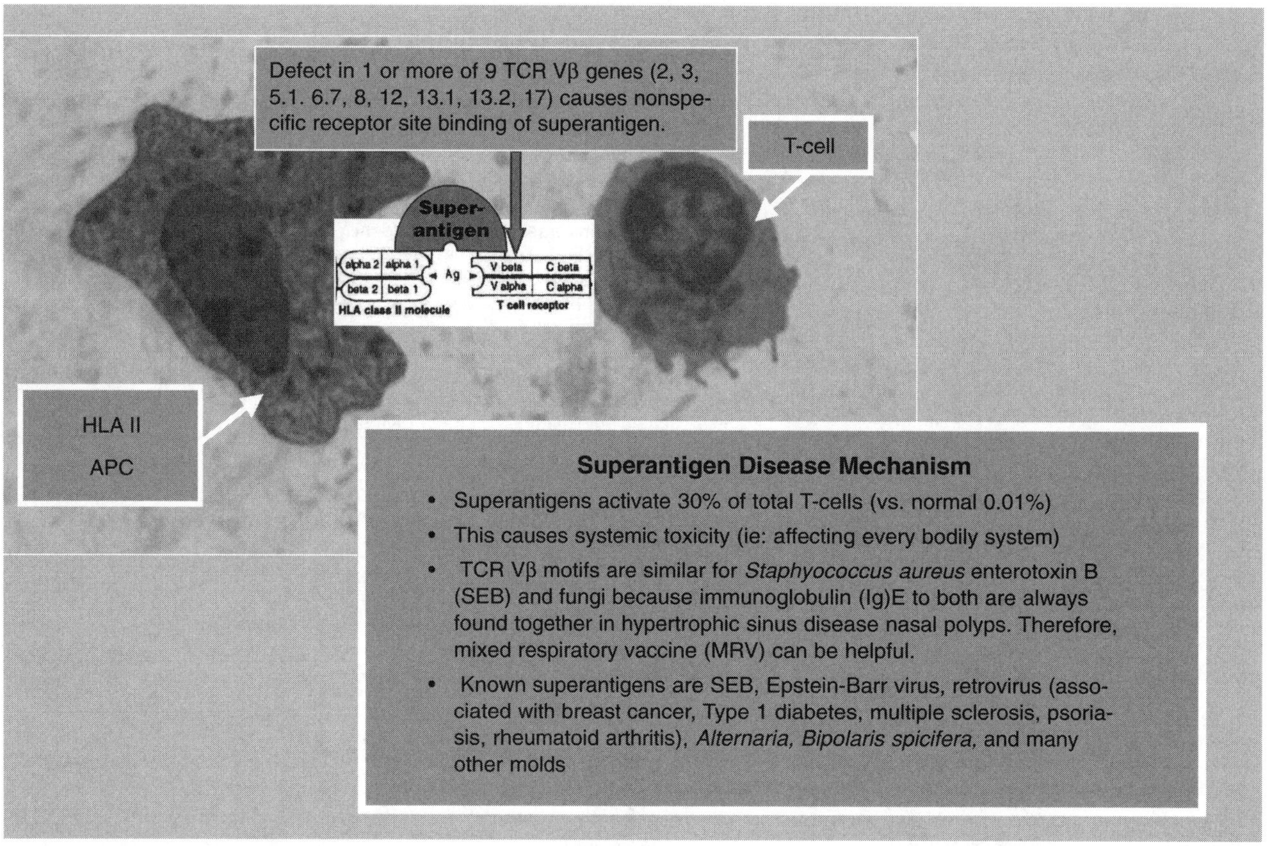

Fig. 2. Superantigen disease mechanism. Fungal superantigens bind to the side of the alpha 1 chain of the antigen-processing cell (APC) and to the side of the variable beta chain helper T-cell receptor (TCR Vβ) site simultaneously, bypassing antigen specificity and causing APC-induced T-cell activation.[8,9,11,12]

tomography (CT) scan. Symptoms must have been present for at least 3 mo and had to include 2 or more of the following: (a) facial pain or pressure, (b) facial congestion or fullness, (c) nasal obstruction or blockage, (d) nasal discharge/purulence/discoloration, (e) postnasal drainage, or (f) hyposmia/anosmia. Sinusitis must have been present for at least 3 mo and been treated with antibiotics for 4–6 wk (or 4 or more sinusitis episodes per year treated with antibiotics for 7–10 days each), with symptoms persisting or recurring after cessation of antibiotic treatment.

Treatment protocols. Both the patients themselves and the air in their home, office, and/or car environments (wherever fungal load was found to be elevated) were treated to reduce fungal load. Treatment of patients consisted of (a) normal saline nasal irrigations twice daily to remove fungi mechanically and (b) 2 antimicrobial nasal sprays (containing 2 structurally different antibiotics, an antifungal, and a steroid), administered as 3 sprays, 4 times daily for 4–10 wk, depending on the time required to clear the sinus mucosa, as verified by endoscopic exam. Our nasal spray protocol is summarized in Table 1.

Environmental remediation consisted of reducing air fungal load by following an environmental treatment protocol that included finding and repairing moisture intrusion and implementing portable high-efficiency particulate air (HEPA) filtration or inline central heating, ventilaton, and air conditioning system (HVAC) filtration (94% efficient at 0.35-μm particle size), in combination with either ionizers or the disbursement of botanical extract by evaporation. The odorless botanical extract was dispersed by the patient or environmental consultant by placing 2–4 foil-covered containers, with wicks perforating the foil, on top of a portable HEPA air filtration unit that exited clean air from the top of the machine. These setups were placed in each room used by the patient. In some cases, Room VI 2500 ionizers were used in combination with portable HEPA air filters, instead of the botanical extract. (Botanical extract and an ionizer could not be used in the same room because the ionizer destroys the activity of the extract.) With the patients' permission, sources of moisture intrusion were identified and controlled. Cars were treated by spraying botanical extract into the HVAC system from the outside suction vent at the windshield on both

Table 1.—Antimicrobial Nasal Spray Protocol

1. Begin clindamycin and garamycin sprays (all antibiotic sprays contain an antifungal, clotrimazole or other) with steroids.
   —Omit steroids if patient has steroid difficulty.
   —Add monteleukast for nasal polyps and for maintenance therapy.
   —If patient is allergic to clindamycin, use cefazolin or trimethoprim/sulfamethoxazole.
   —If patient is allergic to garamycin, use ciprofloxacin.
2. If fungus is visible, or if there is a history of significant fungal exposure, use amphotericin-B, clindamycin, and gentamycin sprays.
   —Use amphotericin 1st (unstable), then the clindamycin and gentamycin sprays.
   —Always use all sprays together: 3 sprays qid.
3. Use sprays for 2–8 wk, or until mucosa is clear by sinus endoscopy or computed tomography sinus scan.
4. Maintain with:
   —Agumax® (4 different citrus seed and citrus skin extracts) mixed with betadine and a steroid.
   —Monteleukast nose sprays: 3 sprays each nostril bid (qid when traveling or after exposure).

Note: All spray formulas, and pharmacist training, provided by PharmaSource Int'l, Inc. (Denver, Colorado).

sides, and on the inside of the car. (Portable HEPA filters, botanical extract containers, and ionizers were supplied by National Allergy Supply, Inc. [Atlanta, Georgia] and PharmaSource Int'l, Inc. [Denver, Colorado]; central inline air filters and moisture intrusion correction were supplied by Mead Indoor EnviroTech, Inc. [Atlanta, Georgia].)

All environmental treatment devices were evaluated for effectiveness by an independent mycology laboratory prior to use in the study. Before each device was tested, a gravity-exposure Sabouraud dextrose agar (SDA) plate (PharmaSource Int'l, Inc. [Denver, Colorado]) was exposed for 1 hr inside a test room documented to contain fungal colonies too numerous to count (TNTC). All devices reduced the test room's fungal colony count from TNTC to 0–2 colonies by day 6.

Monitoring. Prior to treatment of the patient and environment, endoscopic sinus photography was performed, and nose and air fungal cultures were taken. Bacterial cultures were taken when visible pus was present. A Calgi swab on SDA was used for nose culture, and a 1-hr gravity SDA plate exposure for the environment. Plates containing nose swab samples were incubated in a dark area at room temperature for 3 days. If growth was observed, the plate was sent to a mycology laboratory (Quest Diagnostics Mycology Laboratory [Atlanta, Georgia]) for counting and identification.

Air samples were taken by exposing open plates in the areas where the patient spent most time, as well as in areas of likely contamination (e.g., bedroom, kitchen, den, office, car, basement, crawlspace, attic) for 1 hr with the central HVAC fan in the "on" position. The plates were placed at least 0.91 m (3 ft) from any wall. After exposure, the plates were enclosed in foil and sent to a mycology laboratory (Mold Lab Int'l [Knoxville, Tennessee]) for evaluation.

After treatment of the patient and air to reduce fungal load—in accordance with the aforementioned protocols—1-hr SDA plate exposures were repeated for the environmental air, and a 2nd set of endoscopic sinus photographs were taken for each patient. Comparisons were then made between the environmental fungal colony counts and the endoscopic sinus photographs. If nasal and environmental air fungal cultures did not correspond, a different source of sinus infection was suspected.

Laser air particle counts were performed by Mead Indoor EnviroTech (Atlanta, Georgia) before and after treatment of environmental air. A ParticleScan Pro laser airborne particle counter (IQAir, Inc. [Santa Fe Springs, California]) set at 0.3 μm per 1 ft^3 (0.03 m^3) was used to count particles. The resulting values were expressed as number of particles per 0.1 ft^3 for convention. The RAST assay was used to detect serum IgG (delayed mold reaction) and IgE fungus-specific antibody levels for 7 common fungal antigens: *Alternaria alternata/tenuis, Aspergillus fumigatus, Candida albicans, Cephalosporium acre, Cladosporium herbarum, Helminthosporium halodes,* and *Penicillium notatum/chrysogenu.* Antibodies for the same 7 antigens were tested by Esoterix Laboratory Services, Inc. (Austin, Texas) with enzymatic immune assay.

Literature review. A review of current literature was conducted to aid in formulating an hypothesis for the pathogenesis of both AFS and systemic fungal symptoms.

Results

A total of 639 patients with persistent CRS were studied. After treatment of the patients and their environments to reduce fungal load, 343 of the patients, who had abnormal sinus mucosa before treatment, showed normal mucosa without infection. Nasal mucosa failed to normalize in 22 patients. Of these, 3 were found to have lymphoma (1 T-cell and 2 B-cell) and 19 had persistent positive nasal fungal cultures (likely resulting from uncontrolled exposure to environmental fungi). In 219 cases, patients were unable to reduce the mold counts in their environments below 5 colonies and had various degrees of mucosal disease remaining. Fifty-five patients were lost to follow-up. All patients had elevated fungus-specific serum IgG levels to 2 or more of the 7 common fungi tested.

Particle and fungal counts. Laser air particle counts higher than 50,000 particles per 0.1 ft^3 (0.003 m^3) at 0.3-μm particle size were associated with sinus mucosa abnormalities ranging from edema, to polyps and/or pus. However, counts below 50,000 were associated with sinus mucosal health *only* if mold counts were below 5 colonies per 1-hr gravity-plate colony count. Approximately 90% of patients cultured both mold and bacteria. The bacteria types seen were typical and have been described by many researchers. It is generally agreed that chronic antibiotic treatment does not stop CRS.

Literature review. Review of the current literature supported our hypothesis that the pathogenesis of CRS, AFS, and systemic fungal symptoms is a genetic defect at the TCR Vβ site, requiring antigen (fungus) presence. When more than 4 fungus colonies (per 1-hr gravity-plate exposure) are present in these patients' environmental air, chronic sinusitis and/or multiple systemic symptoms can present clinically.

Systemic fungal symptoms. The most common systemic symptoms seen to occur with fungal exposure, to resolve with fungal removal, and to recur with repeated fungal exposure are listed below. When these symptoms occur and recur in this manner, it is logical to hypothesize that fungi are the causative agents. The symptoms vary widely among patients and are general. No symptom is listed unless it began with fungal exposure, was concurrent with positive nasal and environmental fungal cultures, and resolved with fungal removal. The following systemic fungal symptoms were observed in our study subjects:

- Headache in the frontal sinus, radiating to the temples.
- Dizziness—both imbalance and vertigo with spinning—and impaired depth perception (e.g., patient misses stair step going down).
- Hearing loss—both high- and low-frequency, both temporary and permanent.
- Tremors, with severe leg pain and thigh weakness. Inability to walk. Non-neurological muscle weakness, fibromyalgia, lip and hand paresthesias, seizures (rare).
- Pain in some or all joints.
- Cognitive disorder and memory loss (e.g., cannot do serial 7's [mental subtraction: 100 – 7 = 93, 93 – 7 = 86, etc.], cannot remember instructions, and cannot tell directions to common places).
- Other symptoms, including loss of smell and taste, chronic sinus infections, bloating diarrhea, esophageal reflux (specific to hypopharyngeal, esophageal, and gastric candidiasis), severe fatigue.
- Immune suppression IgG subclass deficiency.

See Figure 1 for the likely mechanism of the Type III hypersensitivity reaction.

When patients were treated with the combination of medical therapy (Table 2) and environmental remediation described herein, symptoms of sinus congestion, headache, pressure, and pain, as well as the other systemic symptoms reported, generally tended to resolve within 2–8 wk after 1-hr gravity-plate fungal counts fell below 5 colonies.

Table 2.—Treatment for Systemic Fungal Symptoms, with or without Sinusitis

- Systemic antifungals:
  —Fluconazole (Diflucan®), itraconazole (Sporonox®), terbinafine (Lamisil®), or Voriconazole (VFEND®) × 14–28 days.
  —Grapefruit seed extract (delayed-release [PharmaSource Int'l, Inc. {Denver, Colorado}]) 50–75 mg po bid for 3–6 mo, depending on symptom resolution.
- Nasal sprays:
  —Antifungal: amphotericin-B.
  —Leukotriene inhibitor: montelukast (Singulair®).
  —Antibiotic: clindamycin (gentamycin if infection is visible by endoscopic exam).
- Valdecoxib (Bextra®) 10–20 mg × 14–28 days.
- Saline nasal wash.
- Detoxification (e.g., drink distilled water, sauna, take mega-antioxidant vitamins with thymus).
- Antigen neutralization, and autogenous lymphocytic factor (for immune-compromised).

Discussion

On the basis of 16 yr of clinical experience with 639 patients—using the treatment protocol described herein and monitoring with both endoscopic photography and sinus CT scans—it can be stated that "As long as fungi remain, so will the irritation and the sinusitis." Experience has also shown that 1-hr gravity SDA agar plate exposure is a simple, reliable, and predictable method of assessing environmental fugal levels to determine human health risk. The following scale correlates with the results we obtained by comparing endoscopic sinus photos at various mold colony counts with laser air particle counts (Fig. 3 and 4): (a) a count of 0–4 total fungal colonies per room is within a normal range for sinus health, (b) a count of 5–8 colonies per room is cause for concern—illness is probable in patients with the IgG immune reaction to fungi herein described, and (c) a count of 9+ colonies per room is hazardous—illness is very likely among the CRS patient population.

Laser air particle counts below 50,000 per 0.003 m^3 (0.1 ft^3), at 0.3 μm particle size, were associated with sinus mucosal health. However, environmental air particle counts in this range, with fungal colony counts greater than 4, were associated with sinus mucosal abnormalities. In other words, high fungal counts will overpower an otherwise clean air particle count and cause illness.

Key points in treatment. Removal of the antigen (fungi) from the patient and the air halts the reaction (i.e., CRS and systemic fungal symptoms). Nasal irriga-

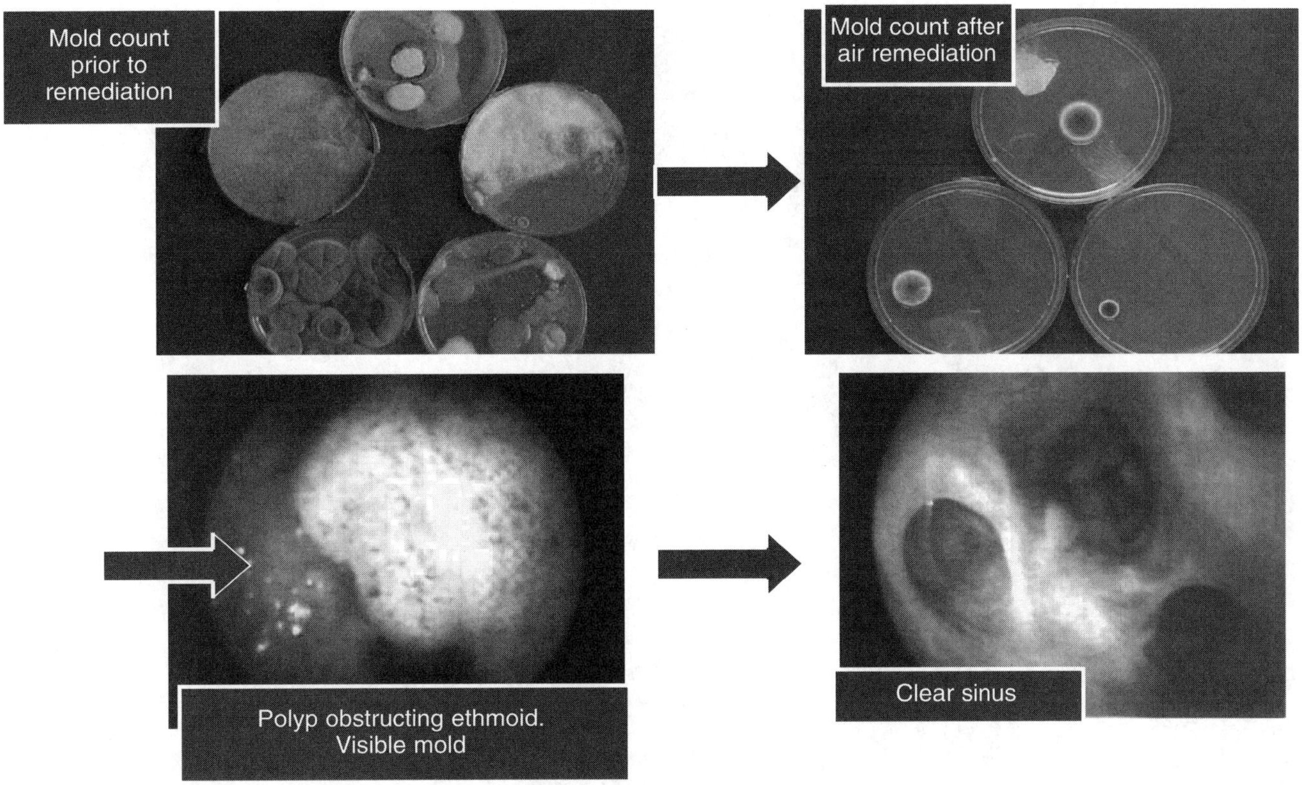

Fig. 3. Resolution of ethmoid polyps 6 wk post-treatment. Once the environmental air fungal count dropped below 5 colonies per 1-hr agar plate exposure, antimicrobial nose sprays cleared nasal polyps in 6 wk.

tion with normal saline is essential for reducing nasal antigen load and antigen-induced inflammation. Because more than 90% of patients cultured both fungi and bacteria, combined antifungal and antibiotic nasal sprays are key to long-term control of CRS. Appropriate use of oral antibiotics, antifungals (fluconazole, delayed-release grapefruit seed extract capsules, terbinafine, itraconazole, voriconazole) and antihistamine/decongestants is also recommended. Clinical experience has shown that immunotherapy for IgE- and IgG-specific molds is beneficial, with neutralization as needed. Mixed respiratory vaccine for chronic *Staphylococcus aureus* is helpful if staph is consistently cultured. Approximately 10% of patients studied had severe nasal polyposis and cultured staph. IgE to both staphylococcal enterotoxin B and dermatiaceous fungi are almost always found together in situ in hypertrophic sinus disease nasal polyps.[22] Endoscopic sinus surgery is essential for patients who fail medical therapy. However, absent removal of fungus from the nose and air, bacterial sinusitis remains, regardless of immunotherapy, vaccine, or surgical intervention.

The 93% of CRS patients in the Mayo study[2] who had positive fungal nasal cultures also exhibited secondary bacterial infections, likely as a result of the mucosal pitting and mucus pooling that occurs when an eosinophil binds to fungi in the nasal mucosa, ruptures, and releases major basic protein into the mucosa. This may explain our observation that antibiotic nose sprays are more effective when combined with an antifungal spray. When fungi were successfully removed from the nose and air, 94% of patients showed endoscopic sinus mucosal improvement to a normal appearance (i.e., no mucosal edema, polyps, or purulence), as depicted in Figures 3 and 4.

Treatment failure strategy. When environmental fungal counts are in the range conducive to good sinus health (0–4 colonies), endoscopic sinus surgery has been performed, all treatment has failed, and a patient still has CRS, the following approach is recommended. The patient should be referred to a hematologist/oncologist for thorough work-ups for lymphoma, all collagen vascular diseases, Immune-Deficiency Syndrome, endocrine disorders, and Chronic Fatigue Syndrome. Because mold contamination previously unidentified by history and environmental testing may be the cause of sinusitis treatment failure, patients should be asked again about potential sources of exposure (e.g., gardening and other hobbies, cleaning, construction, moving of old boxes, or use of old books). Areas involved should be tested (or retested) with SDA agar plates, and an effort should be made to eliminate these exposures.

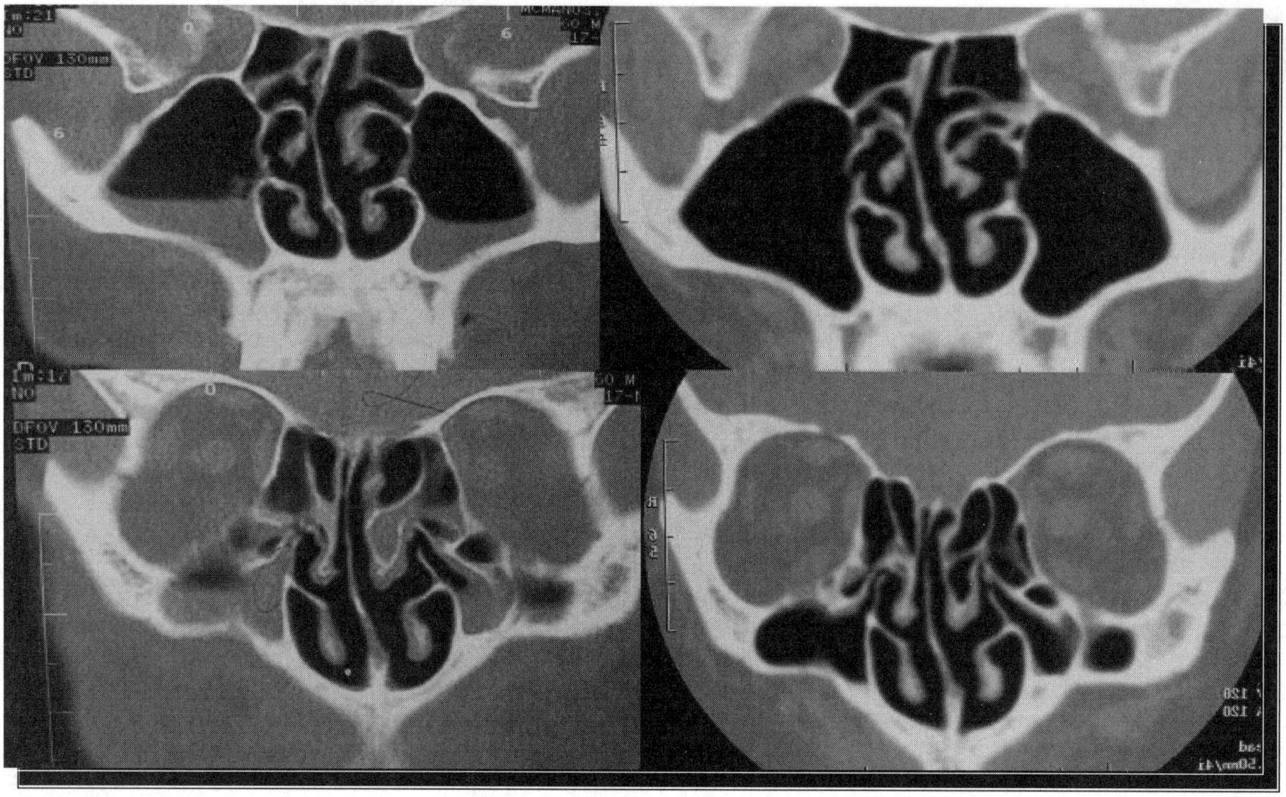

Fig. 4. Computed tomography scans of subject's sinuses pre- and post-treatment. *Left.* 2 scans showing sinusitis prior to treatment. *Right.* 2 scans showing clear sinuses 1 mo after environmental treatment protocol and use of antimicrobial (antibiotic and antifungal) nose sprays.

Conclusions

CRS and systemic fungal symptoms are likely caused by an immune response to fungal antigen. When the antigen is removed from the nose and the environmental air, the immune reaction stops; the sinus mucosal edema, polyps, and/or pus improve or resolve; and systemic symptoms improve or resolve. Fungus is likely responsible for most CRS, and for a complex of systemic symptoms, as a result of a genetic defect on the TCR Vβ site. Exceptions to this are underlying diseases, failure to find the source of the mold exposure, and permanent systemic immune damage resulting from lengthy fungal exposure. Chronic sinusitis patients who have recurring exposure to environmental air containing fungal concentrations in excess of 4 colonies per 1-hr agar plate exposure appear to have an increased risk of persistent chronic sinusitis and/or systemic symptoms, regardless of the medical treatment provided.

* * * * * * * * * *

The author thanks Malcolm Wilkinson, R.Ph., for tirelessly compounding antimicrobial nasal sprays for these patients over the last 20 yr; the patients themselves for teaching him how to help them back to wellness; and his staff for their help in assembling all the data.

Submitted for publication September 20, 2003; revised; accepted for publication November 19, 2003.

Requests for reprints should be sent to Donald Patrick Dennis, M.D., F.A.C.S., Atlanta Center for ENT and Facial Plastic Surgery, 3193 Howell Mill Road, Suite 215, Atlanta, GA 30327.

E-mail: ddennis@mindspring.com

* * * * * * * * * * *

References

1. Kaliner MA, Osguthorpe JD, Fireman P, et al. Sinusitis: bench to bedside. Current findings, future directions. J Allergy Clin Immunol 1997; 99:S829–48.
2. Ponikau JU, Sherris DA, Kern EB, et al. The diagnosis and incidence of allergic fungal sinusitis. Mayo Clin Proc 1999; 74:877–84.
3. Braun H, Buxina W, Freudenschuss K, et al. Eosinophilic fungal rhinosinusitis: a common disorder in Europe? Laryngoscope 2003; 113:264–69.
4. Ponikau JU. Chronic Rhinosinusitis: The War of the Immune System Against Fungi. Bronx, NY: American Rhinologic Society News, January 2001. Available from <http://american-rhinologic.org/cgi-bin/menu.cgi?m=main.menu&state=10012100001000000010000000&citem=4&f=news.0102.ponikau.phtml>

5. Waxman JE, Sale SR, Spector JG, et al. Allergic *Aspergillus* sinusitis: concepts in diagnosis and treatment of a new clinical entity. Laryngoscope 1987; 97:261–66.
6. Nelson S. Gell-Coombs Classification of Hypersensitive Reactions. Geneva, Switzerland: Health on the Net Foundation, 2001. Available from <http://www.hon.ch/Library/Theme/Allergy/Glossary/gcc.html>
7. Hamilos DL. Chronic sinusitis. J Allergy Clin Immunol 2000; 106:213–27.
8. Oswald MR. Basic Immunology. Kansas City, KS: University of Kansas Medical Center, 2000; 4:72. Available from <http://www.kumc.edu/instruction/medicine/pathology/ed/keywords/kw_hypersen1.html>
9. Schubert MS. A superantigen hypothesis for the pathogenesis of chronic hypertrophic rhinosinusitis, allergic fungal sinusitis, and related disorders. Ann Allergy Asthma Immunol 2001; 4:181–87.
10. Kotzin BL, Leung DY, Klapper J, et al. Superantigens and their potential role in human disease. Adv Immunol 1993; 54:99–166.
11. Klapper J, Kotzin B, Herron L, et al. Vβ-specific stimulation of human T cells by staphylococcal toxins. Science 1989; 244:811–13.
12. Choi Y, Kotzin B, Hernon L, et al. Interaction of *Staphylococcus aureus* toxin "superantigen" with human T cells. Proc Natl Acad Sci USA 1989; 86:8941–45.
13. Marrack P, Kappler J. The staphylococcal enterotoxins and their relatives. Science 1990; 248:705–09.
14. Krakauer T. Immune response to staphylococcal superantigens. Immunol Res 1999; 20:163–73.
15. Liu B, Wang Y, Melana SM, et al. Identification of proviral structure in human breast cancer. Cancer Res 2001; 61:1754–59.
16. Perron H, Jouvin-Marche E, Michel M, et al. Multiple sclerosis retrovirus particles and recombinant envelope trigger an abnormal immune response in vitro, by inducing polyclonal Vbeta16 T-lymphocyte activation. Virology 2001; 287(2):321–32.
17. Bour H, Demidem A, Garrigue JL, et al. In vitro T cell response to staphylococcal enterotoxin B superantigen in chronic plaque type psoriasis. Acta Derm Venereol 1995; 75:218–21.
18. Leung DY, Travers JB, Giorno R, et al. Evidence for a streptococcal superantigen-driven process in acute guttate psoriasis. J Clin Invest 1995; 96:2106–12.
19. Valdimarsson H, Sigmundsdottir H, Jonsdottir I. Is psoriasis induced by streptococcal superantigens and maintained by M-protein-specific T cells that cross-react with keratin? Clin Exp Immunol 1997; 107(suppl):21–24.
20. Paliard X, West SG, Lafferty JA, et al. Evidence for the effects of a superantigen in rheumatoid arthritis. Science 1991; 253:325–29.
21. Cole BC, Griffiths MM. Triggering and exacerbation of autoimmune arthritis by the *Mycoplasma arthritidis* superantigen MAM. Arthritis Rheum 1993; 36:994–1002.
22. Bachert C, Gevaert P, Holtappels G, et al. Total and specific IgE in nasal polyps is related to local eosinophilic inflammation. J Allergy Clin Immunol 2001; 107:607–14.

Health Symptoms Caused by Molds in a Courthouse

TANG G. LEE
Faculty of Environmental Design
The University of Calgary
Calgary, Alberta, Canada

ABSTRACT. A majority of occupants of a newly renovated historic courthouse in Calgary, Alberta, Canada, reported multiple (3 or more) health-related symptoms, and several reported more than 10 persistent symptoms. Most required at least 1 day outside of the building to recover from their symptoms. Molds that produce mycotoxins, such as *Stachybotrys chartarum* and *Emericella nidulans,* were identified in the building, along with fungal organisms of the genera *Aspergillus, Penicillium, Streptomyces, Cladosporium, Chaetomium, Rhizopus/Mucor, Alternaria, Ulocladium,* and *Basidiomycetes.* Renovations to this historic had building failed to provide adequate thermal and vapor barriers, thus allowing moist indoor air to migrate into the building enclosure, causing condensation to develop. Mold grew on the condensation and was dispersed throughout the courthouse, including on furniture and files. The courthouse was closed and a new facility was modified with low-offgassing materials, better ventilation and air filtration, and strict building maintenance to accommodate those occupants of the older building who had developed multiple chemical sensitivities.
<Key words: mold, condensation, multiple chemical sensitivity, symptoms>

MORE THAN 500 indoor air quality investigations were conducted during the 1970s and 1980s by the U.S. National Institute for Occupational Safety and Health (NIOSH). Microbial contamination was found to be the primary cause of poor air quality in only 5% of cases; in the remaining 95% of cases, poor air quality was attributed to inadequate ventilation, outdoor air contaminants, contaminants within the building envelope, or contaminants from unknown sources.[1] By the 1990s, however, microorganisms were identified as the primary source of indoor air contamination in 35–50% (depending on building type) of indoor air quality cases.[2] This change has been attributed in part to the adoption of an interdisciplinary investigative approach that combines evaluation of physical, chemical, and microbiological constituents of indoor air environments.

In 1990, occupants of an historic courthouse in Calgary, Alberta, Canada—which housed the Calgary Court of Appeals—reported persistent health-related symptoms almost immediately following renovations to the structure. After several attempts to eliminate suspected air contaminants were unsuccessful, the Alberta ministries of Provincial Justice and Infrastructure retained an interdisciplinary team from the University of Calgary to examine the building and conduct a health survey of occupants. The team was led by an architect and professor of building science (TGL), and consisted of a mechanical engineer, a facilities manager, a building contractor, a microbiologist, a social worker, and several graduate students. A specialist in respiratory diseases (T. M. Matthew van Olm, M.D. [Calgary, Alberta, Canada]) reviewed the draft report on the indoor air quality investigation, including air samples, survey results, and the symptoms of his own patients who occupied this courthouse. He provided a medical opinion supporting the link between reported health symptoms and air contaminants found in the building.

Method

Survey of health symptoms. In January 2001, a survey of health symptoms was administered to 32 of 43 occupants of the Calgary Court of Appeals building, including judges, support staff, security guards, and articling students. An experienced social worker developed the questionnaire and administered the survey. Results of this survey, and of the air quality and building examination conducted by the interdisciplinary team, were presented in a report submitted to Alberta Justice.[3]

Air contaminant sampling. The historic Court of Appeals building is a 2-story structure located in downtown Calgary and bordered by heavily traveled roads. Sampling for bioaerosols was conducted on December 29, 2000, and January 10, 2001, and bulk sampling for bioaerosols was conducted December 29, 2000, and January 2 and 10, 2001. A total of 32 indoor and outdoor air sampling strips and 9 swab/bulk samples were taken. The air samples were collected with a Reuters Centrifugal Sampler (RCS [Biotest Diagnostics Corp. {Denville, New Jersey}]), which was rented from the microbiological laboratory and calibrated by them. The impeller blades were disinfected with alcohol wipes between sample locations, and dried to prevent cross-contamination. The sampler operates by impaction, drawing particles from the air onto rose bengal flexible agar strips inserted into the sampler. The sampler was activated for 4 min running time per sample—the same sampling time used by Health Canada in the development of fungal exposure guidelines.[4]

The samples collected were kept in sterile plastic bags within a closed cooler maintained at approximately about 4 °C for transport. The samples were delivered to an analytical laboratory (Bio-Chem Consulting Ltd [Calgary, Alberta, Canada]) for incubation, analysis, and speciation. Laboratory analysis was conducted with stringent adherence to standard microbiological practices and techniques.

Testing was also conducted for typical contaminants from vehicle emissions (i.e., carbon dioxide and carbon monoxide). Some volatile organic compounds were measured with a MultiWarn II multigas detection instrument (National Draeger, Inc. [Mississauga, Ontario, Canada]). This instrument measured methane (CH_4), chlorine (Cl_2), sulfur dioxide (SO_2), and acetaldehyde (CH_3CHO). Levels of other volatile organic compounds were sampled with carbotrap tubes and assessed with gas chromatography/mass spectrometry by an accredited laboratory. In addition, airborne particles 0.3, 0.5, 1.0, and 5.0 μm in size were counted using a laser APC Plus particle counter (Biotest Diagnostics Corp. [Denville, New Jersey]). Although no standards exist for the maximum number of airborne particles, we compared the levels measured at the courthouse with those tabulated by TGL and his graduate students at the University of Calgary from over 5,000 measurements conducted in various locations throughout North America between 1994 and 2000.

Results

Symptoms reported. Many of the survey respondents reported experiencing more than 1 symptom while

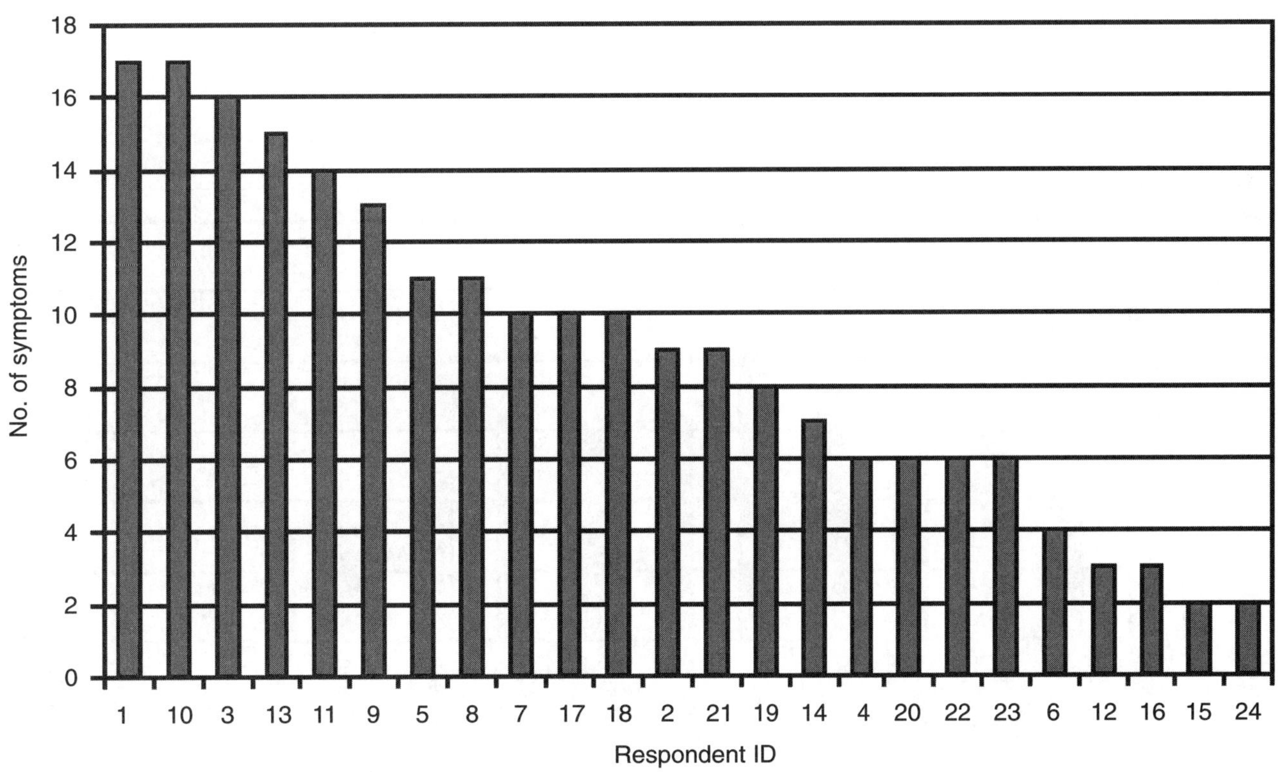

Fig. 1. Numbers of symptoms reported by occupants of the courthouse.

working in the building (Fig. 1). Approximately one-third reported 10 or more symptoms, with 9 symptoms being the average. Multiple symptoms of this magnitude clearly demonstrated significant illness experienced by many occupants of this building.

Symptoms reported included 28 of 31 potential illnesses identified on the survey (excluding menstrual problems, vomiting, and weight loss). More than half of the respondents reported prevalent symptoms such as headaches (71%), sleepiness (66%), and fatigue (66%). Other common symptoms included eye irritation (54%), chapped lips and dry skin (58% and 50%, respectively), dry throat and sore throat (46% and 41%, respectively), stuffiness (45%), and sinus congestion (41%). The most prevalent health symptoms reported by respondents are shown in Figure 2. Several other symptoms, including runny nose, bleeding nose, nausea, and general muscle aches were reported by 4–25% of respondents (Fig. 3).

Time required to recover from symptoms. For 96% of the occupants surveyed, symptoms improved on weekends and during holidays (Fig. 4). Eleven of the respon-

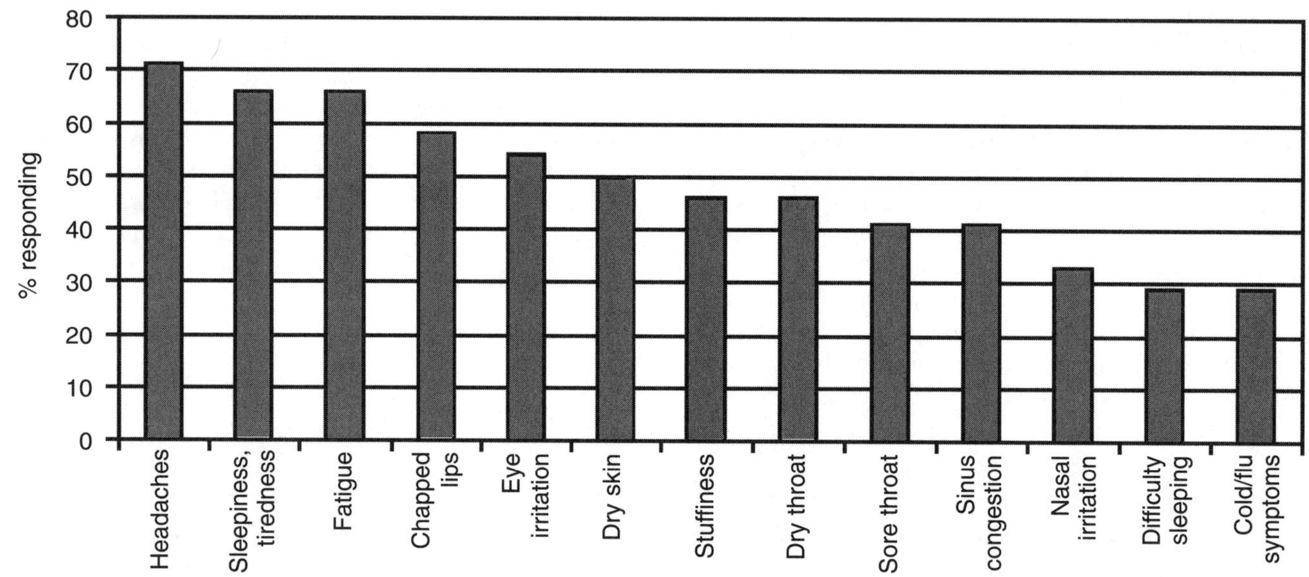

Fig. 2. Primary symptoms experienced while in the building, as reported by occupants.

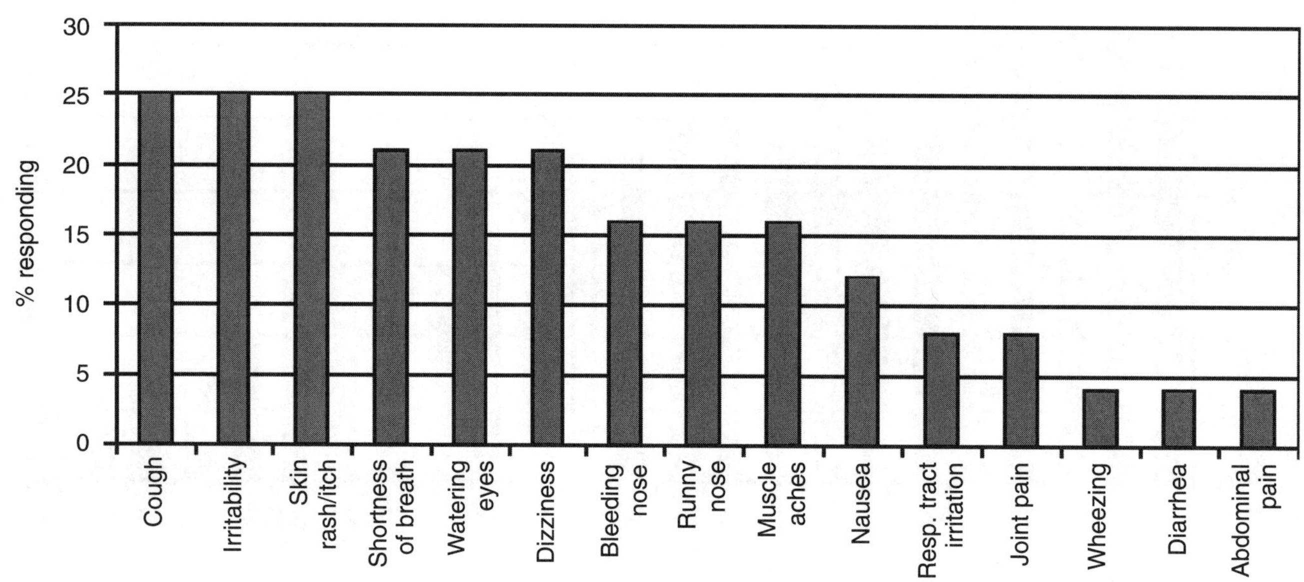

Fig. 3. Secondary symptoms experienced while in the building, as reported by occupants.

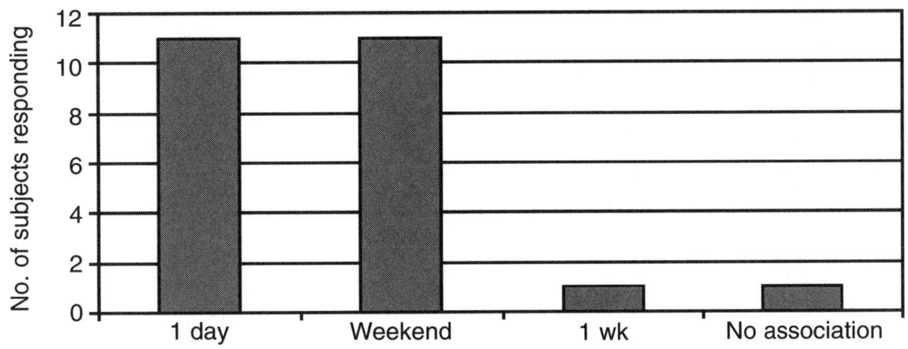

Fig. 4. Time required to recover from reported illnesses.

dents required 1 day outside of the Court of Appeals building to recover from symptoms attributed to the building's environment. Another 11 respondents required a weekend, or at least 2 days, to recover from building-related symptoms.

Some occupants mentioned that they performed the same kind of work in other buildings, but did not experience the reported symptoms. All of the judges also had offices in Edmonton and did not experience these symptoms in that location. Some respondents indicated an immediate reduction in symptoms following their move out of the building. Several staff reported that their asthma and allergy medications had to be increased when they worked at the Calgary building and, more importantly, their medications did not seem to have the same relieving effect or potency as in other environments. Symptoms were reported to be worst at the end of the workday and at the end of the work week (Thursdays and Fridays).

Potential for psychological influence. We also considered the possible influence on reported symptoms of (a) preexisting psychological conditions of individual occupants and (b) mass psychogenic illness among building occupants. However, analysis of the survey and interview results revealed 4 factors that would render these conditions unlikely. First, the building occupants expressed a high degree of job satisfaction. They also expressed satisfaction with the aesthetic quality of the renovated historic building, citing the high ceilings, interior finishes, and amenities. Thus, their symptoms could not be attributed to an unsatisfactory work environment. Second, symptoms were reported soon after individuals commenced work in the renovated building, and occupants did not experience these symptoms when performing the same types of jobs in other buildings. Third, episodes of fainting/dizziness occurred at different times for different occupants. Fourth, 5 occupants experienced sudden-onset dizziness, with at least 1 incident resulting in fainting during work duties. One occupant was taken from the building by ambulance and diagnosed at the hospital with a "respiratory ailment."

Airborne contaminants. Contaminants from vehicle emissions were not elevated in the building; levels of carbon dioxide and carbon monoxide were less than 600 ppm and 5 ppm, respectively. Levels of other air contaminants (i.e., sulfur dioxide, chlorine, methane, formaldehyde, and total volatile organic compounds) were not detectable or were very low; however, acetaldehyde reached 6 ppm, which exceeds the guideline of 5 ppm established by Health Canada.[5] Counts of 0.3-µm airborne particles were high—ranging from 900,000 to over 3.6 million particles per cubic foot (32,000–127,000 particles per liter)—as compared with counts for a typical office (80,000/ft^3 [2,800/l]). The courthouse had as many airborne particles as in a smoky bingo hall.

The fungus *Stachybotrys chartarum* was isolated from the cellulose fireproofing on the ceiling of the basement mechanical room, as well as from some of the furniture in the courthouse. *Aspergillus nidulans/Emericella nidulans* and *Penicillium solitum* were also identified in air. *S. chartarum* is known to cause immediate (Type I) allergenic response. It is also known to produce a variety of mycotoxins (macrocyclic trichothecenes) which have caused toxicosis in both humans and animals. However, the propensity for an individual isolate to produce mycotoxins indoors is unknown.[6,7] *A. nidulans/E. nidulans* is also known to cause immediate (Type I) allergenic responses and is a potential producer of several mycotoxins.[8,9] Other isolates, which were not speciated because of cost constraints, likely included the genera *Aspergillus, Penicillium, Streptomyces, Cladosporium, Chaetomium, Rhizopus, Mucor, Alternaria, Ulocladium,* and *Basidiomycetes.*

Penicillium and *Aspergillus* are 2 genera of fungi that produce toxins as a part of their metabolic cycle. At least 1 of these was found in all 11 samples in which we isolated viable spores. Of significance was the fact that a 12th (bulk) sample taken from the outside sandstone wall of the courthouse did not contain either of these genera, which indicates that they were not drawn into the building but originated within.

Of additional importance is the fact that fungi of the genera *Penicillium, Streptomyces, Cladosporium,* and *Aspergillus* were found in the fiberglass air filters of the wall-mounted heating units used in all perimeter offices. Fungal growth in this location was a likely indicator that moisture was condensing on the filter, or that fungal spores were being drawn into the filter from other locations. With the fan discharging air throughout the room, spores in the air filter were also dispersed.

Discussion

The onset of reported symptoms occurred following renovation of this old courthouse approximately 10 yr ago. The renovation included the replacement of leaky windows and the addition of humidification equipment. Unfortunately, no thermal insulation or proper vapor barrier were installed to prevent air migration across the building envelope. Condensation can form within a wall when warm indoor air with high humidity contacts cold surfaces, such as those prevalent in winter months. This provides an ideal condition for mold growth. An earlier attempt to "clean" the air in the courthouse by improving ventilation actually resulted in even greater mold formation because exfiltration of warm moist air into the walls produced more condensation.[10]

The multidisciplinary investigatory team received input from a physician specializing in respirology, and from the 2 microbiologists who conducted the mycological analyses of airborne and grab samples obtained in the courthouse. Within 2 days of our preliminary report, the courthouse was closed and the court cases transferred to another city (Edmonton), 300 km to the north.

Two years after closing of the courthouse, approximately a dozen staff and judges still experience health-related symptoms. The Court of Appeals was eventually moved to a new facility that had been renovated to minimize air contaminants. Only low-offgassing materials were used, and the new premises has enhanced ventilation and filtration. None of the previous building's contents, including files and equipment, were allowed to be transferred to the new facility. Nontoxic cleaning solutions, and equipment such as high-efficiency particulate air (HEPA) vacuum cleaners, are used in conjunction with a strict housekeeping and building maintenance schedule. Cleaning solutions are chosen for their low toxicity.

Summary

Molds that produce mycotoxins caused multiple health symptoms in occupants of an old courthouse that had been renovated improperly. The lack of adequate thermal and air barriers resulted in condensation that became a mold amplifier. Despite having moved to a new facility, the immunocompromised staff from the old courthouse now requires a work environment with minimal air contaminants to mitigate adverse health effects.

* * * * * * * * * *

Submitted for publication September 20, 2003; revised; accepted for publication October 30, 2003.

Requests for reprints should be sent to Tang G. Lee, M.Arch, MRAIC, Faculty of Environmental Design, The University of Calgary, 2500 University Drive NW, Calgary, Alberta, Canada T2N 1N4.

E-mail: lee@ucalgary.ca

* * * * * * * * * *

References

1. Seitz TA. NIOSH indoor air quality investigations: 1971–1988. In: Weekes DM, Grammage RB (Eds). The Practitioners Approach to Indoor Air Quality Investigations. Proceedings of the Indoor Air Quality International Symposium, May 1989, Cincinnati, Ohio. Akron, OH: American Industrial Hygiene Association; pp 163–71.
2. Lewis FA. Regulating indoor microbes: the OSHA proposed rule on IAQ. In: Johanning E, Yang CS (Eds). Proceedings of the International Conference on Fungi and Bacteria in Indoor Air Environments: Health Effects, Detection and Remediation, Saratoga Springs, New York, October 1994. Albany, NY: Eastern New York Occupational and Environmental Health Center, 1995; pp 5–9.
3. Lee TG, Stooke T. Environmental Assessment of the Alberta Court of Appeals Building. Report for Alberta Justice and Alberta Infrastructure. Edmonton, Alberta: Alberta Justice, March 5, 2001. (E-mail: Lynn.Varty@just.gov.ab.ca)
4. Fungal Contamination in Public Buildings: A Guide to Recognition and Management. A report of the Federal/Provincial Committee on Environmental and Occupational Health, Environmental Health Directorate. Ottawa, Canada: Health Canada, 1995; p 59. Available from <www.hc-sc.gc.ca/hecs-sesc/air_quality/pdf/fungal.pdf>
5. Exposure Guidelines for Residential Indoor Air Quality. A report of the Federal/Provincial Committee on Environmental and Occupational Health, Environmental Health Directorate. Ottawa, Canada: Health Canada, July 1989; p 8.
6. Manville C. Poor drainage leads to fungal infestation, adverse health effects. Indoor Air Quality Update 1996; 3: 11–12.
7. Croft WA, Jarvis BB, Yatawara CS. Airborne outbreak of trichothecene toxicosis. Atmos Environ 1986; 20:549–52.
8. Al-Doory Y, Domson JF. Mould Allergy. Philadelphia, PA: Lea and Febiger, 1984; 287 pp.
9. Crissy JT, Lang H, Parish LC. Manual of Medical Mycology. Cambridge, MA: Blackwell Sciences, 1995; 263 pp.
10. Lee TG, Stooke T. Mould propagation resulting from air pressure differences across the building envelope. Proceedings of the 9th International Conference on Indoor Air Quality and Climate (Indoor Air 2002), Monterey, California, June 30–July 5, 2002. Santa Cruz, CA: Indoor Air 2002 Conference Secretariat; pp 742–47.

Studies on the Role of Fungi in Sick Building Syndrome

DAVID C. STRAUS
J. DANNY COOLEY
WING C. WONG
Department of Microbiology and Immunology
Texas Tech University Health Sciences Center
Lubbock, Texas
CYNTHIA A. JUMPER
Departments of Internal Medicine and Physiology
Texas Tech University Health Sciences Center
Lubbock, Texas

ABSTRACT. Sick Building Syndrome is a term used to describe symptoms in humans which result from problems with indoor air quality. Common complaints include dyspnea, flu-like symptoms, watering eyes, and allergic rhinitis. Although there is likely no single cause for Sick Building Syndrome, fungal contamination in buildings has increasingly been associated with this spectrum of symptoms. The authors describe 2 case studies, and other experimentation, that have investigated the role of fungi in the occurrence of Sick Building Syndrome.
<Key words: fungi, indoor air quality, *Penicillium chrysogenum*, Sick Building Syndrome, *Stachybotrys chartarum*>

SICK BUILDING SYNDROME (SBS) probably originated as a result of the oil embargos that occurred in the mid-1970s. Since that time, buildings in developed countries have increasingly been built "tight" (e.g., with windows that could not be opened by the occupants) to conserve energy and reduce heating and cooling costs. An unexpected consequence of this building style, however, was the trapping of contaminants (e.g., fungi and chemicals) indoors, which led to the SBS phenomenon, first described in 1982.[1] The switch to construction with cellulose-based materials (e.g., pressed particle board, ceiling tile, and sheetrock) in our homes, schools, and commercial buildings undoubtedly contributes to the problem because, when dampened by high humidity or water leakage, such materials provide an ideal substrate for the growth of molds.

Complaints commonly seen in persons suffering from SBS include dyspnea, headache, watering eyes, flu-like symptoms, and allergic rhinitis.[2] The causes of SBS have been exceedingly difficult to elucidate.[3,4] We recently examined the role of fungi in SBS,[5] and other researchers have shown that the presence of fungi indoors can produce allergic reactions in building inhabitants.[6] It is now known that fungi growing inside buildings can make people sick. In 2002, the U.S. Centers for Disease Control and Prevention (CDC) presented a report to Congress which stated, "While there remain many unresolved scientific questions, we do know that exposure to high levels of mold causes some illnesses in susceptible people. Because molds can be harmful, it is important to maintain buildings, prevent water damage and mold growth, and clean up moldy materials."[7] In addition, the report affirmed that mold in buildings can cause 3 categories of human disease: allergic rhinitis, hypersensitivity pneumonitis, and exacerbation of asthma.

Study 1

This study[5] was made possible because of our association with an indoor air quality (IAQ) company, which permitted us access to air and surface microbial samples, personal interviews, and occupant-generated complaints from the buildings studied. The study was a 22-mo examination of 48 schools in the United States that had contracted with the company because of IAQ concerns. Samples of indoor air, outdoor air, and surface microbes were taken with a 2-stage Andersen Air sampler (Model 2000 [Andersen Samplers {Atlanta, Georgia}]). Sabouraud's dextrose agar (pH 5.6) (Difco [Detroit, Michigan]) was used for culturing the air and surface samples. We tabulated the results according to the areas examined: (a) indoor complaint areas, (b) indoor noncomplaint areas, and (c) outdoors. In addition, levels of carbon dioxide, formaldehyde, nitrogen dioxide, hydrogen sulfide, sulfur dioxide, and carbon monoxide; particulates; temperature; and relative humidity were measured at each site. Finally, the IAQ company performed building remediation at all sites where it was deemed necessary. This included removal and replacement of building materials that showed physical deterioration; removal and disinfection of all microbial contamination; and remediation of the heating, ventilation, and air conditioning systems. The latter included removal of visible surface contaminants and cleaning of the air-side surfaces of all internal-air–handling assemblies.

A questionnaire regarding SBS symptoms was distributed to personnel in each of the 48 schools examined. The questionaire[5]—developed using a U.S. National Institute for Occupational Safety and Health survey as a model—was administered by the IAQ company and evaluated by us. As described previously,[5] the questionnaire asked respondents for information that included the nature of their complaints, symptom patterns, and observations about building conditions that might explain the symptoms experienced. A total of 622 occupants (38% of staff members) expressed concerns regarding IAQ in their buildings. The most common symptoms reported were sinus and lung congestion, nasal drainage, and itchy and watering eyes. After remediations were performed, as described above, the same questionnaire was readministered to building occupants. The number of individuals responding to the 2nd questionnaire was comparable to the number who responded to the 1st request. Staff complaints following remediation never exceeded 3%—a significant reduction ($p < 0.001$) in the number of IAQ complaints. We found no correlations between the levels of particulates, carbon dioxide, and other chemicals, and occupant complaints.

Five fungal genera were identified consistently in the outdoor air, and comprised the majority of common outdoor fungi: *Cladosporium* (81.5%), *Penicillium* (5.2%), *Chrysosporium* (4.9%), *Alternaria* (2.8%), and *Aspergillus* (1.1%). In more than 50% of the school buildings, there were discernible reductions in the number of fungi in the indoor air samples from noncomplaint areas vs. the outdoor air samples; however, the fungal profiles were similar. In all of the buildings, the number of *Cladosporium* sp. was significantly lower in the indoor air samples from noncomplaint areas than in the outdoor air samples. At 20 schools, there were significant ($p < 0.001$) increases in the number of *Penicillium* sp. in the indoor air samples from complaint areas, compared with the outdoor air samples and with the indoor air samples from noncomplaint areas. Five schools had higher concentrations of *Penicillium* spores in the air samples from complaint areas. In 11 schools, the indoor air complaint areas appeared microbially similar to outdoor air; however, *Stachybotrys chartarum* was isolated from swab samples of visible growth. *S. chartarum* requires a great deal of water to be able to colonize surfaces. It is also known to be a potent producer of trichothecene mycotoxins.[5] In addition, it is very difficult to isolate viable *S. chartarum* spores from air. In the remaining 12 schools, fungal profiles were not significantly different ($p = 0.10$) in the areas examined. Following building remediation, fungal air profiles indoors generally returned to what was seen in the air outside the building.

Study 2

In another study,[8] we examined the fungal air ecology inside and outside of a multilevel hotel in the southwestern United States over a 6-hr period to determine whether conditions of the inside air and outside air changed differently or dramatically over time. We asked the question, "Is an indoor air sample just a "snapshot" of the fungal ecology of the air, or does it actually tell the investigator something about the 'health' of the building?" A hotel room with visible mold growth on the walls was made available to us for a 6-hr period. Samples of the inside air and outside air were taken in triplicate, cultured, and evaluated to determine fungal profiles, as described above.[5] Our results demonstrated that, whereas the fungal profile of the outside air changed continually, on an hourly basis, the fungal profile of air within a building with indoor fungal growth (in this case, *Penicillium* sp.) did not change dramatically during the 6-hr test period. This study reinforced our belief that "sick" buildings tend to remain sick and do not "cure" themselves over time. In comparison, the fungal profile of the outdoor air was constantly changing because the wind blew new fungal spores into the sampling area, and old fungal spores out of the area.

Additional Studies

Why, then, does exposure to spores of *Penicillium* and *Stachybotrys* species result in the phenomenon known as Sick Building Syndrome? Results from our laboratory, and the work of others, suggest possible mechanisms. Licorish et al.[9] found that inhalation of *Penicillium* sp. spores in "natural state" numbers was immunopathogenic for asthma. *Penicillium* spores range in size from 1 to 3 μm and are easily inhaled into the lungs of humans and animals. In a series of studies, we examined the effects of inhalation of *Penicillium* sp. spores in C57BL/6 mice (Harlan Sprague Dawley [Indianapolis, Indiana]).

In the 1st study of the series,[10] we administered viable *Penicillium chrysogenum* spores to mice intranasally. We determined that the spores could be isolated from the lungs of mice as early as 25 min, and as late as 1.5 days, after intranasal instillation of 10^6 spores with a conidial population that was 25% viable. We found that 18% of the viable spores actually made their way to the lungs, where they remained alive for up to 36 hr—sufficient time to produce potentially allergenic compounds.

In the next study,[11] we examined the effects of the inhalation of *Penicillium chrysogenum* spores on the rodent immune system. We found that inhalation of *P. chrysogenum* spores (10^6 spores with 25% viability) resulted in increased levels of serum immunoglobulin (Ig) E and increased interleukin 4 and 5 levels in mouse lung. The results suggest that long-term exposure to, and inhalation of, viable *P. chrysogenum* spores induces Type 2 T-helper cell-mediated inflammatory reactions.

In the final study in this series,[12] we used the same animal system (C57BL/6 mice) and organism (*Penicillium chrysogenum*). We found that mice that inhaled these organisms did not develop allergenic reactions (e.g., IgE production) when challenged with low doses of viable *P. chrysogenum* spores administered intranasally. We used doses in this study that were consistent with those found in outside air. We also demonstrated that protease antigens from viable spores induced allergic responses in the mice when instilled intranasally. These data illustrated the significance of the inhalation of *P. chrysogenum* spores in allergic sensitization to this fungus.

Conclusions

It appears likely that the health consequences observed in exposure to *Stachybotrys chartarum* are attributable, at least in part, to mycotoxin production by this organism. *S. chartarum* produces very potent mycotoxins known as trichothecenes,[13] which have been demonstrated to be produced when this fungus grows inside of buildings.[14,15] These mycotoxins are found on the conidia (vegetative spores) of *S. chartarum*,[16] which can be inhaled by humans.[17] Therefore, it is logical to conclude that it is potentially dangerous to have *S. chartarum* growing in a building inhabited by people or animals. Indeed, it has been concluded that *S. chartarum* meets, ". . . in a regulatory sense, all of the conditions necessary to make a public health recommendation . . . in terms of hazard identification. The allergens and toxins of *S. chartarum* can result in disease in humans, as well as in domestic and laboratory animals."[18] We believe that the inhalation of *Penicillium* species spores (as well as the spores of other fungi) is responsible for the allergic rhinitis, hypersensitivity pneumonitis, and exacerbation of asthma observed in individuals working, living, or attending school in mold-infested buildings, as described in the CDC report.[7]

* * * * * * * * * *

The authors thank Assured IAQ® of Dallas, Texas, for financial support and access to the buildings examined in Studies 1 and 2. They also acknowledge the following individuals who helped to generate the data reported herein: Trevor Brasel, Suzanne Graham, Jim Hutson, Jared Martin, James McGrath, Chris Schwab, and Jim Willams.

Submitted for publication September 20, 2003; revised; accepted for publication October 30, 2003.

Requests for reprints should be sent to David C. Straus, Ph.D., Department of Microbiology and Immunology, Texas Tech University Health Sciences Center, Lubbock, TX 79430.

E-mail: david.straus@ttuhsc.edu

* * * * * * * * * *

References

1. Hodgson M. Field studies on the sick building syndrome. Ann NY Acad Sci 1992; 641:21–36.
2. Mishra SK, Ajello L, Ahearn DG, et al. Environmental mycology and its importance to public health. J Med Vet Mycol 1992; 30:287–305.
3. Finnigan MS, Pickering CAC, Burge PS. The sick building syndrome: prevalence studies. Br Med J 1984; 289: 1573–75.
4. Feder G. Sick building syndrome. Br Med J 1985; 290:322.
5. Cooley JD, Wong WC, Jumper CA, et al. Correlation between the prevalence of certain fungi and sick building syndrome. Occup Environ Med 1998; 55:579–84.
6. Salvaggio J, Aukrust L. Mold-induced asthma. J Allergy Clin Immunol 1981; 68:327–46.
7. Redd SC. State of the Science on Molds and Human Health. Statement for the Record before the Subcommittees on Oversight and Investigations and Housing and Community Opportunity Committee on Financial Services, U.S. House of Representatives, July 18, 2002; pp 1–11. Available from <http://www.cdc.gov/nceh/airpollution/images/moldsci.pdf>
8. McGrath JJ, Wong WC, Cooley JD, et al. Continually measured fungal profiles in sick building syndrome. Curr Microbiol 1999; 38:33–36.
9. Licorish K, Novey HS, Kozak P, et al. Role of *Alternaria* and *Penicillium* spores in the pathogenesis of asthma. J Allergy Clin Immunol 1985; 76:819–25.
10. Cooley JD, Wong WC, Jumper CA, et al. Cellular and humoral responses in an animal model inhaling *Penicillium chrysogenum* spores. In: Johanning E (Ed). Proceedings of the 3rd International Conference on Bioaerosols, Fungi, and Mycotoxins: Health Effects, Assessment, Prevention and Control, Saratoga Springs, New York, September 1998. Albany, NY: Eastern New York Occupational and Environmental Health Center, 1999; pp 403–10.
11. Cooley JD, Wong WC, Jumper CA, et al. An animal model for allergic penicilliosis induced by the intranasal instillation of viable *Penicillium chrysogenum* conidia. Thorax 2000; 55:489–96.
12. Schwab CJ, Cooley JD, Brasel T, et al. Characterization of

exposure to low levels of viable *Penicillium chrysogenum* conidia and allergic sensitization induced by a protease allergen extract from viable *Penicillium chrysogenum* conidia in mice. Int Arch Allergy Immunol 2003; 130:200–08.

13. Jarvis BB, Sorenson WG, Hintikka E-L, et al. Study of toxin production by isolates of *Stachybotrys chartarum* and *Memnoniella echinata* isolated during a study of pulmonary hemosiderosis in infants. Appl Environ Microbiol 1998; 64:3620–25.
14. Croft WA, Jarvis BB, Yatawara CS. Airborne outbreak of trichothecene toxicosis. Atmos Environ 1986; 20:549–52.
15. Nikulin M, Pasanen A-L, Berg S, et al. *Stachybotrys atra* growth and toxin production in some building materials and fodder under different relative humidities. Appl Environ Microbiol 1994; 60:3421–24.
16. Sorenson WG, Fraxer DG, Jarvis BB, et al. Trichothecene mycotoxins in aerosolized conidia of *Stachybotrys atra*. Appl Environ Microbiol 1987; 53:1370–75.
17. Elidemir O, Colasurdo, GN, Rossman SN, et al. Isolation of *Stachybotrys* from the lungs of a child with pulmonary hemosiderosis. Pediatrics 1999; 104:964–66.
18. Miller JD, Rand TG, Jarvis BB. *Stachybotrys chartarum*: cause of disease or media darling? Med Mycol 2003; 41: 271–91.

Analysis for Mycotoxins: The Chemist's Perspective

BRUCE B. JARVIS
Department of Chemistry and Biochemistry
University of Maryland
College Park, Maryland

ABSTRACT. Mycotoxins are fungal metabolites that pose a health risk to exposed animals and humans. In recent years, concern has mounted regarding human exposure to mycotoxins via inhalation of mold spores produced in damp buildings and homes. Although mycotoxins can be detected in such buildings, reliable means for measuring an occupant's level of exposure to most mycotoxins are lacking. The author briefly reviews the chemical methods currently available for mycotoxin analysis, outlining accepted practices and discussing the limitations of these measurements.
<Key words: biomarkers, chromatographic methods, fungi, mycotoxins, reliability, trichothecenes>

MOST READERS of the *Archives of Environmental Health* have taken at least 2 yr of college chemistry (perhaps sometime in the distant past), but may still lack the experience to judge the quality of the many reports on the presence, or absence, of mycotoxins in environmental samples which are currently flooding the desks of lawyers and doctors who deal with the indoor "toxic" mold problem. This review is intended to provide some background that will aid readers in making reasoned judgments about the reliability of such reports and encourage them to seek additional opinions about a report's validity.

There are literally thousands of mycotoxins—small-molecular–weight fungal metabolites which pose a health risk to humans and animals that (most often) ingest contaminated food or feed, or (less often) are exposed via inhalation of toxin-bearing spores.[1] Of these mycotoxins, only 5 (i.e., deoxynivalenol [DON], aflatoxin, fumonisin, zearalenone, and ochratoxin) are known to commonly affect human health or to result in significant losses in animal productivity.[2] Thousands of other fungal metabolites are toxic to various animal, plant, bacterial, or fungal cells.[3] Because these compounds comprise hundreds of different structural types, chemical analyses must be tailored to the individual mycotoxin. In addition, the biological activities of these toxins vary tremendously, even within a specific class of mycotoxin. Thus, reliance on a bioassay (e.g., cytotoxicity) of an environmental sample as a measure of a specific mycotoxin is hampered by both false-positive and false-negative results. This brief review emphasizes both the power and the limitations of chemical analyses for mycotoxins in environmental samples.[4]

The class of fungal secondary metabolites referred to as mycotoxins is formed somewhat arbitrarily in that mycotoxins are characterized as fungal metabolites that pose a risk (most often through contamination of food) to humans or livestock. Other fungal metabolites may be just as toxic (or more so) as mycotoxins, but are not considered mycotoxins because humans and animals are not generally exposed to them. Furthermore, members of a known class of mycotoxins produced by fungi that have no environmental impact on humans or livestock are still considered mycotoxins because they are related closely in structure to other members of the same class. For example, many trichothecene mycotoxins—produced by various fungi, and even found in some plants[5]—which have never been reported to pose a risk to humans or animals are still considered mycotoxins because their structure is similar to that of other trichothecenes that do constitute a risk.

Isolation and Characterization of Mycotoxins

Before analytical methods can be applied, mycotoxins must be isolated and characterized. One always starts with a pure culture of the fungus. Ideally, the fungus is grown in liquid culture, but sometimes toxin production can be realized only from solid media cultures. The culture is extracted with an organic solvent, and the crude extract is partitioned among various solvents according to the acid–base and polar properties of the desired metabolite(s). This crude fractionation often can concentrate the desired compound(s) into a fraction that has less than 30% of the original weight of the crude extract. Next, the active fraction is processed by a series of increasingly sophisticated (and expensive) chromatographic techniques that eventually yield the pure compound(s). The compounds are then thoroughly characterized using a combination of techniques, the most important of which are nuclear magnetic resonance (NMR) spectroscopy and x-ray diffraction analysis. Mass spectrometry is also employed; however, this highly sensitive technique tends to be of far greater importance in the search for known metabolites in environmental samples, ranging from tissues to dust. *No completely dependable chemical analysis for mycotoxins can be conducted without the use of standards for the compounds being analyzed.*

Chemical Methods for Mycotoxin Analysis

As in any general method of analysis, chemical analyses range from very crude and uncertain to highly precise and sensitive—often resembling the search for a "needle in a haystack." Environmental samples are *very* dirty. For example, one of the signature compounds for fungi is ergosterol, the cholesterol counterpart found in the cell walls of all fungi. Air samples from indoor environments typically show considerably more cholesterol from the skin scales of human and animal inhabitants than ergosterol from fungi, even in homes and buildings with fungal contamination (J.D. Miller, Professor of Biology and Chemistry, Carleton University, Ottawa, Ontario, Canada; personal communication).

Many of the standard procedures for the analysis of mycotoxins are described in the Association of Analytical Chemists' publication *Official Methods of Analysis of AOAC International*.[6] However, these methods have been developed for specific circumstances and, in particular, for specific matrices in which the mycotoxins are found. In the United States, these matrices are almost exclusively agricultural products (principally grains), because the U.S. Food and Drug Administration has regulatory responsibilities in the commercial markets for these important products.

It is essential to appreciate the limitations of chemical analyses. Experience gained during development of the AOAC standardized procedures (in which many laboratories participated in establishing the protocols) has revealed the following limitations: at the 1 ppm level (1 μg of analyte per gram of sample), the precision of analysis is ±16%; at 1 ppb, precision drops to ±45%; and at levels below 1 ppb (1 ng/gm), the precision of analysis quickly becomes unacceptable.[7]

Chemical analysis for mycotoxins consists of 4 steps: (1) sample preparation, (2) extraction, (3) cleanup of the extract, and (4) measurement of the specific toxins. Although techniques for obtaining agricultural samples can be somewhat complicated,[8] sampling of the indoor environment is more straightforward. Three general methods are employed: (1) air sampling, with collection on filters; (2) dust sampling; and (3) bulk sampling of visibly contaminated areas. Of these, bulk sampling is the most likely to yield positive results because there is a clear indication of the presence of fungus. It is important that the fungi in these samples be characterized *down to the species level*, because the results of any subsequent chemical analysis must correspond with the presence of mycotoxin-producing fungi—a point that is often overlooked. Establishment of this close connection between a specific mycotoxin and a specific fungal species is necessary because mycotoxins are restricted in their distribution to only a small number of fungal species.[9] For example, many species of fungi produce trichothecenes, but only a small fraction of these species actually produce T-2 toxin. The satratoxin trichothecenes produced by *Stachybotrys chartarum* are also produced by a very small number of *Myrothecium roridum* isolates[5] and, interestingly, by a mushroom,[10] but these instances are so idiosyncratic that, for all practical purposes, the satratoxins of interest in human health are those produced by *Stachybotrys* sp. fungi.

Claims that trichothecene mycotoxin T-2 toxin was found in an indoor environment from which no species of T-2 toxin-producing *Fusarium* species could be isolated are most certainly wrong in their chemical analysis.[11,12] In these cases, the analyses were performed with thin-layer chromatrography (TLC)—a method with low sensitivity and specificity. Researchers who find aflatoxin B_1 in the blood of individuals who live or work in buildings (other than occupational environments such as grainaries) with no detectible levels of aflatoxin-producing *Aspergillus* species (e.g., *A. flavus*) should look to another source (such as diet) for this mycotoxin, or check the credibility of their analysis. Although *Aspergillus* grows on various crop plants,[2] dietary aflatoxin exposure in the United States is modest.[13]

Given that levels of mycotoxins in indoor environments are low in most circumstances, sample sizes should be large. Ideally, the extraction solvent will extract the desired fungal metabolites with a minimum of extraneous compounds. What occurs most often in practice is a balance between these 2 factors such that more than 50% of the mycotoxins are extracted, accompanied by only a small amount of contaminating material, so that an accurate analysis can still be conducted. There are no rigid rules governing the choice of extraction solvent, but most extraction systems consist of alcohols (methanol or ethanol) with varying amounts of water (0–20%). Choice is frequently dictated by the nature of the material being extracted, and often involves trial and error to establish conditions that give reproducible and accurate results.

Environmental extracts are complex mixtures and nearly always require cleaning prior to analysis. One particularly effective type of cleanup column is an immunoaffinity column (IAC), wherein antibodies specific to a given mycotoxin are immobilized on a packing material such as Sepharose™. Passing the crude extract (or body fluid such as milk or plasma) through the column concentrates the mycotoxin on the column, allowing most of the interfering material to pass through. The mycotoxin then can be eluted from the column by washing with an organic solvent such as methanol or acetonitrile. Use of this technology, however, is constrained by cost and availability, as only a limited number of IACs are available commercially for mycotoxins (e.g., aflatoxins, DON, and ochratoxin A).

A more typical cleanup procedure involves passing the crude extract through a short chromatography column that may contain a variety of packing material (usually silica), either normal or reversed-phase; washing the column with 2 or more mixtures of solvents; and collecting the desired analyte in a particular fraction. Again, the extraction procedure is often delineated through laboratory optimization experiments. For example, the procedure for *Stachybotrys chartarum* samples calls for a polyethyleneimine cleanup column that can effectively concentrate 2 classes of mycotoxins—macrocyclic trichothecenes (satratoxins) and spirocyclic phenolics—into 2 separate fractions that can then be analyzed effectively with high-performance liquid chromatography (HPLC).[14] Without this cleanup, it is very difficult to detect the satratoxins on HPLC analysis because they are obscured by the more prominent spirocyclic phenolics.

Chromatographic procedures are central to mycotoxin assays. Chromatography separates (resolves) the various components within a mixture and provides a measure of the identity of the components. There is, however, great variation among chromatographic methods. TLC (with spots observed visually) has low resolution and sensitivity, whereas gas-liquid chromatography–mass spectrometry (GLC-MS) and HPLC-MS have high resolution and sensitivity. In addition, the MS-coupled techniques enable one to confirm with certainty the identity of the analyte.

The TLC technique is in wide use because of its speed and cost-effectiveness, but it should never be used to establish with certainty the presence or absence of a mycotoxin in the indoor environment. It is instructive to quote from an early paper by Croft et al.,[15] who found trichothecenes in a *Stachybotrys chartarum*-contaminated home in Chicago, Illinois. Regarding the extract, following solvent partitions: "after several column chromatographies (silica gel, methanol in dichloromethane eluate) yielded several fractions which upon TLC (Takitani and Asabe, 1983) *suggested* the presence of macrocyclic trichothecenes" (emphasis added). Further purification led to the isolation of pure toxins that were then characterized by NMR spectroscopy and shown to be identical to the standards. Thus, Croft et al. found TLC to be helpful as a *guide*, but certainly not to be relied on as a tool for identification of the toxins.

Currently, the best and most sensitive analytical methods for the chemical analysis of mycotoxins are GLC-MS[16] and HPLC-MS.[17] The latter method can incorporate an additional mode of analysis—diode array detection (DAD)—that allows one to detect the ultraviolet spectrum of the analyte as well. The combination of HPLC-DAD-MS with MS/MS detection (in which the MSs of fragment ions can be acquired) is so powerful (but expensive!) that, in many cases, analyses can be conducted reliably without the need for quality control through comparison with standards data.[17] However, in cases that involve putative human intoxications, it is advisable to have the relevant standards on hand to ensure the quality of the analysis. The limits of detection by HPLC-DAD-MS for common mycotoxins range from 1 pg to 3 ng per injection[17]—values several orders of magnitude lower than those found with TLC.

One of the basic problems in assessing human exposure to mycotoxins is the need to establish a reliable means for measuring the level of exposure—that is, a biomarker that will give a reliable measure of a specific mycotoxin to which an individual has been exposed. Only a few such biomarkers exist for mycotoxins (e.g., for aflatoxins,[18] ochratoxin A,[19] and DON[20]). In some cases, the biomarker is the toxin (or protein-bound toxin) itself (e.g., ochratoxin A[21] and aflatoxin B_1[18] in blood). In other cases, it may be a well-characterized metabolite such as the glucuronide of DON in urine,[20] or a deoxyribonucleic acid (DNA) conjugate of aflatoxin B_1.[18] However, for the vast majority of mycotoxins, we lack not only characterized biomarkers, but also information on the pharmacokinetics or metabolism of these compounds in animals, especially humans.

Another important factor is that some mycotoxins (or the adducts they form with biopolymers) are fairly long-lived and can be detected in animal tissue or blood products many weeks after exposure; others are rapidly metabolized to water-soluble glucuronides and excreted in urine within a matter of hours or a few days. Ochratoxin A and aflatoxins are examples of the former; simple trichothecenes such as DON and T-2 toxins are examples of the latter. Pharmacokinetic data are available regarding the metabolic fate of several of the simple trichothecenes (including T-2 toxin and DON in laboratory and domestic animals),[22] as well as diacetoxyscirpenol in humans.[20,23] Such data have contributed to the development of standards for tolerable daily intakes of these compounds in food.[24] However, no such data are available for the macrocyclic trichothecenes, such as the satratoxins.

Claims by some researchers that specific immunoglobulins (e.g., IgG) to mycotoxins can be measured in patients' blood are, in my opinion, without substance. It is theoretically possible for a low-molecular–weight compound to haptenize (leading to the formation of human antibodies), as occurs in people with penicillin allergy and in patients with Acquired Immune Deficiency Syndrome (AIDS) who take many drugs over an extended period of time. However, there has been no demonstration in peer-reviewed literature that humans exposed to environmental levels of mycotoxins generate such specific antibodies. Both monoclonal and polyclonal antibodies to mycotoxins can be produced and used to measure mycotoxin levels in environmental samples (including human blood and tissue). Such antibodies form the basis for many commercially available immunoassays.[25,26] Antibodies to the satratoxins have been prepared,[27] and may prove useful in the future for measuring macrocyclic trichothecene satratoxins in individuals exposed to *Stachybotrys chartarum*.[28]

Future advances in analyses for mycotoxins are likely to be in the areas of biosensors and increased sensitivity/selectivity of immunoassays. The chemist probably will always want to see confirmatory data (e.g., MS data) but, at the very least, will demand that these newer assays be shown to provide numbers that can be confirmed as reliable by several laboratories and certified by organizations such as AOAC International.

* * * * * * * * * *

Submitted for publication September 20, 2003; revised; accepted for publication October 30, 2003.

Requests for reprints should be sent to Bruce B. Jarvis, Dept. of Chemistry and Biochemistry, University of Maryland, College Park, MD 20742-2021.

E-mail: bj6@umail.umd.edu

* * * * * * * * * *

References

1. Bennett JW, Klich M. Mycotoxins. Clin Microbiol Rev 2003; 16:497–516.
2. Miller JD. Mycotoxins. In: Francis FJ (Ed). Encyclopedia of Food Science and Technology. New York: Wiley, 1999; pp 1698–706.
3. Cole RJ, Schweikert MA, Jarvis BB. Handbook of Secondary Fungal Metabolites. Vol. 1–3. New York: Academic Press, 2003.
4. Cole RJ (Ed). Modern Methods in the Analysis and Structural Elucidation of Mycotoxins. New York: Academic Press, 1986.
5. Jarvis BB. Macrocyclic trichothecenes. In: Sharma RP, Salunkhe DK (Eds). Mycotoxins and Phytoalexins. Boca Raton, FL: CRC Press, 1991; pp 361–421.
6. Association of Analytical Chemists (AOAC). Official Methods of Analysis of AOAC International. 17th ed, Rev. 2. Gaithersburg, MD: AOAC 2003; 2,200 pp. Available for purchase from <http://www.aoac.org/pubs/oma_revised.htm>
7. Pohland AE, Trucksess MW. Mycotoxin method evaluation: an introduction. In: Trucksess MW, Pohland AE (Eds). Methods in Molecular Biology. Vol 157, Mycotoxin Protocols. Totowa, NJ: Humana Press, 2000; pp 3–10.
8. Whitaker TB. Sampling techniques. In: Trucksess MW, Pohland AE (Eds). Methods in Molecular Biology. Vol 157, Mycotoxin Protocols. Totowa, NJ: Humana Press, 2000; pp 11–24.
9. Frisvad JC, Thrane U, Filtenborg O. Role and use of secondary metabolites in fungal taxonomy. In: Frisvad JC, Bridge PD, Arora DK (Eds). Chemical Fungal Taxonomy. New York: Marcel Dekker, 1998; pp 289–319.
10. Saikawa Y, Okamoto H, Inui T, et al. Toxic principles of a poisonous mushroom *Podostroma cornu-damae*. Tetrahedron 2001; 57:8277–81.
11. Smoragiewicz W, Cossette B, Boutard A, et al. Trichothecene mycotoxins in the dust of ventilation systems in office buildings. Int Arch Occup Environ Health 1993; 65: 113–17.
12. Croft WA, Jastromski B, Croft AL, et al. Clinical confirmation of trichothecene mycotoxicosis in patient urine. J Environ Biol 2002; 23:301–20.
13. International Agency for Research on Cancer (IARC). IARC Monographs on the Evaluation of Carcinogenic Risks to Humans. Vol. 82, Some Traditional Herbal Medicines, Some Mycotoxins, Naphthalene and Styrene. Lyon, France: IARC Press, 2002.
14. Hinkley SF, Jarvis BB. Method for *Stachybotrys* toxins. In: Trucksess MW, Pohland AE (Eds). Methods in Molecular Biology. Vol 157, Mycotoxin Protocols. Totowa, NJ: Humana Press, 2000; pp 173–94.
15. Croft WA, Jarvis BB, Yatawara CS. Airborne outbreak of trichothecene toxicosis. Atmos Environ 1986; 20: 449–552.
16. Nielsen KF, Thrane U. Fast method for detection of trichothecenes in fungal culture using gas chromatography–tandem mass spectrometry. J Chromatogr A 2001; 929:75–87.
17. Nielsen KF, Smedsgaard J. Fungal metabolite screening: database of 474 mycotoxins and fungal metabolites for dereplication by standardised liquid chromatography–UV mass spectrometry methodology. J Chromatogr A 2003; 1002:111–36.
18. Sabbioni G, Sepai O. Determination of human exposure to aflatoxins. In: Sinha KK, Bhatnagar D (Eds). Mycotoxins in Agriculture and Food Safety. New York: Marcel Dekker, 1994; pp 183–226.
19. Usleber E, Dietrich R, Schneider E, et al. Immunochemical method for ochratoxin A. In: Trucksess MW, Pohland AE (Eds). Methods in Molecular Biology. Vol 157, Mycotoxin Protocols. Totowa, NJ: Humana Press, 2000; pp 81–94.
20. Meky FA, Turner PC, Ashcroft AE, et al. Development of a urinary biomarker of human exposure to deoxynivalenol. Food Chem Toxicol 2003; 41:265–73.
21. Hagelberg S, Hult K, Fuchs R. Toxicokinetics of ochratoxin A in several species and its plasma-binding properties. J Appl Toxicol 1989; 9:91–96.
22. Prelusky DB, Rotter BA, Rotter RG. Toxicology of mycotoxins. In: Miller JD, Trenholm HL (Eds). Mycotoxins in Grain: Compounds Other than Aflatoxin. St. Paul, MN: Eagan Press, 1994; pp 359–404.
23. Ueno Y. Trichothecenes: Chemical, Biological, and Toxi-

cological Aspects. Amsterdam: Elsevier, 1983.

24. Joint Food and Agriculture Organization of the United Nations (FAO)/World Health Organization (WHO) Expert Committee on Food Additives (JECFA). 56th meeting of the JECFA, Mycotoxins. Geneva, Switzerland: Feb 2001. <http://www.who.int/pcs/jecfa/Summary56.pdf>
25. Chu FS. Recent studies on immunoassays for mycotoxins. In: Beier RC, Stanker LH (Eds). Immunoassays for Residue Analysis. Washington, DC: American Chemical Society, 1996; pp 294–313.
26. Trucksess MW, Pohland AE (Eds). Methods in Molecular Biology. Vol 157, Mycotoxin Protocols. Totowa, NJ: Humana Press, 2000.
27. Chung Y-J, Jarvis BB, Tak H, et al. Immunochemical assay for satratoxin G and other macrocyclic trichothecenes associated with indoor air contamination by *Stachybotrys chartarum*. Toxicol Mech Meth 2003, 13:247–53.
28. Miller JD, Jarvis BB, Rand TG. *Stachybotrys chartarum*: cause of human disease or media darling? Med Mycol 2003; 41:271–91.

Organophosphorus Ester-Induced Chronic Neurotoxicity

MOHAMED B. ABOU-DONIA
Department of Pharmacology and Cancer Biology
Duke University Medical Center
Durham, North Carolina

ABSTRACT. Organophosphorus compounds are potent neurotoxic chemicals that are widely used in medicine, industry, and agriculture. The neurotoxicity of these chemicals has been documented in accidental human poisoning, epidemiological studies, and animal models. Organophosphorus compounds have 3 distinct neurotoxic actions. The primary action is the irreversible inhibition of acetylcholinesterase, resulting in the accumulation of acetylcholine and subsequent overstimulation of the nicotinic and muscarinic acetylcholine receptors, resulting in cholinergic effects. Another action of some of these compounds, arising from single or repeated exposure, is a delayed onset of ataxia, accompanied by a Wallerian-type degeneration of the axon and myelin in the most distal portion of the longest tracts in both the central and peripheral nervous systems, and is known as organophosphorus ester-induced delayed neurotoxicity (OPIDN). In addition, since the introduction and extensive use of synthetic organophosphorus compounds in agriculture and industry half a century ago, many studies have reported long-term, persistent, chronic neurotoxicity symptoms in individuals as a result of acute exposure to high doses that cause acute cholinergic toxicity, or from long-term, low-level, subclinical doses of these chemicals. The author attempts to define the neuronal disorder that results from organophosphorus ester-induced chronic neurotoxicity (OPICN), which leads to long-term neurological and neurobehavioral deficits. Although the mechanisms of this neurodegenerative disorder have yet to be established, the sparse available data suggest that large toxic doses of organophosphorus compounds cause acute necrotic neuronal cell death in the brain, whereas sublethal or subclinical doses produce apoptotic neuronal cell death and involve oxidative stress.
<Key words: acetylcholinesterase, apoptosis, brain, necrosis, nervous system, neurobehavioral, neuropathological, neuropsychological, OPICN, OPIDN, organophosphorus compounds>

ORGANOPHOSPHORUS COMPOUNDS are chemicals that contain both carbon and phosphorus atoms.[1] They are derivatives of phosphoric (H_3PO_4), phosphorus or phosphonic (H_3PO_3), and phosphinic (H_3PO_2) acids (Fig. 1). The biological action of organophosphorus compounds is related to their phosphorylating abilities. This is dependent on the electrophilicity (positive character) of the phosphorus atom, which is determined by its substituent groups. Steric factors of substituents also play a major role in determining the biological activity of these chemicals. Lipid solubility is important because it enhances the ability of these compounds to cross biological membranes and the blood-brain barrier, leading to increased biological activity. Organophosphorus compounds are an economically important class of chemical compounds with numerous uses, such as in pesticides, industrial fluids, flame retardants, therapeutics, and nerve gas agents.

Most modern synthetic organophosphorus compounds are tailor-made to inhibit acetylcholinesterase (AChE), an enzyme essential for life in humans and other animal species. Tetraethylpyrophosphate was the 1st organophosphate synthesized as an AChE inhibitor in 1854.[2] Later, dimethyl and diethyl phosphorofluoridates were synthesized.[2] During World War II, organophosphorus compounds were developed primarily as agricultural insecticides, and later as chemical warfare agents. The majority of organophosphorus insecticides are organophosphorothioates; nerve agents are organophosphonates or organophosphonothioates; industrial chemicals are typically organophosphates[3] (Table 1).

Biologically, organophosphorus compounds are neu-

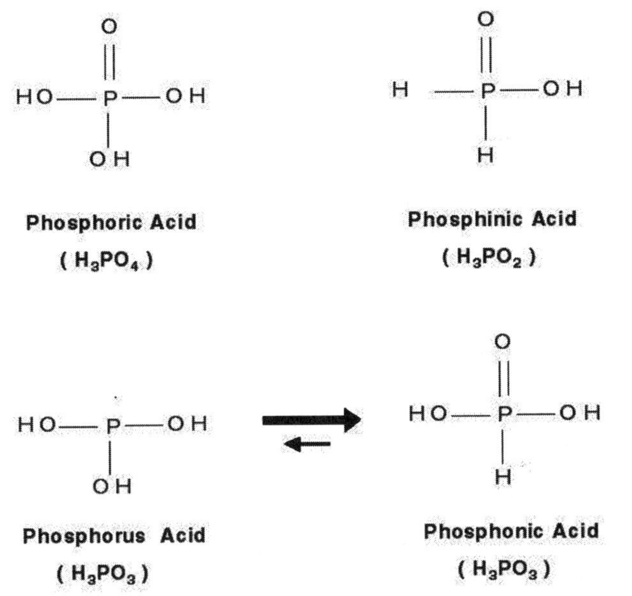

Fig. 1. Chemical structures of phosphoric acid, phosphinic acid, phosphorus acid, and phosphonic acid.

rotoxic to humans and other animals via 3 distinct actions: (1) cholinergic neurotoxicity, (2) organophosphorus ester-induced delayed neurotoxicity (OPIDN), and (3) organophosphorus ester-induced chronic neurotoxicity (OPICN).

Cholinergic Neurotoxicity of Organophosphorus Compounds

Organophosphorus compounds cause cholinergic neurotoxicity by disrupting the cholinergic system that includes AChE and its natural substrate, the neurotransmitter acetylcholine.[3] Acetylcholine is released in response to nerve stimulation and binds to post-synaptic acetylcholine receptors, resulting in muscle contraction or gland secretions. Its action is rapidly terminated by hydrolysis with AChE via the serine hydroxyl in the catalytic triad of AChE.[2] The 3-dimensional structure of AChE reveals an active center located at the base of a narrow gorge about 20 Å in depth.[4] The active center includes the following sites (Fig. 2): (a) the catalytic triad: Glu 334, His 447, and Ser 203; (b) an aceyl pocket: Phe 295 and Phe 297; (c) a choline subunit: Trp 86, Glu 202, and Tyr 337; and (d) a peripheral site: Trp 286, Tyr 72, Tyr 124, and Asp 74.

Organophosphorus ester inhibition of AChE. Organophosphorus esters inhibit AChE by phosphorylating the serine hydroxyl group at the catalytic triad site (Fig. 3). The phosphoric or phosphonic acid ester formed with the enzyme is extremely stable and is hydrolyzed very slowly.[1] If the phosphorylated enzyme contains methyl or ethyl groups, the enzyme is regenerated in several hours by hydrolysis. On the other hand, virtually no hydrolysis occurs with an isopropyl group (e.g., sarin) and the return of AChE is dependent upon synthesis of a new enzyme. Phosphorylated AChE undergoes aging—a process that involves the loss of an alkyl group, resulting in a negatively charged monoalkyl enzyme (Fig. 4).[3] Organophosphorus compounds undergo detoxification by binding to other enzymes that contain

Table 1.—Compounds Cited in the Text and Their IUPAC Designations

Compound	IUPAC designation
Chlorpyrifos	*O,O*-diethyl *O*-3,5,6-trichloro-2-pyridyl phosphorothioate
Cyclosarin (GF)	*O*-cyclohexyl methylphosphonofluoridate
DFP	*O,O*-diisopropyl phosphorofluoridate
DEET	*N,N*-diethyl-m-toluamide
Diazinon	*O,O*-diethyl *O*-2-isopropyl-6-methylpyrimidin-4-yl phosphorothioate
Fenthion	*O,O*-dimethyl *O*-4 methylthio-m-tolyl phosphorothioate
Malathion	*S*-1,2-*bis*(ethoxycarbonyl)ethyl *O,O*-dimethyl phosphorodithioate
Methamidophos	*O,S*-dimethyl phosphoramidothioate
Permethrin	3-phenoxybenzl (1*RS*)-*cis-trans*-3-(2,2-dichlorovinyl)-2,2-dimethylcyclo-propanecarboxylate
Quinalphos	*O,O*-diethyl *O*-quinoxalin-2-yl phosphorothioate
Tabun (GA)	*O*-ethyl *N,N*-dimethylphosphoamido-cyanidate
Sarin (GB)	*O*-isopropyl methylphosphonofluoridate
Soman (GD)	*O*-2,2-trimethypropyl methylphosphono-fluoridate
TOCP	Tri-ortho-cresylphosphate
VX	*O*-ethyl *S*-2-diisopropylaminoethyl methylphosphonothioate
VR	*O*-isobutyl *S*-2-diethylaminoethyl methylphosphonothioate

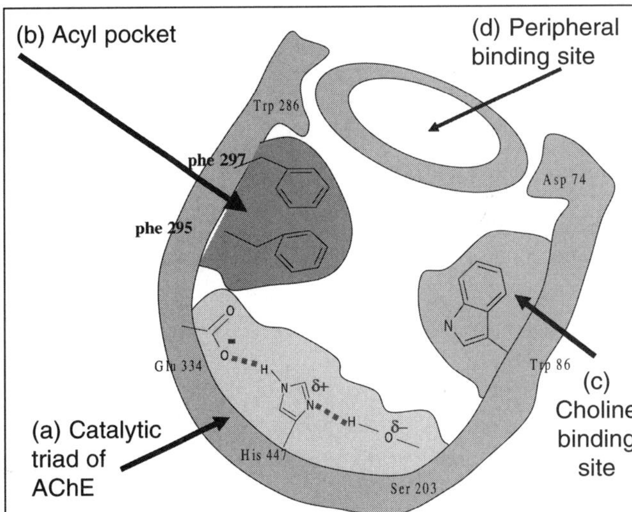

Fig. 2. Active center gorge of mammalian acetylcholinesterase (AChE).

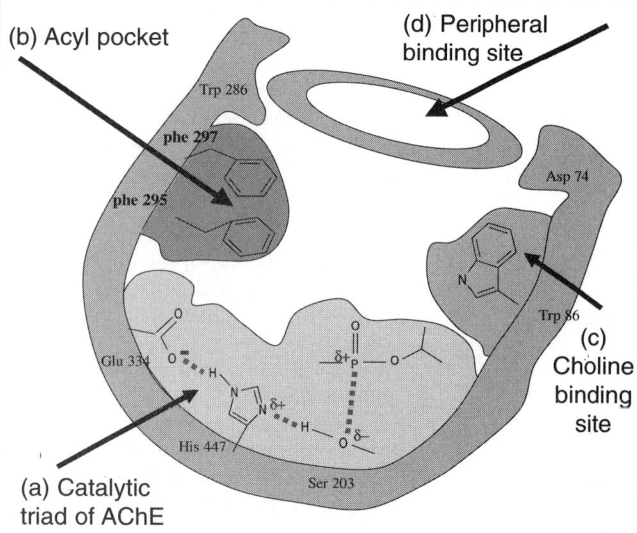

Fig. 3. Sarin phosphonyl enzyme: isopropyl methylphosphonyl acetylcholinesterase (AChE).

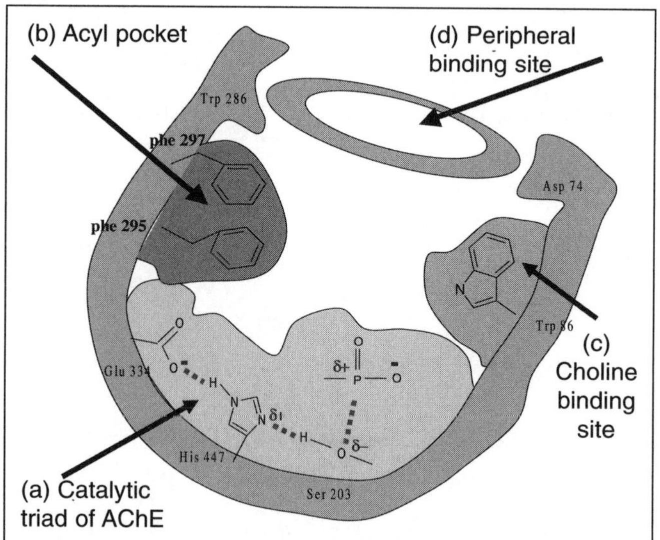

Fig. 4. Sarin aged phosphonyl enzyme: methylphosphonyl acetylcholinesterase (AChE).

the amino acid serine. These enzymes include plasma butyrylcholinesterase (BChE)[5,6] and paraoxonase.[7,8]

Inhibition of AChE results in the accumulation of acetylcholine at both the muscarinic and nicotinic receptors in the central nervous system (CNS) and the peripheral nervous system (PNS). Excess Acetylcholine initially causes excitation, and then paralysis, of cholinergic transmission, resulting in some or all of the cholinergic symptoms, depending on the dose size, frequency of exposure, duration of exposure, and route of exposure, as well as other factors such as combined exposure to other chemicals and individual sensitivity and susceptibility.

Human exposure. Human exposure—mostly via inhalation—to the organophosphorus nerve agent sarin was recently documented in 2 terrorist incidents in Japan. At midnight on June 27, 1994, sarin was released in Matsumoto City.[9] Of the 600 persons who were exposed, 58 were admitted to hospitals, where 7 died. Although miosis was the most common symptom, severely poisoned patients developed CNS symptoms and cardiomyopathy. A few victims complained of arrhythmia and showed cardiac contraction. The 2nd terrorist attack by sarin was in the Tokyo subway trains, at 8:05 A.M. on March 20, 1995, when a total of 5,000 persons were hospitalized and 11 died.[10] Patients with high exposure to sarin in the Tokyo subway incident exhibited marked muscle fasciculation, tachycardia, high blood pressure (nicotinic responses), sneezing, rhinorrhea, miosis, reduced consciousness, respiratory compromise, seizures, and flaccid paralysis.[11] Patients with mild exposure complained of headaches, dizziness, nausea, chest discomfort, abdominal cramps, and miosis. Interestingly, patients had pupillary constriction, even when their cholinesterase activity was normal. Furthermore, inhibition of red blood cell AChE activity was a more sensitive indicator of exposure than serum BChE activity.[12] The absence of bradycardia and excessive secretions—which are common in dermal or ingestion exposures—suggested that the major route of exposure to the sarin gas in these instances was via inhalation. The patients were treated with atropine eye drops for marked miosis, and with pralidoxime iodide (2-PAM).

Organophosphorus Ester-Induced Delayed Neurotoxicity (OPIDN)

Characteristics of OPIDN. OPIDN is a neurodegenerative disorder characterized by a delayed onset of prolonged ataxia and upper motor neuron spasticity as a result of a single or repeated exposure to organophosphorus esters.[13–16] The neuropathological lesion is a central-peripheral distal axonopathy caused by a chemical transection of the axon (known as Wallerian-type degeneration), followed by myelin degeneration of distal portions of the long and large-diameter tracts of the CNS and PNS.[17] Incidents of OPIDN have been documented for over a century. The earliest recorded cases were attributed to the use of tri-*o*-cresyl phosphate (TOCP)-containing creosote oil for treatment of pulmonary tuberculosis in France in 1899.[13,15,16] In 1930, TOCP was identified as the chemical responsible for an estimated 50,000 cases of OPIDN in the Southern and Midwestern regions of the United States.[13,15,16] More recently, Himuro et al.[18] reported that a 51-yr–old man who was exposed to sarin during the Tokyo subway incident and survived its acute toxicity, then died 15 mo later. Neuropathological alterations and neurological deficits observed in this individual were consistent with the dying-back degeneration of the nervous system characteristic of OPIDN. This incident indicated that humans are more sensitive than experimental animals to sarin-induced OPIDN, inasmuch as it required

26–28 daily doses of LD_{50} (25 μg/kg, i.m.) sarin to produce OPIDN in the hen.[19] OPIDN has been divided into 3 classes: *Type I* is caused by the pentavalent phosphates and phosphonates, as well as their sulfur analogs; *Type II* is produced by the trivalent phosphites;[20] and *Type III* is induced by phosphines.[21,22] All 3 OPIDN types are produced by organophosphorus compounds and characterized by central-peripheral distal axonopathy. Type II differs from Type I in terms of the susceptibility of rodents and the presence of neuropathological lesions in neuronal cell bodies.[20] Type III OPIDN is not accompanied by inhibition of the neurotoxicity target esterase (NTE), thus casting further doubt on this enzyme as the target for OPIDN.[21,22]

Mechanisms of OPIDN. Early studies on the mechanisms of OPIDN centered on the inhibition of the esterases AChE[23] and BChE[24] by organophosphorus esters. Subsequent studies eliminated both enzymes as targets for OPIDN.[25] Johnson[26] proposed an NTE—an enzymatic activity preferentially inhibited by organophosphorus compounds capable of producing OPIDN as its target. Despite numerous studies since the introduction of this concept 35 yr ago, the NTE hypothesis has not advanced our understanding of the mechanism of OPIDN because: (a) evidence for the involvement of NTE in the development of OPIDN is only correlative; (b) it has not been shown how inhibition and aging of NTE leads to axonal degeneration; (c) NTE, which is present in neuronal and non-neuronal tissues and in sensitive and insensitive species, has no known biochemical or physiological function; (d) some organophosphorus pesticides that produce OPIDN in humans do not inhibit or age NTE;[27–30] and (e) phosphines that produce Type III OPIDN do not inhibit NTE.[21,22] However, the most convincing evidence against this hypothesis is the recent finding that NTE-knockout mice are sensitive to the development of OPIDN,[31–33] indicating that this enzyme is not involved in the mechanisms of OPIDN.

Protein kinases as targets for OPIDN. Because research on esterases did not increase our understanding of the mechanisms of OPIDN, we have been studying the involvement of protein kinase-mediated phosphorylation of cytoskeletal proteins in the development of OPIDN. These studies were prompted by the following observations: (a) Since organophosphorus compounds are effective phosphorylating agents, it is reasonable to expect that they would interfere with normal kinase-mediated phosphorylation of a serine or threonine group at the target protein. (b) The earliest ultrastructural alterations in OPIDN are seen mostly as aggregation and accumulation of cytoskeletal proteins, microtubules, and neurofilaments, followed by their dissolution and disappearance. (c) The structural and functional status of cytoskeletal proteins are affected significantly by protein kinase-mediated phosphorylation.

Anomalous hyperphosphorylation of cytoskeletal elements is associated with OPIDN, a neurodegenerative disorder characterized by distally located swellings in large axons of the CNS and PNS, with subsequent axonal degeneration. Central to our hypothesis is the observation that increased aberrant protein kinase-mediated phosphorylation of cytoskeletal proteins could result in the destabilization of microtubules and neurofilaments, leading to their aggregation and deregulation in the axon.[34] Protein kinases are able to amplify and distribute signals because a single protein kinase can phosphorylate many different target proteins. Several protein kinases are turned on by 2nd messengers. For example, calcium/calmodulin-dependent protein kinase II (CaM kinase II) is inactive until it is bound by the calcium-calmodulin complex that induces conformational changes and causes the enzyme to unfold an inhibitory domain from its active site.[35] We have demonstrated substantial increases in the autophosphorylation,[36,37] enzymatic activity,[38] protein levels, and mRNAs of CaM kinase II in hens treated with diisopropylphosphorofluoridate (DFP).[39] These aberrant alterations have resulted in increased phosphorylation of the following cytoskeletal proteins: tubulin, neurofilaments, microtubule associated proteins-2 (MAP-2), and tau proteins.[39–42] Increased activity of CaM kinase II can affect the stability of cytoskeletal proteins through post-translational modification. Phosphorylation of these proteins interrupts their interaction, polymerization, and stabilization, leading to their degeneration.[39,43,44]

On the other hand, early studies identified transcription factors as critical phosphoproteins in signaling cascades. Immediate early genes control gene expression and therefore affect long-term cellular responses. We have demonstrated that the transcription of c-fos is elevated in OPIDN, perhaps through the activation of cAMP (adenosine monophosphate) response element binding (CREB), which is phosphorylated by CaM kinase II. Subsequent to c-fos activation,[45] we observed altered gene expression of CaM kinase II,[43] neurofilaments,[46] glial fibrillary acidic protein (GFAP), and vimentin.[47] Our results also showed an increase in medium (NF-M) and a decrease in low (NF-L) and high (NF-H) molecular weight neurofilaments in the spinal cords of hens treated with DFP.[46] This imbalance in the stoichiometry of neurofilament proteins interferes with their interaction with microtubules and promotes neurofilament dissociation from microtubules, leading to the aggregation of both cytoskeletal proteins.[48] Immunohistochemical studies in nervous system tissues from TOCP- and DFP-treated hens demonstrated aberrant aggregation of phosphorylated neurofilament, tubulin, and CaM kinase II.[49]

Organophosphorus Ester-Induced Chronic Neurotoxicity (OPICN)

Various epidemiological studies have demonstrated that individuals exposed to a single large toxic dose, or to small subclinical doses, of organophosphorus compounds have developed a chronic neurotoxicity that persists for years after exposure and is distinct from both cholinergic and OPIDN effects.[50] This disorder has been variously referred to in the literature as: "chronic neurobehavioral effects,"[11] "chronic organophosphate-induced neuropsychiatric disorder (COPIND),[50] "psychiatric sequelae of chronic exposure,"[51] "central nervous system effects of chronic exposure,"[52] "psychological and neurological alterations,"[53] "long-term effects,"[54] "neuropsychological abnormalities,"[55] "central cholinergic involvement in behavioral hyperreactivity,"[56] "chorea and psychiatric changes,"[57] "chronic central nervous effects of acute organophosphate pesticide intoxication,"[58] "chronic neurological sequelae,"[59,60] "neuropsychological effects of long-term exposure,"[61] "neurobehavioral effects,"[62] and "delayed neurologic behavioral effects of subtoxic doses."[63] Our review of the literature indicated that these studies describe a nervous system disorder induced by organophosphorus compounds which involves neuronal degeneration and subsequent neurological, neurobehavioral, and neuropsychological consequences. We will next define and describe this disorder, and refer to it as "organophosphorus ester-induced chronic neurotoxicity" or OPICN.

Characteristics of OPICN. OPICN is produced by exposure to large, acutely toxic—or small subclinical—doses of organophosphorus compounds. Clinical signs, which continue for a prolonged time ranging from weeks to years after exposure, consist of neurological and neurobehavioral abnormalities. Damage is present in both the PNS and CNS, with greater involvement of the latter. Within the brain, neuropathological lesions are seen in various regions, including the cortex, hippocampal formation, and cerebellum. The lesions are characterized by neuronal cell death resulting from early necrosis or delayed apoptosis. Neurological and neurobehavioral alterations are exacerbated by concurrent exposure to stress or to other chemicals that cause neuronal cell death or oxidative stress. Because CNS injury predominates, improvement is slow and complete recovery is unlikely.

OPICN following large toxic exposure to organophosphorus compounds. Several studies have reported that some individuals who were exposed to large toxic doses of organophosphorus compounds, and who experienced severe acute poisoning and subsequent recovery, eventually developed the long-term and persistent symptoms of OPICN. Many of the adverse effects produced by organophosphorus compounds are not related to AChE inhibition.[64] Individuals with a history of acute organophosphate exposure reported an increased incidence of depression, irritability, confusion, and social isolation.[65] Such exposures resulted in decreased verbal attention, visual memory, motoricity, and affectivity.[66] Rosenstock et al.[58] reported that even a single exposure to organophosphates requiring medical treatment was associated with a persistent deficit in neuropsychological functions. A study of long-term effects in individuals who experienced acute toxicity with organophosphorus insecticides indicated dose-dependent decreases in sustained visual attention and vibrotactile sensitivity.[59] In another study, one-fourth of the patients who were hospitalized following exposure to methamidophos exhibited an abnormal vibrotactile threshold between 10 and 34 mo after hospitalization.[67]

Callender et al.[60] have described a woman with chronic neurological sequelae following acute exposure to a combination of an organophosphorothioate insecticide, pyrethrin, piperonyl butoxide, and petroleum distillates. Initially, she developed symptoms of acute cholinergic toxicity. One month after exposure, she experienced severe frequent headaches, muscle cramps, and diarrhea. After 3.5 mo, she developed numbness in her legs, tremors, memory problems, anxiety-depression, and insomnia. One year following exposure, she developed weakness, imbalance, and dizziness, and was confined to a wheelchair. Her symptoms were all characteristic of OPIDN. Twenty-eight months after exposure, she developed "delayed sequelae of gross neurologic symptoms," consisting of coarse tremors, intermittent hemiballistic movements of the right arm and leg, flaccid fasciculations of muscle groups, muscle cramps, and sensory disturbances.

Some victims of the Tokyo subway sarin incident, who developed acute cholinergic neurotoxicity, also developed long-term, chronic neurotoxicity characterized by CNS neurological deficits and neurobehavioral impairments.[11] Six to 8 mo after the Tokyo poisoning, some victims showed delayed effects on psychomotor performance, the visual nervous system, and the vestibule-cerebellar system.[68] It is noteworthy that females were more likely than males to exhibit delayed effects on the vestibular-cerebellar system. Three years after the Matsumoto attack in Japan, some patients complained of fatigue, shoulder stiffness, weakness, and blurred vision.[9] Others complained of insomnia, bad dreams, husky voice, slight fever, and palpations. Colosio et al.[62] reviewed the literature on the neurobehavioral toxicity of pesticides, and reported that some individuals who were acutely poisoned with organophosphorus compounds developed long-term impairment of neurobehavioral performance. They also concluded that these effects were only "an aspecific expression of damage and not of direct neurotoxicity."

OPICN following subclinical exposures to organophosphorus compounds. Reports on OPICN in individuals following long-term, subclinical exposures, without previous acute poisoning, have been inconsistent, mostly because of difficulty in the quantitative determination of exposure levels, but also because of problems with selection of controls. Several studies of workers exposed to low subclinical doses of organophosphorus insecticides failed to show neurobehavioral alterations between pre- and post-exposure measurements.[69–75] It has been suggested that the levels of exposure of subjects in these reports might have been below the threshold level needed to cause neurobehavioral deficits, and that studies of prolonged low-level exposures may eventually reveal neurobehavioral deficits.[62] Consistent with this opinion are the reports of impairment in neurobehavioral performance in individuals exposed to low-levels of organophosphorus insecticides. Professional pesticide applicators and farmers who had been exposed to organophosphorus pesticides showed elevated levels of anxiety, impaired vigilance and reduced concentration.[76] Kaplan et al.[77] reported persistent long-term cognitive dysfunction and defects in concentration, word finding, and short-term memory in individuals exposed to low subclinical levels of the organophosphorus insecticide chlorpyrifos. A significant increase in hand vibration threshold was reported in a group of pesticide applicators,[78] and significant cognitive and neuropsychological deficits have been found in sheep dippers who had been exposed to organophosphorus insecticides.[72,73] Male fruit farmers who were chronically exposed to organophosphorus insecticides showed significant slowing of their reaction time.[79] Female pesticide applicators exhibited longer reaction times, reduced motor steadiness, and increased tension, depression, and fatigue compared with controls.[74] Workers exposed to the organophosphorus insecticide quinalphos during its manufacture exhibited alterations in CNS function that were manifested as memory, learning, vigilance, and motor deficits, despite having normal AChE activity.[80] Rescue workers and some victims who did not develop any acute neurotoxicity symptoms nevertheless complained of a chronic decline in memory 3 yr and 9 mo after the Tokyo attack.[81] Pilkington et al.[55] reported a strong association between chronic low-level exposure to organophosphate concentrates in sheep dips and neurological symptoms in sheep dippers—suggesting that long-term health effects may occur in at least some sheep dippers exposed to these insecticides over their working lives.

Neurological and neurobehavioral alterations. Although the symptoms of OPICN are a consequence of damage to both the PNS and CNS, they are related primarily to CNS injury and resultant neurological and neurobehavioral abnormalities. Studies on the effects of exposure to organophosphorus compounds over the past half century have shown that chronic neurological and neurobehavioral symptoms include headache, drowsiness, dizziness, anxiety, apathy, mental confusion, restlessness, labile emotions, anorexia, insomnia, lethargy, fatigue, inability to concentrate, memory deficits, depression, irritability, confusion, generalized weakness and tremors.[51,52,82,83] Respiratory, circulatory, and skin problems may be present as well in cases of chronic toxicity.[1] It should be noted that not every patient exhibits all of these symptoms. Gershon and Shaw[51] reported that most of the symptoms that develop after organophosphate exposure resolve within 1 yr. Jamal[50] conducted an extensive review of the health effects of organophosphorus compounds and concluded that either acute or long-term, low-level exposure to these chemicals produces a number of chronic neurological and psychiatric abnormalities that he called "chronic organophosphate-induced neuropsychiatric disorder," or COPIND. Jamal recommended a multifaceted approach to the evaluation of the toxic effects of chronic, subclinical, repeated, low-level exposures to organophosphorus compounds; included were structural and quantitative analyses of symptoms and clinical neurological signs. In the present article, our concept of OPICN encompasses structural, functional, physiological, neurological, and neurobehavioral abnormalities, including neuropsychiatric alterations or COPIND. OPICN may be caused by an acute exposure that results in cholinergic toxicity, or by exposure to subclinical doses that do not produce acute poisoning.

Neuropathological alterations. Petras[84] investigated the neuropathological alterations in rat brains 15–28 days following intramuscular injections of large, acutely toxic doses (79.4–114.8 μg/kg) of the nerve agent soman. He reported that the brain damage in all 4 animals that developed seizures was comparable to damage present in 3 of the 4 animals that exhibited only limb tremor. Neuropathological lesions were characterized by axonal degeneration, seen in the cerebral cortex, basal ganglia, thalamus, subthalamic region, hypothalamus, hippocampus, fornix, septum, preoptic area, superior colliculus, pretectal area, basilar pontine nuclei, medullary tegmentum, and corticospinal tracts. Although the mechanism of soman-induced brain injury was not known, Petras noted that the lesions did not resemble those present in experimental fetal hypoxia[85] or OPIDN.[17] These results are consistent with later findings obtained after acute soman exposure,[86,87] exposure to the nerve agent sarin,[88] and neuronal necrosis induced by the organophosphorus insecticide fenthion.[89] Petras also indicated that soman-treated rats did not need to experience a seizure to develop lesions. Abdel-Rahman et al.[90] demonstrated neuropathological alterations in rat brain 24 hr after administration of an intramuscular LD_{50} dose (100 μg/kg) of sarin. Neuronal degeneration was present in the cerebral cortex, dentate gyrus, CA1 and CA3 subfields of the hippocampal formation, and the in Purkinje cells of the cerebellum. Neuronal degeneration of hippocampal cells is consistent with organophosphorus compound-induced alterations in behavior, and cognitive deficits such as impaired learning and memory.[91–94] Furthermore, chronic exposure to organophosphorus compounds resulted in long-term cognitive deficits, even in the absence of clinical signs of acute cholinergic toxicity.[95,96] Shih et al.[97] demonstrated that lethal doses ($2 \times LD_{50}$) of all tested nerve agents (i.e., tabun, sarin, soman, cyclosarin, VR, and VX) induced seizures accompanied by neuropathological lesions in the brains of guinea pigs, similar to those lesions reported for other organophosphorus compounds in other species.[98–103] Recent reports have indicated that anticonvulsants protected guinea pigs against soman- and sarin-induced seizures and the development of neuropathological lesions.[104,105] Time-course studies also have reported that sarin-induced brain lesions exacerbated over time and extended into brain areas that were not initially affected.[88,106] Kim et al.[107] found that that an intraperitoneal injection of 9 mg/kg ($1.8 \times LD_{50}$) DFP in rats protected with pyridostigmine bromide and atropine nitrate caused tonic-clonic seizures, followed by prolonged mild clonic epilepsy accompanied by early necrotic and delayed apoptotic neuronal degeneration. Early necrotic brain injury in the hippocampus and piriform/entorhinal cortices was seen between 1 and 12 hr after dosing. On the other hand, typical apoptotic terminal deoxynucleotidyl transferase-mediated dUTP-X nick end labeling (TUNEL)-positive cell death began to appear at 12 hr in the thalamus. Daily dermal administration of $0.01 \times LD_{50}$ of malathion for 28 days caused neuronal degeneration in the rat brain that was exacerbated by combined exposure to the insect repellent DEET and/or the insecticide permethrin.[108]

Correlation between neuropathological lesions and neurological and neurobehavioral alterations. Neuropathological changes—the hallmark of OPICN—could explain the neurological, neurobehavioral, and neuropsychological abnormalities reported in humans and animals exposed to organophosphorus compounds. A subcutaneous dose of 104 µg/kg soman induced status epilepticus in rats, followed by degeneration of neuronal cells in the piriform cortex and CA3 of the hippocampus.[103] Similar results have been reported in a variety of species.[98,109,110] In another study, only those mice treated with a subcutaneous dose of 90 µg/kg of soman which developed long-lasting convulsive seizures exhibited the neuropathological alterations.[111] Twenty-four hours after dosing, numerous eosinophilic cells and deoxyribonucleic acid (DNA) fragmentation (TUNEL-positive) cells were observed in the lateral septum, the endopiriform and entorhinal cortices, the dorsal thalamus, the hippocampus, and the amygdala. Animals that had only slight tremors and no convulsions did not show any lesions.[111] Guinea pigs given a subcutaneous dose of 200 µg/kg soman ($2 \times LD_{50}$) developed seizures and exhibited neuropathological lesions in the amygdala; the substantia nigra; the thalamus; the piriform, entorhinal, and perirhinal cortices; and the hippocampus between 24–48 hr following injection.[104] Male guinea pigs developed epileptiform seizures after receiving $2 \times LD_{50}$ subcutaneous doses of the following nerve agents: tabun (240 µg/kg), sarin (84 µg/kg), soman (56 µg/kg), cyclosarin (114 µg/kg), VX (16 µg/kg), or VR (22 µg/kg). The seizures were accompanied by necrotic death of neuronal cells, with the amygdala having the most severe injury, followed by the cortex and the caudate nucleus.[97]

An intraperitoneal injection of 9 mg/kg ($1.8 \times LD_{50}$) DFP caused severe early (15–90 min) tonic-clonic limbic seizures, followed by prolonged mild clonic epilepsy.[107] Necrotic cell death was seen 1 hr after DFP administration, primarily in the CA1 and CA3 subfields of the hippocampus and piriform/entorhinal cortices, and manifest as degeneration of neuronal cells and spongiform of neuropils. Whereas the severity of hippocampal injury remained the same for up to 12 hr, damage to the piriform/entorhinal cortices, thalamus, and amygdala continued to increase up to 12 hr. Furthermore, apoptotic death of neuronal cells (TUNEL-positive) was seen in the thalamus at 12 hr, and peaked at 24 hr. Rats that survived $1 \times LD_{50}$ sarin (95 µg/kg) exhibited persistent lesions, mainly in the hippocampus, piriform cortex, and thalamus.[88] Furthermore, brain injury was exacerbated over time; at 3 mo after exposure, other areas that were not initially affected became damaged. A recent study has described the early neuropathological changes in the adult male rat brain 24 hr after exposure to a single intramuscular dose of 1.0, 0.5, 0.1, or 0.01 $\times LD_{50}$ (100 µg/kg) sarin.[90] Sarin at $1.0 \times LD_{50}$ caused extensive severe tremors, seizures, and convulsions accompanied by damage involving mainly the cerebral cortex, the hippocampal formation (dentate gyrus, and CA1 and CA3 subfields) and the cerebellum. Damage was evidenced by (a) a significant inhibition of plasma BChE, brain region AChE, and M2 M-acetylcholine receptor ligand binding; (b) an increase in permeability of the blood-brain barrier; and (c) diffuse neuronal cell death coupled with decreased MAP-2 expression within the dendrites of surviving neurons. The $0.5 \times LD_{50}$ sarin dose did not cause motor convulsions, and only moderate Purkinje neuron loss. The 0.1 and $0.01 \times LD_{50}$ doses of sarin caused no alterations at 24 hr after dosing. These results indicate that sarin-induced acute brain injury is dose-dependent.

In animals treated with $1 \times LD_{50}$ sarin, both superficial layers (I–III) and deeper layers (IV–V) of the motor cortex and somatosensory cortex showed degeneration of neurons. In the deeper layers of the cortex, neuron degeneration was seen in layer V. Pyramidal neurons in layers III and V of the cortex are the source of the axons of the corticospinal tract, which is the largest descending fiber tract (or motor pathway) from the brain controlling movements of various contralateral muscle groups. Thus, sarin-induced death of layers III and V neurons of the motor cortex could lead to considerable motor and sensory abnormalities, ataxia, weakness, and loss of strength. Furthermore, disruption of the hippocampal circuitry because of the degeneration of neurons in different subfields can lead to learning and memory deficits. Lesions in the cerebellum could result in gait and coordination abnormalities. Because the severely affected areas (e.g., the limbic system, corticofugal system, and central motor system) are associated with mood, judgment, emotion, posture, locomotion, and skilled movements, humans exhibiting acute toxicity symptoms following exposure to large doses of organophosphates may also develop psychiatric and motor deficits. Inasmuch as the damaged areas of the brain do not regenerate, these symptoms are expected to persist long-term.[112–114] These findings are in agreement with a recent study by Kilburn[115] which evaluated the neurobehavioral effects of chronic low-level exposure to the organophosphorus insecticide chlorpyrifos in 22 patients. Kilburn demonstrated, for the 1st time, an association between chlorpyrifos sprayed inside homes and offices and neurophysiological impairments in balance, visual fields, color discrimination, hearing reaction time, and grip strength. These patients also had psychological impairment of verbal recall and cognitive function, and two-thirds of them had been prescribed antidepressant drugs. In addition, the patients exhibited severe respiratory symptoms, accompanied by airway obstruction. Other chlorpyrifos-induced neurotoxicity incidents in humans have been reported.[116] These results are consistent with the report that daily dermal application of 0.1 mg/kg chlorpyrifos to adult rats resulted in sensorimotor deficits.[117] Also, maternal exposure to a daily dermal dose of 0.1 mg/kg chlorpyrifos during gestational days 4–20 caused an increased expression of GFAP in the cerebellum and hippocampus of offspring on postnatal day 30.[118] A major component of astrocytic intermediate neurofilaments, GFAP is up-regulated in response to reactive gliosis resulting from insults such as trauma, neurodegenerative disease, and exposure to neurotoxicants.[119]

Mechanisms of OPICN. Recent studies have shown that large toxic doses of organophosphorus compounds cause early convulsive seizures and subsequent encephalopathy, leading to the necrotic death of brain neuronal cells, whereas small doses produce delayed apoptotic death. Pazdernik et al.[103] have proposed the following 5 phases that result in organophosphorus compound-induced cholinergic seizures: (1) initiation, (2) limbic status epilepticus, (3) motor convulsions, (4) early excitotoxic damage, and (5) delayed oxidative stress. The mechanisms of neuronal cell death in OPICN that appear to be mediated through necrosis or apoptosis, which may involve increased AChE gene expression, are discussed below.

Necrosis. The large toxic doses of organophosphorus compounds which induce early seizures activate the glutamatergic system and involve the Ca^{2+}-related excitotoxic process,[120,121] possibly mediated by the N-methyl-D-aspartate (NMDA) subtype of glutamate receptors.[122,123] De Groot[104] hypothesized that accumulated acetylcholine, resulting from acute inhibition of AChE by organophosphorus compounds, leads to activation of glutamatergic neurons and the release of the excitatory L-glutamate amino acid neurotransmitter. This in turn produces increased depolarization and subsequent activation of the NMDA subtype of glutamate receptors—and the opening of NMDA ion channels—resulting in massive Ca^{2+} fluxes into the post-synaptic cell and causing neuronal degeneration. Thus, glutamate-induced neuronal degeneration during seizures may occur as a result of lowering of the threshold for glutamate excitation at NMDA receptor sites.

Activation of nitric oxide synthase, following stimulation of NMDA receptor sites, increases the level of nitric oxide, which functions as a signaling or cytotoxic molecule responsible for neuronal cell death.[124] As a retrograde messenger, nitric oxide induces the release of several neurotransmitters, including excitatory amino acid L-glutamate[125] which alters neurotransmitter balance and affects neuronal excitability. The production of nitric oxide is enhanced in AChE-inhibitor–induced seizures.[126,127] Kim et al.[107] demonstrated the involvement of nitric oxide in organophosphate-induced seizures and the effectiveness of nitric oxide synthesis inhibitors in preventing such seizures.

Apoptosis. Neuronal degeneration caused by apoptosis or programmed cell death may have physiologic or pathologic consequences. Elimination of precancerous, old, or excess cells is carried out by apoptosis without injury to surrounding cells as seen in necrosis.[128] Small doses of organophosphorus compounds cause delayed neuronal cell death that involves free radical generation (i.e., reactive oxygen species [ROS]). Organophosphates that cause mitochondrial damage/dysfunction also cause depletion of ATP and increased generation of ROS, which results in oxidative stress.[129,130] ROS can cause fatal depletion of mitochondrial energy (ATP), induction of proteolytic enzymes, and DNA fragmentation, leading to apoptotic death.[129,131,132] These results are consistent with the DNA damage detected in the lymphocytes in peripheral blood in 8 individuals, following residential exposure to the organophosphorus insecticides chlorpyrifos and diazinon.[133]

The brain is highly susceptible to oxidative stress-induced injury for several reasons: (a) its oxygen requirements are high; (b) it has a high rate of glucose consumption; (c) it contains large amounts of peroxidizable fatty acids; and (d) it has relatively low antioxidant capacity.[131,132] A single sublethal dose of $0.5 \times LD_{50}$ sarin, which did not induce seizures, nevertheless caused delayed apoptotic death of rat brain neurons in the cerebral cortex, hippocampus, and Purkinje cells of the cerebellum 24 hr after dosing.[90,134] Furthermore, rats treated with a single $0.1 \times LD_{50}$ dose of sarin, and which did not exhibit brain histopathological alterations 1, 7, or 30 days after dosing, nevertheless showed apoptotic death of brain neurons in the same areas mentioned above, 1 yr after dosing.[90,135] These results are consistent with the sensorimotor deficits exhibited by sarin-treated animals 3 mo after exposure; the animals showed continued deterioration when tested 6 mo after dosing.

Increased AChE gene expression. Recent studies have suggested that AChE may play a role in the pathogenesis of OPICN similar to that reported for Alzheimer's Disease.[136,137] We have demonstrated that sarin induced the AChE gene in the same regions of the brain that underwent neuronal degeneration.[138] AChE has been shown to be neurotoxic in vivo and in vitro; it accelerates assembly of amyloid peptide in Alzheimer's fibrils, leading to cell death via apoptosis.[139] Some studies have demonstrated increased AChE expression in apoptotic neuroblastoma SK-N-SH cells after long-term culturing.[139] Brain AChE has been shown to be toxic to neuronal (Neuro 2a) and glial-like (B12) cells.[137] There are also reports that transgenic mice overexpressing human AChE in brain neurons undergo progressive cognitive deterioration.[140] These results suggest that sarin may provoke an endogenous cell suicide pathway cascade in susceptible neurons (e.g., in the caspase-3 pathway), resulting in the release of AChE into adjacent brain tissues. The aggregation of AChE initiates more apoptotic neuronal death. Amplification of this cascade thus may result in the progressive neuronal loss that is the hallmark of sarin-induced chronic neurotoxicity. It is noteworthy that a common symptom of both OPICN and Alzheimer's Disease is memory deficit, suggesting that the aging process may be accelerated following exposure to organophosphorus compounds in OPICN.

Other factors. The occurrence and severity of OPICN is influenced by factors such as environmental exposure to other chemicals, stress, or individual genetic differences. For example, cholinotoxicants such as organophosphates or carbamates—which do not have a positive charge and are capable of crossing the blood-brain barrier—act at the same receptors and thus exacerbate OPICN. Individuals with low levels of the plasma enzymes BChE[5,6,141] or paraoxonase[7,8] that act as the 1st line of defense against neurotoxicity (by removing organophosphates from circulation through scavenging or hydrolysis) are vulnerable to the development of persistent OPICN. All of these factors may be involved in development of the phenomenon known as "chemical sensitivity" or "multiple chemical sensitivity."[142] Thus, prior chemical exposure, stress, or genetic factors might make individuals predisposed or susceptible to CNS injury upon subsequent exposure to other chemicals.

Combined exposure to other chemicals that cause oxidative stress can intensify OPICN which results from exposure to organophosphorus compounds.[108] Furthermore, stress that also causes oxidative stress decreases the threshold level required to produce neuronal damage and results in increased OPICN following combined exposure to stress and organophosphates. Thus, OPICN may explain the reports that Persian Gulf War veterans showed a higher than normal propensity toward persistent neurological complaints such as memory and attention deficits, irritability, chronic fatigue, muscle and joint pain, and poor performance on cognition tests.[143–147] A large number of these personnel were exposed to low levels of sarin during the demolition of Iraqi munitions at Khamisiya,[148,149] as well as to other chemicals and to stress.[150,151] Also, OPICN may explain the recent report that Persian Gulf War veterans are at an almost 2-fold greater risk of developing amyotrophic lateral sclerosis (ALS) than other veterans,[152] which is consistent with Haley's[153] suggestion that the increase in ALS is "a war-related environmental trigger." Furthermore, OPICN induced by low-level inhalation of organophosphates present in jet engine lubricating oils and the hydraulic fluids of aircraft[154] could explain the long-term neurologic deficits consistently reported by crewmembers and passengers, although organophosphate levels may have been too low to produce OPIDN.[155]

Prognosis

Previous reports have indicated that, subsequent to exposure to organophosphorus compounds, an individual could develop acute cholinergic neurotoxicity, followed by OPICN. In a few cases, OPIDN may occur with or without the development of cholinergic neurotoxicity, with OPICN developing later. Furthermore, OPICN may occur following long-term, low-level exposure to organophosphorus compounds, and without the development of acute neurotoxicity. Because the long-term, persistent effects of OPICN result from neuronal degeneration of the PNS and CNS, induced by organophosphates, it is unlikely that improvement is the consequence of the regeneration of brain neurons, inasmuch as such repair is not typical of the CNS. Clinical improvement may occur, however, through repair of the PNS. Also, reversible changes in the CNS that might be present initially (e.g., edema) could later subside, giving the *appearance* of repair. Furthermore, if damage is not too extensive, other neurons having the same function could meet the added demands and maintain normal activity. When the CNS is severely damaged, however, neither of these repair mechanisms is possible and some loss of function will likely occur.

Conclusions

Herein we have described the long-term, persistent neurodegenerative disorder induced by exposure to organophosphorus compounds. We define this effect as organophosphorus ester-induced chronic neurotoxicity, or OPICN. Numerous cases documenting this disorder have been reported since the extensive use of these chemicals in industry and agriculture began more than 50 yr ago. Although largely characterized by chronic neurobehavioral alterations, OPICN involves other molecular, neurochemical, neurophysiological, neuropathological, neuropsychological, and neurological changes. The term "neurotoxicity" encompasses all of these, and adequately describes this neurodegenerative disorder.

* * * * * * * * * *

The author would like to express his appreciation for the artwork of Anjelika Dechkovskaia.

Submitted for publication May 14, 2004; revised; accepted for publication May 30, 2004.

Requests for reprints should be sent to Mohamed B. Abou-Donia, Ph.D., Department of Pharmacology and Cancer Biology, Box 3813, Duke University Medical Center, Durham, NC 27710.

* * * * * * * * * *

References

1. Abou-Donia MB. Organophosphorus pesticides. In: Chang LW, Dyer RS (Eds). Handbook of Neurotoxicology. New York: Marcel Dekker, 1994; pp 419–47.
2. Koelle GB. Protection of cholinesterase against inevitable inactivation by diisopropyl fluorophosphate in vitro. J Pharmacol Exp Ther 1946; 88:232–37.
3. Abou-Donia MB. Neurotoxicology. Boca Raton, FL: CRC Press, 1992; pp 3–24.
4. Sussman JL, Harel M, Frolow F. Atomic structure of acetylcholinesterase from Torpedo californica. A prototypic acetylcholine-binding protein. Science 1992; 253:872–78.
5. Whittaker M. The pseudocholinesterase variants: esterase levels and increased resistance to fluoride. Acta Genet Basel 1994; 14:281–85.
6. Lockridge O. Genetic variants of human serum cholinesterase influence metabolism of the muscle relaxant succinylcholine. Pharmacol Ther 1990; 47:35–60.
7. Mackness B, Mackness MI, Arrol S, et al. Effect of the molecular polymorphisms of human paraoxonase (PONI) on the rate of hydrolysis of paraoxon. Br J Pharmacol 1997; 122:265–68.
8. Davies HG, Richter RJ, Keifer M, et al. The effect of the human serum paraoxonase polymorphism is reversed with diazoxon, soman, and sarin. Nat Genet 1996; 14: 334–36.
9. Morita H, Yanagisawa N, Nakajima T, et al. Sarin poisoning in Matsumoto, Japan. Lancet 1995; 346:290–93.
10. Okumura T, Takasu N, Ishimatsu S, et al. Report on 640 victims of the Tokyo subway sarin attack. Ann Emerg Med 1995; 28:129–35.
11. Yokoyama K, Araki S, Murata K, et al. Chronic neurobehavioral effects of Tokyo subway sarin poisoning in relation to posttraumatic stress disorder. Arch Environ Health 1998; 53:249–56.
12. Masuda N, Takatsu M, Morinari H. Sarin poisoning in Tokyo subway. Lancet 1995; 345:1446–47.
13. Smith MI, Elvove E, Frazier WH. The pharmacological action of certain phenol esters with special reference to the etiology of so-called ginger paralysis. Public Health Rep 1930; 45:2509–24.
14. Johnson MK. The delayed neuropathology caused by some organophosphorous esters: mechanism and challenge. Crit Rev Toxicol 1975; 2:289–316.
15. Abou-Donia MB. Organophosphorous ester-induced delayed neurotoxicity. Annu Rev Pharmacol Toxicol 1981; 21:511–48.
16. Abou-Donia MB, Lapadula LM. Mechanisms of organophosphorus ester-induced delayed neurotoxicity: type I and type II. Annu Rev Pharmacol Toxicol 1990; 30:405–40.
17. Cavanagh JB, Patangia GN. Changes in the central nervous system in the cat as the result of tri-*o*-cresylphosphate poisoning. Brain 1995; 88:165–80.
18. Himuro K, Murayama S, Nishiyama K, et al. Distal sensory axonopathy after sarin intoxification. Neurology 1998; 51:1195–97.
19. Davies OR, Holland PR. Effect of oximes and atropine upon the development of delayed neurotoxic signs in chickens following poisoning by DFP and sarin. Biochem Pharmacol 1972; 21:3145–51.
20. Abou-Donia MB. Triphenyl phosphite: a type II organophosphorus compound-induced delayed neurotoxic agent. In: Chambers JE, Levi PE (Eds). Organophosphates: Chemistry, Fates, and Effects. Part IV: Toxic Effects – Organismal. San Diego, CA: Academic Press, 1992; pp 327–51.
21. Abou-Donia MB, Wilmarth KR, Jansen KF, et al. Triphenylphosphine (TPP): a type III organophosphorous compound-induced delayed neurotoxic agent (OPIDN). Toxicologist 1996; 30:311.
22. Abdel-Rahman AA, Jensen KF, Farr CH, et al. Daily treatment of triphenylphosphine (TPP) produces organophosphorous-induced delayed neurotoxicity (OPIDN). Toxicologist 1997; 36:19.
23. Bloch H, Hottinger A. Uber die spezifitat der cholinesterase-hemmung durch tri-o-kresyl phosphat. Int Z Vitaminforsch 1943; 13:90.
24. Earl CJ, Thompson RHS. The inhibitory action of tri-ortho-cresyl phosphate and cholinesterases. Br J Pharmacol 1952; 7:261–69.
25. Aldridge WN, Barnes JM. Further observations on the neurotoxicity of organophosphorus compounds. Biochem Pharmacol 1966; 15:541–47.
26. Johnson MK. The delayed neurotoxic effect of some organophosphorus compounds. Br Med Bull 1969; 114:711–17.
27. Lotti M. The pathogenesis of organophosphate delayed neuropathy. Crit Rev Toxicol 1992; 21:465–87.
28. Curtes JP, Develay P, Hubert JP. Late peripheral neuropathy due to acute voluntary intoxication by organophosphorous compounds. Clin Toxicol 1981; 18:1453.
29. de Jager AE, van Weerden TW, Houthoff HJ, et al. Polyneuropathy after massive exposure to parathion. Neurology 1981; 31:603–05.
30. Stamboulis E, Psimaras A, Vassilopoulos D. Neuropathy following acute intoxication, with mercarbam (Opester). Acta Neurol Scand 1991; 83:198.
31. Winrow CJ, Hemming ML, Allen DA, et al. Loss of neuropathy target esterase in mice links organophosphate exposure to hyperactivity. Nat Genet 2003; 33:477–85.
32. O'Callahan JP. Neurotoxic esterase: not so toxic? Nat Genet 2003; 33:1–2.

33. Bus J, Maurissen J, Marable B, et al. Association between organophosphate exposure and hyperactivity? Nat Genet 2003; 34(3):235.
34. Abou-Donia MB. Involvement of cytoskeletal proteins in the mechanisms of organophosphorous ester delayed neurotoxicity. Clin Exp Pharmacol Physiol 1995; 22:358–59.
35. Schulman H. Advances in Second Messenger and Phosphorylation Research. New York: Raven Press, 1988; pp 39–111.
36. Patton SE, Lapadula DM, Abou-Donia MB. Relationship of tri-o-cresyl phosphate-induced delayed neurotoxicity to enhancement of in vitro phosphorylation of hen brain and spinal cord. J Pharmacol Exp Ther 1986; 239: 597–605.
37. Suwita E, Lapadula DM, Abou-Donia MB. Calcium and calmodulin-enhanced in vitro phosphorylation of hen brain cold-stable microtubules and spinal cord neurofilament triplet proteins after a single oral dose of tri-o-cresyl phosphate. Proc Natl Acad Sci USA 1986; 83: 6174–78.
38. Abou-Donia MB, Viana ME, Gupta RP, et al. Enhanced calmodulin binding concurrent with increased kinase-dependent phosphorylation of cytoskeletal proteins following a single subcutaneous injection of diisopropyl phosphorofluoridate in hens. Neurochem Res 1993; 22:165–73.
39. Gupta RP, Abou-Donia MB. Tau proteins-enhanced Ca2+/calmodulin (CaM)-dependent phosphorylation by the brain supernatant of diisopropyl phosphorofluoridate (DFP)-treated hen: tau mutants indicate phosphorylation of more amino acids in tau by CaM kinase II. Brain Res 1998; 813:32–43.
40. Gupta RP, Abou-Donia MB. In vivo and in vitro effects of diisopropyl phosphorofluoridate (DFP) on the rate of hen brain tubulin polymerization. Neurochem Res 1994; 19:435–44.
41. Gupta RP, Abou-Donia MB. Neurofilament phosphorylation and [125I] calmodulin binding by Ca2+ /calmodulin-dependent protein kinase in the brain subcellular fractions of diisopropyl phosphorofluoridate (DFP)-treated hen. Neurochem Res 1995; 20(9):1095–105.
42. Gupta RP, Abou-Donia, MB. Diisopropyl phosphorofluoridate (DFP) treatment alters calcium-activated proteinase activity and cytoskeletal proteins of the hen sciatic nerve. Brain Res 1995; 677:162–66.
43. Gupta RP, Bing G, Hong JS, et al. cDNA cloning and sequencing of Ca2+ /calmodulin-dependent protein kinase _ subunit and its mRNA expression in diisopropyl phosphorofluoridate (DFP)-treated hen central nervous system. Mol Cell Biochem 1998; 181:29–39.
44. Gupta RP, Abou-Donia MB. Tau phosphorylation by diisopropyl phosphorofluoridate (DFP)-treated hen brain supernatant inhibits its binding with microtubules: role of Ca2+/calmodulin-dependent protein kinase II in tau phosphorylation. Biochem Pharmacol 1999; 53: 1799–1806.
45. Gupta RP, Damodaran TV, Abou-Donia MB. C-fos mRNA induction in the central and peripheral nervous systems of diisopropyl phosphorofluoridate (DFP)-treated hens. Neurochem Res 2000; 25(3):327–34.
46. Gupta RP, Abdel-Rahman AA, Jensen KF, et al. Altered expression of neurofilament subunits in diisopropyl phosphorofluoridate (DFP)-treated hen spinal cord and their presence in axonal aggregates. Forthcoming.
47. Damodaran TV, Abou-Donia MB. Alterations in levels of mRNAs coding for glial fibrillary acidic protein (GFAP) and vimentin genes in the central nervous system of hens treats with diisopropyl phosphorofluoridate (DFP). Neurochem Res 2000; 25:809–16.
48. Jensen KF, Lapadula DM, Anderson JK, et al. Anomalous phosphorylated neurofilament aggregations in central and peripheral axons of hens treated with tri-ortho-cresyl phosphate (TOCP). J Neurosci Res 1992; 33:455–60.
49. Abou-Donia MB, et al. Unpublished results, 2004.
50. Jamal G. Neurological syndromes of organophosphorus compounds. Adverse Drug React Toxicol Rev 1997; 16:133–70.
51. Gershon S, Shaw FB. Psychiatric sequelae of chronic exposure to organophosphorous insecticides. Lancet 1961; 1:1371–74.
52. Dille JR, Smith PW. Central nervous system effects of chronic exposure to organophosphate insecticides. Aerosp Med 1964; 35:475–78.
53. Metcalf DR, Holmes JH. EEG, psychological and neurological alterations in humans with organophosphorous exposure. Ann NY Acad Sci 1969; 160:357–65.
54. Duffy FH, Burchfield JL, Bartels PH, et al. Long-term effects of an organophosphate upon the human electroencephalogram. Toxicol Appl Pharmacol 1979; 47: 161–76.
55. Pilkington A, Buchanan D, Jamal GA, et al. An epidemological study of the relations between exposure to organophosphate pesticides and indices of chronic peripheral neuropathy and neuropsychological abnormalities in sheep farmers and dippers. Occup Environ Med 2001; 58(11):702–10.
56. Russell R, Macri J. Central cholinergic involvement hyper-reactivity. Pharmocol Biochem Behav 1979; 10: 43–48.
57. Joubert J, Joubert PH. Chorea and psychiatric changes in organophosphate poisoning. A report of 2 further studies. S Afr Med J 1988; 74:32–34.
58. Rosenstock L, Keifer M, Daniell WE, et al. Chronic central nervous system effects of acute organophosphate pesticide intoxication. Lancet 1991; 338:223–27.
59. Steenland K, Jenkins B, Ames RG, et al. Chronic neurological sequelae to organophosphate pesticide poisoning. Am J Public Health 1994; 84:731–36.
60. Callender TJ, Morrow L, Subramanian K. Evaluation of chronic neurological sequelae after acute pesticide exposure using SPECT brain scans. J Toxicol Environ Health 1994; 41:275–84.
61. Stephens R, Spurgeon A, Calvert IA, et al. Neuropsychological effects of long-term exposure to organophosphates in sheep dip. Lancet 1995; 345:1135–39.
62. Colosio C, Tiramani M, Maroni M. Neurobehavioral effects of pesticides: state of the art. Neurotoxicology 2003; 24:577–91.
63. Scremin O, Shih TM, Huynh L, et al. Delayed neurologic and behavioral effects of subtoxic doses of cholinesterase inhibitors. J Pharmacol Exp Ther 2003; 304:1111–19.
64. Echbichon DJ, Joy RM. Pesticides and Neurological Diseases. 2nd ed. Boston and London: CRC Press, 1995.
65. Savage EP, Keefe TF, Mounce LM, et al. Chronic neurological sequelae of acute organophosphates pesticide poisoning. Arch Environ Health 1988; 43:38–45.
66. Maroni M, Jarvisalo J, La Ferla L. The WHO-UNDP edidemiological study on the health effects of exposure to organophosphorous pesticides. Toxicol Lett 1986; 33:115–23.

67. McConell R, Keifer M, Rosenstock L. Elevated quantitative vibrotactile threshold among workers previously poisoned with methamidophos and other organophosphate pesticides. Am J Ind Med 1994; 25:325–34.
68. Yokoyama K, Araki K, Murata K, et al. A preliminary study on delayed vestibulo-cerebellar effects of Tokyo Subway sarin poisoning in relation to gender difference: frequency analysis of postural sway. J Occup Environ Med 1998; 40(1):17–21.
69. Rodnitzky RL, Levin HS, Mick DL. Occupational exposure to organophosphate pesticides. Arch Environ Health 1975; 30:98–103.
70. Maizlish N, Schenker M, Weisskopf C, et al. A behavioral evaluation of pest control workers with short-term, low-level exposure to the organophosphate diazinon. Am J Indust Med 1987; 12:153–72.
71. Daniell W, Barnhart S, Demers P, et al. Neuropsychological performance among agricultural pesticide applicators. Environ Res 1992; 59(1):217–28.
72. Beach JR, Spurgeon A, Stephens R, et al. Abnormalities on neurological examination among sheep farmers exposed to organophosphate pesticides. Occup Environ Med 1996; 53(8):520–25.
73. London L, Myers JE, Neil V, et al. An investigation into neurological and neurobehavioral effects of long-term agrochemical use among deciduous fruit farm workers in the Western Cape, South Africa. Environ Res 1997; 73(1–2):132–45.
74. Bazylewicz-Walckzak B, Majzakova W, Szymczak M. Behavioral effects of occupational exposure to organophosphorous pesticides in female greenhouse plantin workers. Neurotoxicology 1999; 20(5):819–26.
75. Steenland M. Chronic neurological effects of organophosphate pesticides. Br Med J 1996; 312: 1311–12.
76. Levin HS, Rodnitzsky RL, Mick DL, et al. Anxiety associated with exposure to organophosphorous compounds. Arch Gen Psychiatry 1976; 33:225–28.
77. Kaplan JG, Kessler J, Rosenberg N, et al. Sensory neuropathy associated with Dursban (chlorpyrifos) exposure. Neurology 1993; 43:2193–96.
78. Stokes L, Stark A, Marshall E, et al. Neurotoxicity among pesticide applicators exposed to organophosphates. Occup Environ Health 1995; 52:648–53.
79. Fielder N, Feldman RG, Jacobson J, et al. The assessment of neurobehavioral toxicity: SOGOMSEC joint report. Environ Health Perspect 1996; 104(Suppl 2): 179–91.
80. Srivastava AK, Gupta BN, Bihari V, et al. Clinical, biochemical and neurobehavioral studies on workers engaged in the manufacture of quinalphos. Food Chem Toxicol 2000; 38(1):65–69.
81. Nishiwaki Y, Maekawa K, Ogawa Y, et al. Effects of sarin on the nervous system in rescue team staff members and police officers 3 years after the Tokyo subway sarin attack. Environ Health Perspect 2001;109: 1169–73.
82. Durham WF, Wolfe HR, Quinby GE. Organophosphorus insecticides and mental alertness. Arch Environ Health 1965; 10:55–66.
83. Tabershaw IR, Cooper WC. Sequelae of acute organic phosphate poisoning. J Occup Med 1966; 8:5–20.
84. Petras JM. Soman neurotoxicity. Fundam Appl Toxicol 1981; 1:242–49.
85. Faro MD, Windle WF. Transneuronal degeneration in brains of monkeys asphyxiated at birth. Exp Neurol 1969; 24:38–53.
86. McDonough JH, Dochterman LW, Smith CD, et al. Protection against nerve agent-induced neuropathology, but not cardiac pathology, is associated with the anticonvulsant action of drug treatment. Neurotoxicology 1995; 15:123–32.
87. Kadar T, Cohen G, Sarah R, et al. Long-term study of brain lesions following soman, in comparison to DFP and metrazol poisoning. Hum Exp Toxicol 1992; 11: 517–23.
88. Kadar T, Shapira S, Cohen G, et al. Sarin induced neuropathology in rats. Hum Exp Toxicol 1995; 14: 252–59.
89. Veronesi B, Jones K, Pope C. The neurotoxicity of subchronic acetylcholinesterase (AChE) inhibition in rat hippocampus. Toxicol Appl Pharmacol 1990; 104: 440–56.
90. Abdel-Rahman A, Shetty AK, Abou-Donia MB. Acute exposure to sarin increases blood brain barrier permeability and induces neuropathological changes in the rat brain: dose-response relationships. Neuroscience 2002; 113(3):721–41.
91. McDonald BE, Costa LG, Murphy SD. Spatial memory impairment and central muscarinic receptor loss following prolonged treatment with organophosphates. Toxicol Lett 1988; 40:47–56.
92. Rafaelle K, Olton D, Annau Z. Repeated exposure to diisopropylfluorophosphate (DFP) produces increased sensitivity to cholinergic antagonists in discrimination retention and reversal. Psychopharmacology (Berl) 1990; 100:267–74.
93. Bushnell PJ, Padilla SS, Ward T, et al. Behavioral and neurochemical changes in rats dosed repeatedly with diisopropyl fluorophosphate. J Pharmacol Exp Ther 1991; 256:741–50.
94. Kassa J, Koupilova M, Vachek J. The influence of low-level sarin inhalation exposure on spatial memory in rats. Pharmacol Biochem Behav 2001; 70:175–79.
95. Prendergast MA, Terry AV Jr, Buccafusco JJ. Chronic, low-level exposure to diisopropyl fluorophosphate causes protracted impairment of spatial navigation learning. Psychopharmacology (Berl) 1997; 130: 276–84.
96. Prendergast MA, Terry AV Jr, Buccafusco JJ. Effects of chronic low-level organophosphate exposure on delayed recall, discrimination and spatial learning in monkeys and rats. Neurotoxicol Teratol 1998; 20: 115–22.
97. Shih TM, Duniho SM, McDonough JH. Control of nerve agent-induced seizures is critical for neuroprotection and survival. Toxicol Appl Pharmacol 2003; 188:69–80.
98. Lemercier G, Carpentier P, Setenac-Roumanou H, et al. Histological and histochemical changes in the central nervous system of the rat poisoned by an irreversible anticholinesterase organophosphorous compound. Acta Neuropathol 1983; 61:123–29.
99. Petras JM. Neurology and neuropathology of soman-induced brain injury: an overview. J Exp Anal Behav 1994; 61:319–29.
100. Carpentier P, Delamanche IS, Lebert M, et al. Seizure-related opening of the blood brain barrier induced by soman: possible correlation with the acute neuropathology observed in poisoned rats. Neurotoxicology 1990; 11:493–508.
101. Clement JG, Broxup B. Efficacy of diazepam and avizafone against soman-induced neuropathology in brain of rats. Neurotoxicology 1993; 14:485–504.

102. Shih TM, Koviak TA, Capacio BR. Anticonvulsants for poisoning by the organophosphorous compound Soman: pharmacological mechanisms. Neurosci Biobehav Rev 1991; 15:349.
103. Pazdernik TL, Emerson MR, Cross R, et al. Soman-induced seizures: limbic activity, oxidative stress, and neuroprotective proteins. J Appl Toxicol 2001; 21: S87–S94.
104. deGroot DMG, Bierman EPB, Bruijnzeel PLB, et al. Beneficial effects of TCP on soman intoxication in guinea pigs: seizures, brain damage, and learning behavior. J Appl Toxicol 2001; 21:S57–S65.
105. Taysee L, Calvet JH, Buee J, et al. Comparative efficacy of diazepam and avizafone against sarin-induced neuropathology and respiratory failure in guinea pigs: influence of atropine dose. Toxicology 2003; 188: 197–209.
106. Abou-Donia MB, et al. Unpublished results, 2004.
107. Kim YB, Hur GH, Shin S, et al. Organophosphate-induced brain injuries: delayed apoptosis mediated by nitric oxide. Environ Toxicol Pharmacol 1999; 7:147–52.
108. Abdel-Rahman AA, Dechkovskaia AM, Goldstein LB, et al. Neurological deficits induced by malathion, DEET and permethrin, alone or in combination in adult rats. J Toxicol Environ Health A 2004; 67:331–56.
109. McLeod CG, Singer AW, Harrington DG. Acute neuropathology in soman-poisoned rats. Neurotoxicology 1984; 5:53–58.
110. Churchill L, Pazdernik TL, Jackson JL. Soman-induced brain lesions demonstrated by muscarinic receptor autoradiography. Neurotoxicology 1985; 6:81–90.
111. Baille V, Dorandeu F, Carpentier P, et al. Acute exposure to a low or mild dose of soman: biochemical, behavioral and histopathological effects. Pharmacol Biochem Behav 2001; 69:561–69.
112. Sidell FR. Soman and sarin: clinical manifestations and treatment of accidental poisoning by organophosphates. Clin Toxicol 1974; 7(1):1–17.
113. West I. Sequelae of poisoning from phosphate ester pesticides. Ind Med Surg 1968; 37(11):832.
114. Namba T, Nolte CT, Jackrel J et al. Poisoning due to organophosphate insecticides. Am J Med 1971; 50:475.
115. Kilburn KH. Evidence for chronic neurobehavioral impairment from chlorpyrifosa, and organophosphate insecticide (Dursban) used indoors. Environ Epidemiol Toxicol 1999; 1:153–62.
116. Blondell J, Dobozy VA. Review of Chlorpyrifos Poisoning Data. Washington, DC: U.S. Environmental Protection Agency, 14 Jan 1997.
117. Abou-Donia MB, Abdel-Rahman AA, Goldstein LB, et al. Sensorimotor deficits and increased brain nicotinic acetylcholine receptors following exposure to chlorpyrifos and/or nicotine in rats. Arch Toxicol 2003; 77: 452–58.
118. Abdel-Rahman AA, Dechkovskaia AM, Mehta-Simmons H, et al. Increased expression of glial fibrillary acidic protein in cerebellum and hippocampus: differential effects on neonatal brain regional acetylcholinesterase following maternal exposure to combined chlorpyrifos and nicotine. J Toxicol Environ Health A, 2003; 66:2047–66.
119. Eng LF, Ghirnikar RS. GFAP and astrogliosis. Brain Pathol 1994; 4:229–37.
120. Olney JW, de Gubareff T, Labruyere J. Seizure-related brain damage induced by cholinergic agents. Nature 1983; 301:520–22.
121. Dawson VL, Dawson TM, Lonedon ED, et al. Nitric oxide mediates glutamate neurotoxicity in primary cortical cultures. Proc Natl Acad Sci USA 1991; 88: 6368–71.
122. Solberg Y, Belkin M. The role of excitotoxicity in organophosphorus nerve agents central poisoning. Trends Pharmacol Sci 1997; 8:183–85.
123. Raveh L, Chapman S, Cohen G, et al. The involvement of the NMDA receptor complex in the protective effect of anticholinergic drugs against soman poisoning. Neurotoxicology 1999; 20:551–60.
124. DawsonTM, Dawson VL, Snyder SH. A novel neuronal messenger molecule in brain: the free radical, nitric oxide. Ann Neurol 1992; 32:297–31.
125. Montague PR, Gancayco CD, Winn MJ., et al. Role of NO production in NMDA receptor-mediated neurotransmitter release in cerebral cortex. Science 1994; 263(5149):973–77.
126. Bagetta G, Massoud R, Rodino P, et al. Systematic administration of lithium chloride and tacrine increases nitric oxide synthase activity in the hippocampus of rats. Eur J Pharmacol 1993; 237:61–64.
127. Kim YB, Hui GH, Lee YS, et al. A role of nitric acid oxide in organophosphate-induced convulsions. Environ Toxicol Pharmacol 1997; 1(3):53–56.
128. Thompson CB. Apoptosis in the pathogenesis and treatment of disease. Science 1995; 267:1456–62.
129. Tsujimoto Y. Apoptosis and necrosis: intracellular ATP level as a determinant for cell death modes. Cell Death Differ 1997; 4:429–34.
130. Murphy AN, Fiskum G, Beal MF. Mitochondria in neurodegeneration: cell life and death. J Cereb Blood Flow Metab 1999: 19:231–45.
131. Floyd RA. Antioxidants, oxidative stress, and degenerative neurological disorders. Proc Soc Exp Biol Med 1999; 222:236–45.
132. Gupta RP, Milatovic D, Dettbarn WD. Depletion of energy metabolites following acetylcholinesterase inhibitor-induced status epilepticus: protection by antioxidants. Neurotoxicology 2001; 22:271–82.
133. Lieberman AD, Craven MR, Lewis HA, et al. Genotoxicity from domestic use of organophosphate pesticides. J Occup Environ Med 1998; 40(11):954–57.
134. Abou-Donia MB, et al. Unpublished data, 2004.
135. Abou-Donia MB, et al. Unpublished results, 2004.
136. Sberna G, Saez-Valero J, Li QX, et al. Acetylcholinesterase is increased in the brains of transgenic mice expressing the C-terminal fragment (CT100) of the B-amyloid protein precursor of Alzheimer's Disease. J Neurochem 1998; 71:723–31.
137. Calderon FH, von Bernhardi R, De Ferrari G, et al. Toxic effects of acetylcholinesterase on neuronal and glial-like cells in vitro. Mol Psychiatry 1998; 3:247–55.
138. Damodaran TV, Jones KH, Patel AG, et al. Sarin (nerve agent GB)-induced differential expression of mRNA coding for the acetylcholinesterase gene in the rat central nervous system. Biochem Pharmacol 2003; 65: 2041–47.
139. Yang L, Heng-Yi H, Zhang XJ. Increased expression of intranuclear AChE involved in apoptosis of SK-N-SH cells. Neurosci Res 2002; 42:261–68.
140. Andres C, Seidman S, Beeri R, et al. Transgenic acetylcholinesterase induces enlargement of murine neuromuscular junctions but leaves spinal cord synapses intact. Neurochem Int 1998; 32:449–56.
141. Abou-Donia MB, Wilmarth KR, Jensen KF. Neurotoxicity resulting from coexposure to pyridostigmine bro-

mide, DEET, and permethrin: implications of Gulf War chemical exposures. J Toxicol Environ Health 1996; 48: 35–56.
142. Rae WJ. Chemical Sensitivity. Vol 1. Boca Raton, FL: Lewis Publishers, 1992; pp 47–154.
143. Kurt TL. Epidemiological association in U.S. veterans between Gulf War illiness and exposure to anti-cholinesterases. Toxicol Lett 1998; 11:1–5.
144. Anger WK, Storzbach D, Binder LM. Neurobehavioral deficits in Persian Gulf veterans: evidence from a population-based study. J Int Neuropsychol Soc 1999; 5:203–12
145. McCauley LA, Rischitelli G, Lambert WE. Symptoms of Gulf War vertrans possibly exposed to organophosphate chemical warfare agents at Khamisiyah, Iraq. Int J Occup Environ Health 2001; 7:3170–75.
146. Storzbach D, Campbell KA, Binder LM. Psychological differences between veterans with and without Gulf War unexplained symptoms. Psychosom Med 2000; 726–35.
147. White RF, Proctor SP, Heeren T. Neuropshchological functions in Gulf War veterans: relationship to self-reported toxicant exposures. Amer J Ind Med 2001; 40: 42–44.
148. Institute of Medicine of the National Academies. Health Consequences of Service during the Persian Gulf War: Initial Findings and Recommendation for Immediate Action. Washington, DC: National Academies Press, 1995.
149. Augerson WS. A Review of The Scientific Literature as it Pertains to Gulf War Illnesses. Vol 5. Chemical and Biological Warfare Agents. Santa Monica, CA: Rand Corporation, 2000.
150. Hyams KC, Wignall FS, Roswell R. War syndromes and their evaluation: from the U.S. Civil War to the Persian Gulf War. Ann Intern Med 1995; 125:398–405.
151. Baker DG, Mendenhall CL, Simbart LA. Relationship between posttraumatic stress disorder and self-reported physical symptoms in Persian Gulf War veterans. Arch Intern Med 1997; 157:2076–78.
152. Horner RD, Kamins KG, Feussner JR, et al. Occurrence of amyotrophic lateral sclerosis among Gulf War veterans. Neurology 2003; 61:742–49.
153. Haley RW. Excess incidence of ALS in young Gulf War veterans. Neurology 2003; 61:750–56.
154. Freudenthal RL, Rausch L, Gerhart JK, et al. Subchronic neurotoxicity of oil formulations containing either tricresyl phosphate or tri-orthocresyl phosphate. J Am Coll Toxicol 1993; 12:409–16.
155. Daughtrey W, Biles R, Jortner B. Subchronic delayed neurotoxicity evaluation of jet engine lubricants containing phosphorus additives. Fundam Appl Toxicol 1996; 32:244–49.

Atmospheric Transport of Mold Spores in Clouds of Desert Dust

EUGENE A. SHINN
DALE W. GRIFFIN
U.S. Geological Survey
St. Petersburg, Florida
DOUGLAS B. SEBA
Academy of Marine Sciences
Alexandria, Virginia

ABSTRACT. Fungal spores can be transported globally in clouds of desert dust. Many species of fungi (commonly known as molds) and bacteria—including some that are human pathogens—have characteristics suited to long-range atmospheric transport. Dust from the African desert can affect air quality in Africa, Europe, the Middle East, and the Americas. Asian desert dust can affect air quality in Asia, the Arctic, North America, and Europe. Atmospheric exposure to mold-carrying desert dust may affect human health directly through allergic induction of respiratory stress. In addition, mold spores within these dust clouds may seed downwind ecosystems in both outdoor and indoor environments.

<Key words: African dust, arsenic, asthma, fungal spores, health effects, mold, public health>

THE DRAMATIC INCREASE in the transport of soil dust across the Atlantic and Pacific oceans during the past 30 yr has been well documented.[1–3] Although geological evidence indicates that African and Asian dusts have been crossing the oceans for eons, the flux of dust has more than doubled since 1970.[4] The timing and increase in volume of African dust flux can be attributed to the ongoing drought in North Africa that began around 1970.[1,5,6] Year-to-year variations in this flux are associated with cyclic latitudinal changes in the North Atlantic Oscillation (NAO), which is a large-scale, multiyear fluctuation in atmospheric pressure between the subtropical high-pressure system located near the Azores in the Atlantic Ocean and the subpolar low-pressure system near Iceland. Positive phases of the NAO—when the system is located at higher latitudes over the North Atlantic—result in increased aridity in the Sahara and Sahel regions of North Africa. These 1- to 3-yr shifts lead to greater dust generation and transport (Fig. 1).[7] On a global scale, regional weather and dust transport are also influenced by a periodic weather phenomenon known as the El Niño Southern Oscillation (ENSO, a large-scale climatic fluctuation of the tropical Pacific Ocean). Since 1970, the highest quantities of transatlantic African dust movement have coincided approximately with ENSO events. The current and conservative estimate for the quantity of desert dust that is transported in Earth's atmosphere each year is approximately 2 billion metric tons.[8] Moving in association with this dust are an estimated 2 quadrillion microorganisms.[9]

Declines in the health of Caribbean coral reefs have been attributed to the NAO-related increase in African dust transport to the Americas.[10–12] We have also focused on the potential health effects of dust on terrestrial ecosystems in the Caribbean and southeastern United States.[9] Approximately half of the airborne particles 1- to 2-μm in aerodynamic diameter ($PM_{2.5}$) inhaled in South Florida during the summer months originated in Africa.[1] Equally significant is Asian dust movement in the Pacific Northwest, especially in Alaska and Western Canada. During large Asian dust events, an estimated 4,000 metric tons of soil per hour can affect the Arctic environment.[13] Asian dust can commingle with industrial aerosols from the burgeoning industrial activities in China, Korea, and Japan before crossing the Pacific Ocean. Research conducted in Korea by Park et al.[14] established a correlation between

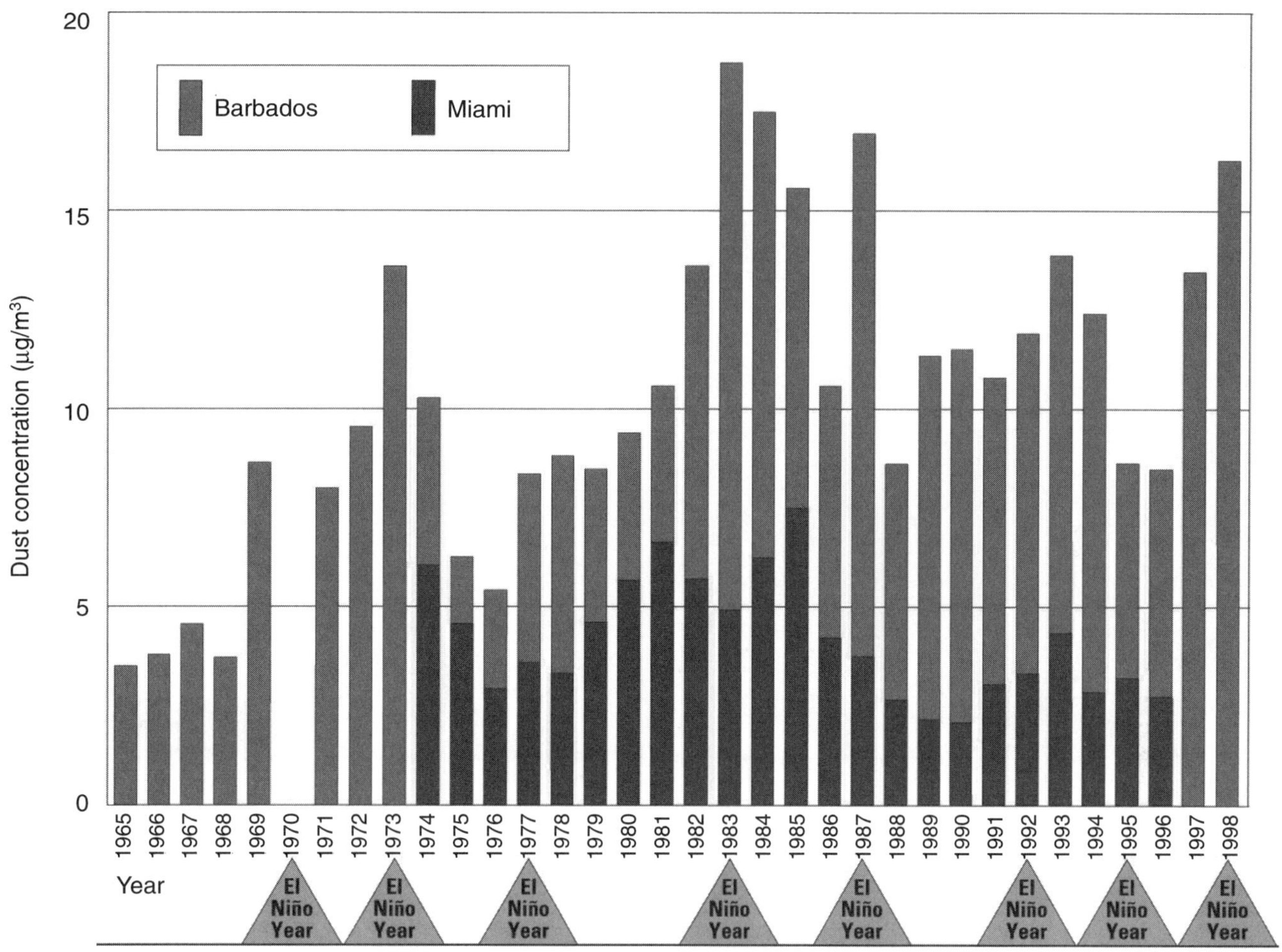

Fig. 1. Average annual flux of Sahelian and Saharan soil dust to Barbados, the Bahamas, and Miami, Florida, 1965–1998. Note the coincidence of African dust flux with the onset of El Niño. (Data courtesy Joseph Prospero, University of Miami. 1997–1998 data for Miami not available.)

dust storms and respiratory-stress–induced apnea. Freye et al.[15] also found an increase in respiratory stress (e.g., asthma, allergic rhinitis, and sinusitis) as a result of higher than normal pollen and fungi counts associated with the ENSO event of 1997–1998 (Fig. 1).

Asthma, a worldwide problem, has increased dramatically in the Caribbean.[16] The Caribbean Asthma Association reported a 17-fold increase in asthma on the island of Barbados since 1973, which corresponds to a prevalence rate of approximately 1 in 4 inhabitants.[17] The resulting increase in the use of asthma medications throughout the Caribbean has had an economic impact on health-care resources.[18] Gyan et al.[19] found a positive correlation between pediatric hospital admissions for respiratory distress in Trinidad and African dust events. This dramatic rise in cases of asthma in the Caribbean parallels the increase in dust flux from Africa to Barbados and Miami (Fig. 1).[1,2]

Anecdotal information from residents of the Caribbean Basin indicates that—in addition to asthma—sinus stress, migraine headache, eye irritation, and chest pain are not uncommon during and after dust events. One study has shown that, once a person is sensitized to an allergen such as fungal spores, even low concentrations can induce an allergic response.[20] In the Middle East, a survey of asthma and allergic rhinitis prevalence identified dust storm exposure as a significant predictor of these illnesses.[21]

In the United States, few people suspect that breathing distress and headache may be caused by exposure to African or Asian dust. Atmospheric dust in the United States often goes unrecognized because the population is accustomed to the hazy conditions caused by anthropogenic emissions. However, the existence of transoceanic dust over the United States is readily apparent in satellite images (Figs. 2–4), which reveal a generalized flux of global dust. Perry et al.[22] found African dust as far north as Maine and as far west as Carlsbad, New Mexico. Approximately half of the dust collected in Carlsbad originated in Africa.

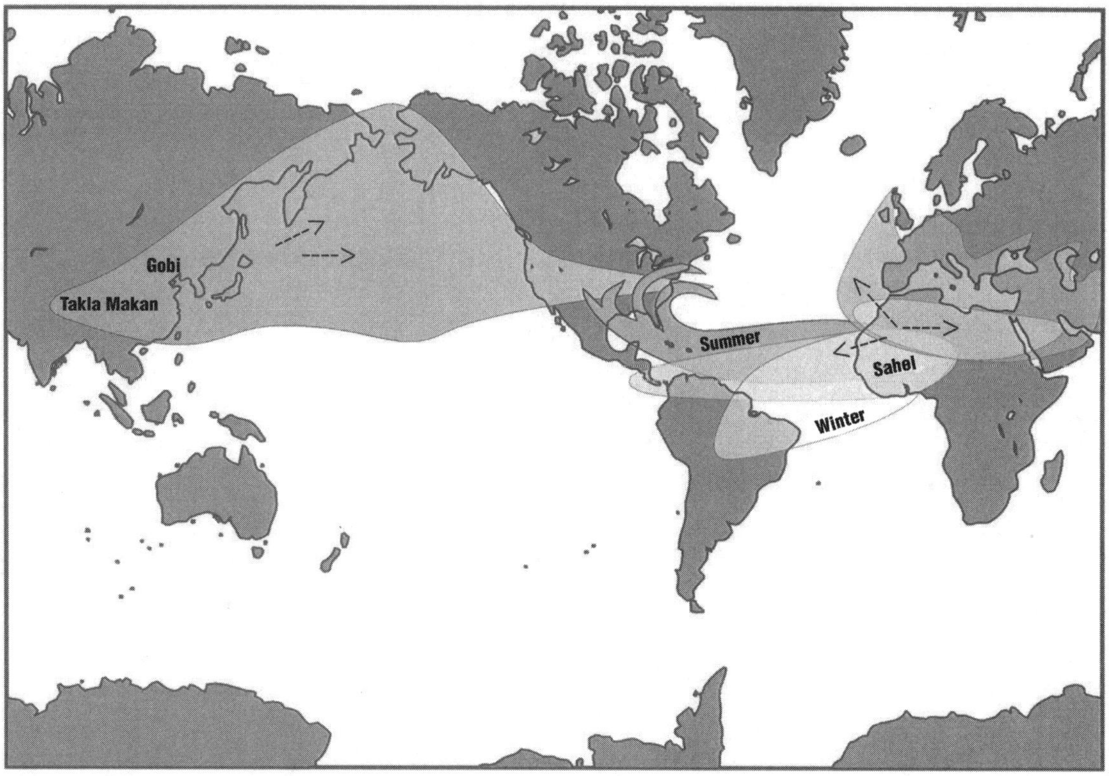

Fig. 2. Generalized flux of dust into the Caribbean and United States. ***Sources:*** **Satellite images and dust sampling data (U.S. Environmental Protection Agency/Nonpoint Source Pollution [EPA/NPS] IMPROVE visibility and particle monitoring program) from Perry et al.[22]; Asian dust flux across the Pacific from Garrison et al.[10]**

Fig. 3. NASA's SeaWiFS (Sea-Viewing Wide Field-of-View Sensor) satellite image showing African dust clouds traversing the Atlantic. (Information on SeaWiFS is available from <http://seawifs.gsfc.nasa.gov/SEAWIFS.html>)

In 2000, the U.S. Geological Survey (USGS) in St. Petersburg, Florida, began the USGS Global Dust Program (initially funded by the U.S. National Aeronautics and Space Administration [NASA]) to investigate the potential health effects of various African dust components. To date, the major focus has been on the identification of cultivatable microorganisms that are transported along with other organic and inorganic constituents of transoceanic dust clouds. In addition to microbiological testing, dust samples are screened for the presence of heavy metals, pesticides, and radiogenic elements. Preliminary work has shown that arsenic, in association with iron, averages about 20 parts per million (ppm) in African dust particulate matter that reaches South and Central Florida.[23] Analyses of these samples have also shown that concentrations of mercury up to 2 ppm, and concentrations of beryllium and lead isotopes (specifically ^{7}Be and ^{210}Pb), are greater than those found in local rocks and soils.

Our previous microbial study identified numerous

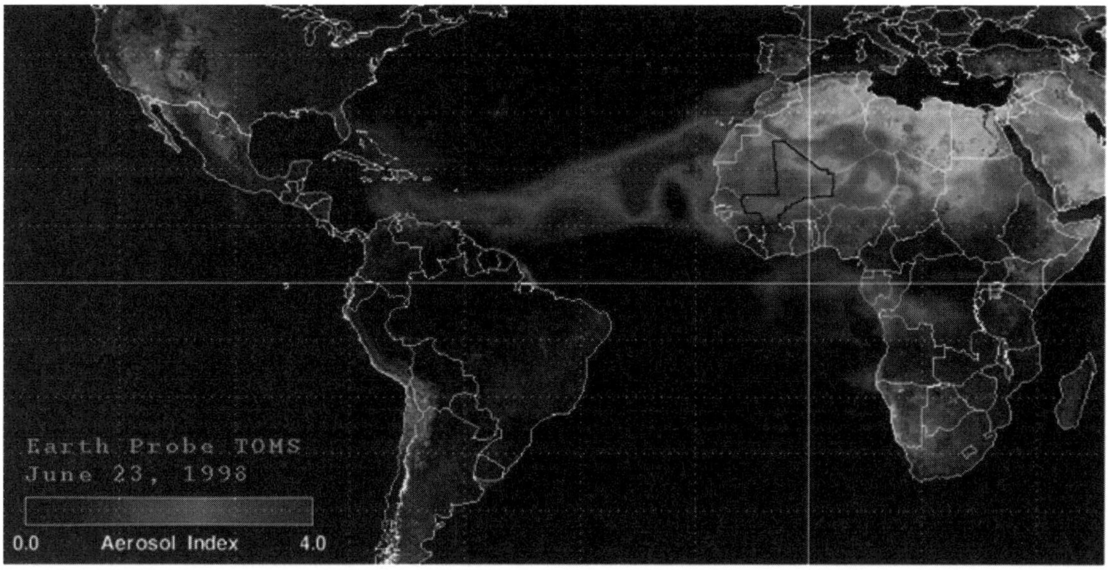

Fig. 4. NASA's Earth Probe TOMS (Total Ozone Mapping Spectrometer) satellite image showing a continuous stream of dust crossing the Atlantic. The North African nation of Mali is outlined in black. (Information on TOMS is available from <http://jwocky.gsfc.nasa.gov>)

species of bacteria and fungi that survive the 5- to 7-day transatlantic transport.[11] The bacteria-to-fungi ratio is variable between dust events, but samples at a site in the U.S. Virgin Islands were usually dominated by bacteria (76% of dust-event isolates).[11] Approximately 30% of the microbes cultured and identified thus far are capable of causing disease in plants and animals, and 10% are opportunistic human pathogens.[11] To date, we have identified fungal species of *Acremonium, Alternaria, Aspergillus, Aureobasidium, Bipolaris, Cladosporium, Coccodinium, Fusarium, Gibberella, Microsporum, Nigrospora, Paecilomyces, Penicillium, Pleospora, Scopulariopsis,* and *Trichophyton.* With the exception of *Pleospora* and *Gibberella* (both of which contain plant pathogens), and *Coccodinium,* all the genera identified include species that are known to cause allergic reactions, pulmonary infections, or skin infections (Table 1).[24,25] These data indicate that any fungus (or spore-forming microorganism) is capable of surviving long-range atmospheric transport, including the common house mold *Stachybotrys chartarum.*

House Mold and the Potential Dust Connection

An eyewitness to the 1930 American Dust Bowl event stated:

> *Since breathing is unpleasant even indoors while a dust storm is at its height, people sought relief through wearing wet cloths over their faces.... But the practice soon proved to be a dangerous one.... The constant breathing of damp air, aggravated by the dust, frequently led to pneumonia, with many deaths resulting....The dust I had labored in all day began to show its effects on my system. My head ached, my stomach was upset, and my lungs were oppressed and felt as if they must contain a ton of fine dirt.*
>
> —Lawrence Svobida, *An Empire of Dust*[27]

The adverse health effects of molds in homes—and the associated multibillion-dollar home insurance crisis—began in the 1970s and are well documented by other contributors to this, and the previous,[28] *Archives of Environmental Health* special issues on molds and mycotoxins. Precise causes of the mold problem are difficult to discern, as indicated by the increase in litigation associated with mold-related insurance claims.[29] In general, the causes can be linked with changes in home and office construction methods, materials, and design, and with the increasing dependence on central air conditioning. Air conditioning not only reduces the circulation of fresh air within a dwelling, but also induces thermal gradients within homes and offices that create localized cool and moist areas conducive to mold growth.[30] In addition, the porous ducting materials increasingly used in central air conditioning systems create a favorable culture medium for microbes when cool and damp.[31,32] The dark and cryptic conditions in the air conditioning ducts and elevator shafts of large buildings are notorious promoters of mold growth.[33,34]

The most frequently mentioned cause of house mold is flooding, especially in homes with gypsum and cardboard interior sheetrock walls. After flood water recedes, mold can grow unseen on standard wooden studs and in porous insulating materials between wall panels.[35] Detection of such mold is difficult, often requiring costly dismantling of structures. The health

Table 1.—Fungal Species Capable of Affecting Humans and Plants, Cultured and Identified in Samples of Atmospheric Desert Dust

Genus	Pathogen type[24,25]	Distribution[24,26]	Disease potential[24,25]
Acremonium	Human	Wide distribution. Common in soil, on plants, and indoors.	Mycetoma (colonization of tissue/bone), onychomycosis (colonization of nail), mycotic keratitis, allergen. Can cause a number of different disease types in immunocompromised individuals.
Alternaria	Human and plant	Wide distribution. Common in soil, on plants, and indoors.	Cutaneous phaeohyphomycosis (colonization of skin), potent allergen (common cause of extrinsic asthma). Can cause deep tissue infections in immunocompromised individuals.
Aspergillus	Human, animal, and insect	Wide distribution. Common in soil, organic detritus, and indoors.	Aspergilliosis (pulmonary [allergic and colonizing], disseminated, central nervous system, cutaneous, nasal–orbital, and iatrogenic), potent allergen. Causes a number of different disease types in immunocompromised individuals.
Aureobasidi-	Human and plant	Found in temperate areas. Common on plant tissue and indoors.	Cutaneous phaeohyphomycosis, invasive disease in immunocompromised individuals.
Bipolaris	Human and plant	Wide distribution. Common on plants and indoors.	Pansinusitis, meningoencephalitis, chronic pulmonary disease.
Cladosporium	Human and plant	Wide distribution. Most commonly isolated fungi in outdoor studies.	Cutaneous phaeohyphomycosis, chromoblastomycosis (subcutaneous skin infections), mycotic keratitis, potent allergen.
Coccodinium	Unknown	Found in tropical air samples.[11]	No information.
Fusarium	Human and plant	Wide distribution. Common isolate in soil and indoors.	Invasive cutaneous infection (erythematous lesions and nodules), systemic granulamatous disease, allergen.
Gibberella	Plant	Wide distribution.	Stalk rot, corn ear rot, Bakanae disease.
Microsporum	Human and animal	Wide distribution.	Dermatophytosis (i.e., ringworm).
Nigrospora	Human	Wide distribution. Common in soil, organic detritus, and indoors.	Allergen.
Pae-	Human and insect	Wide distribution. Common in soil, organic detritus, and indoors.	Mycotic keratitis paecilomycosis , pneumonia, allergen.
Penicillium	Human	Wide distribution. Very common in temperate regions. Common in soils and indoors.	Bronchopulmonary penicilliosis, potent allergen (hypersensitivity and allergic alveolitis).
Pleospora	Plant	Wide distribution.	Leaf spot. Species can produce mycotoxins toxic to certain plants (e.g., the opium poppy).
Scopulariopsis	Human and insect	Wide distribution. Common in soils and indoors.	Pneumonia in immunocompromised individuals, rare subcutaneous and pulmonary cases, associated with Type III allergy.
Trichophyton	Human and animal	Wide distribution. Common in soils and indoors.	Dermatophytosis, allergen.

problems associated with house mold have spawned high-tech mold detection and litigation industries, including a new cottage industry that uses specially trained dogs. Although no single source of fungi is implicated in the house mold problem, it is possible that fungal spores transported in desert dust may affect indoor human health both directly (by initiating an allergic response) and indirectly (as seed for indoor growth of mold). This relationship may be especially true in areas around the globe where desert dust settles and accumulates in warm and damp climates, such as the southeastern United States.

Although molds identified in houses, and in African dust samples, are also common in U.S. soils, it is suspected that microbes that survive transoceanic transport and exposure to ultraviolet radiation—as well as their hot and arid sources—may be more capable of withstanding extreme conditions than native species.[36] The study of airborne fungi and bacteria was initiated in the late 19th century and was later considered important for understanding the worldwide spread of rust fungi in agricultural crops during the 1930s.[37–39] The availability of new technologies (e.g., the polymerase chain reaction test), combined with agricultural and human security issues, are driving a revival in studies of aerobiology and the various pathways through which microbes can spread across the planet. Our atmospheric microbiology research group—one of the few in the United States—is expanding its activities, and we hope to contribute to a better understanding of both the causes and prevention of human and plant disease epidemics.

* * * * * * * * * *

The authors thank Christina Kellogg, Virginia Garrison, and Charles Holmes for technical assistance; Betsy Boynton for computer images; and Barbara Lidz for editorial assistance.

Submitted for publication September 20, 2003; revised; accepted for publication November 24, 2003.

Requests for reprints should be sent to Eugene A. Shinn, U.S. Geological Survey, 600 4th Street South, St. Petersburg, FL 33701.

E-mail: eshinn@usgs.gov

* * * * * * * * * *

References

1. Prospero JM. Long-term measurements of the transport of African mineral dust to the southeastern United States: implications for regional air quality. J Geophys Res 1999; 104(D13):15917–27.
2. Jaffe D, McKendry I, Anderson T, et al. Six "new" episodes of trans-Pacific transport of air pollutants. Atmos Environ 2003; 37(3):391–404.
3. Middleton NJ. A geography of dust storms in southwest Asia. J Climatol 1986; 6(2):183–96.
4. Prospero JM, Lamb PJ. African droughts and dust transport to the Caribbean: climate change implications. Science 2003; 302(5647):1024–27.
5. Prospero JM, Nees RT. Dust concentration in the atmosphere of the Equatorial North Atlantic: possible relationship to the Sahelian drought. Science 1977; 196:1196–98.
6. Prospero JM, Nees RT. Impact of the North African drought and El Niño on mineral dust in the Barbados trade winds. Nature 1986; 320(6064):735–38.
7. Moulin C, Lambert CE, Dulac F, et al. Control of atmospheric export of dust from North Africa by the North Atlantic Oscillation. Nature 1997; 387:691–94.
8. Perkins S. Dust, the thermostat. Sci News 2001; 160: 200–01.
9. Griffin DW, Kellogg CA, Garrison VA, et al. The global transport of dust. Amer Sci 2002; 90(3):228–35.
10. Garrison VH, Shinn EA, Foreman WT, et al. African and Asian dust: from desert soils to coral reefs. Bioscience 2003; 53:469–80.
11. Griffin DW, Garrison VH, Herman JR, et al. African desert dust in the Caribbean atmosphere: microbiology and public health. Aerobiologia 2001; 17(3):203–13.
12. Shinn EA, Smith GW, Prospero JM, et al. African dust and the demise of Caribbean coral reefs. Geophys Res Let 2000; 27(19):3029–32.
13. Rahn KA, Boyrs RD, Shaw GE, et al. Long-range impact of desert aerosol on atmospheric chemistry: two examples. In: Fenner F (Ed). Saharan Dust: Mobilization, Transport, and Deposition. Chichester, U.K.: John Wiley & Sons, 1977; 243–66.
14. Park J-W, Lim YH, Kyung SY, et al. Effects of Asian dust events on peak expiratory flow and respiratory symptoms in subjects with bronchial asthma. Eur Respir J 2003; 22 (45):S301.
15. Freye HB, King J, Litwin CM. Variations of pollen and mold concentrations in 1998 during the strong El Niño event of 1997–1998 and their impact on clinical exacerbations of allergic rhinitis, asthma, and sinusitis. Allergy Asthma Proc 2001; 22(4):239–47.
16. Monteil MA, Juman S, Hassanally R, et al. Descriptive epidemiology of asthma in Trinidad, West Indies. J Asthma 2000; 37(8):677–84.
17. Howitt ME. Asthma management in the Caribbean—an update. Postgrad Doctor—Caribbean 2000; 16(2):86–104.
18. Howitt ME, Naidu R. The economics of asthma treatment in Barbados—a review of the drug cost. West Indian Med J 2000; 49(2):19.
19. Gyan K, Henry W, Lacaille S, et al. African dust clouds are associated with increased paediatric asthma Accident and Emergency admissions on the Caribbean island of Trinidad. Lancet 2002. <www.thelancet.com/era/search?searchtext=african+dust&search.x=39&search.y=6>
20. Gravesen S. Fungi as a cause of allergic disease. Allergy 1979; 34:135–54.
21. Bener A, Abdulrazzaq YM, Al-Mutawwa J, et al. Genetic and environmental factors associated with asthma. Hum Biol 1996; 68(3):405–14.
22. Perry KD, Cahill TA, Eldred RA, et al. Long-range transport of North African dust to the eastern United States. J Geophys Res 1997; 102:11225–38.
23. Holmes CW, Miller R. Atmospherically transported metals and deposition in the southeastern United States: local or transoceanic? Appl Geochem 2003 (forthcoming).
24. St-Germain G, Summerbell R. Identifying Filamentous Fungi. Belmont, CA: Star Publishing, 1996.
25. Rippon JW. Medical Mycology: The Pathogenic Fungi and the Pathogenic Actinomycetes, 3rd ed. Philadelphia, PA: W.B. Saunders, 1988.
26. Ren P, Jankun TM, Belanger K, et al. The relation between fungal propagules in indoor air and home characteristics. Allergy 2001; 56:419–24.
27. Svobida L. An Empire of Dust. Caldwell, ID: Caxton Printers, 1940.
28. Molds and mycotoxins: special issue, Pt 1. Arch Environ Health 2003 Jul; 58(7):385–448.
29. Dotzour MG, Bravenec E. Insurance at a premium: Tierra Grande 2002; 9(2).
30. Eggleston PA. Environmental control for fungal allergen exposure. Curr Allergy Asthma Rep 2003; 3(5):424–29.
31. Ezeonu IM, Noble JA, Simmons RB, et al. Effect of relative humidity on fungal colonization of fiberglass insulation. Appl Environ Microbiol 1994; 60(6):2149–51.
32. Foarde KK, Menetrez MY. Evaluating the potential efficacy of three antifungal sealants of duct liner and galvanized steel as used in HVAC systems. J Ind Microbiol Biotechnol 2002; 29(1):38–43.
33. Samson RA. Occurrence of moulds in modern living and working environments. Eur J Epidemiol 1985; 1(1):54–61.
34. Lutz BD, Jin J, Rinaldi MG, et al. Outbreak of invasive *As-*

pergillus infection in surgical patients, associated with a contaminated air-handling system. Clin Infect Dis 2003; 37(6):786–93.
35. U.S. Environmental Protection Agency (EPA), Office of Radiation and Indoor Air (6609J). Flood cleanup—avoiding indoor air quality problems. Washington, DC: Indoor Air Quality Information Clearinghouse, 2003; EPA Publication No. 402-F-93-005.
36. Imshenetsky AA, Lysenko SV, Kozlova TM, et al. Resistance of mesospheric microorganisms to periodic freezing—thawing. Mikrobiologiia 1983; 52(6):902–08.
37. Bowden J, Gregory PH, Johnson CG. Possible wind transport of coffee rust across the Atlantic Ocean. Nature 1971; 229:500–01.
38. Nagarajan S, Singh DV. Long-distance dispersion of rust pathogens. Annu Rev Phytopathol 1990; 28:139–53.
39. Meier FC, Lindbergh CA. Collecting microorganisms in the arctic atmosphere. Arctic Monthly 1935; 40:5–20.

The Way We Build Now

GEOFFREY H. HUTTON
Hutton + Rostron
Guildford, Surrey
United Kingdom

ABSTRACT. Vernacular building for humans evolved in much the same way that shelter evolved for birds and animals: largely by trial and error, and experience. Healthy accommodation is critical for the survival of all species, and the existence of molds and fungal spores in the air and food supply has been part of the evolutionary equation. With human intervention, the environment has been changed by the use of chemical products and the availability of cheap energy, at a rate outpacing evolutionary adaptation. The interaction of these challenges with building design and human health is significant. Relevant factors should be explored, and their effects understood, so that solutions in building can be developed before the use of fossil fuels must be abandoned altogether.

<Key words: building design, environmental health, evolution, *Serpula lacrymans*, vernacular housing>

BUILDING DESIGN is a multifactorial equation with thousands of unknowns and few knowns, which interacts with health at every level. The factors involved have been optimized in the evolution of vernacular building, but not met so successfully in concept-based design. Water, with its contribution to decay and the growth of molds within structures, is a constant in the equation.

Birds are great builders.[1] They compete with other animals to exploit local resources for their nests, and they take great care in selecting the nest's location and structure. Too high in the tree, and the wind will destroy the nest; too low and a predator will take the eggs; too strong or elaborate construction wastes otherwise valuable breeding or feeding time. The nest must be defensible, inconspicuous, protect the eggs from breakage, and not harbor pests or support fungal growth; it must retain sufficient heat for insulation and be well drained and well ventilated. The initial owner may decide to clear out after a year and build a new "hygienic model" next season, or it may invest in a family home to be refurbished yearly, or may even be a squatter, such as the cuckoo. Each creature makes a home, or creates shelter for itself, in balance with its environment, and passes on the benefits to its descendents. Every climate and context has residents who have solved their accommodation problems sufficiently well to have survived and prospered over many generations. In nature, the animal, home, and environment are clearly interdependent.

Humans, too, have solved the problems of climate and context with the building methods and materials available to them, whether as nomads or in settled communities. Possible building solutions were understood by the users, and the technology was within their grasp and was shared by the community. Differentiation between building functions was slight—the essentials of shelter and enclosure were met by common methods for spanning, loadbearing, and enclosing structures. Properties inherent in the available materials were relied on for weather resistance, strength, and insulation (Fig. 1). The building interior was heated when necessary by combustion of the limited fuel available, which was important primarily for cooking. Thus, buildings evolved and served satisfactorily for millennia, and are still in use, providing the ultimate barrier against the environment. The technology that was developed incorporated the accumulated experience of generations for practical survival and prosperity. Empirical knowledge included solving such issues as ventilation, temperature, and evaporation with an elegance that is often overlooked in analysis (Fig. 2).

Fig. 1. Building a hut in Nigeria involves erection of a simple frame to be covered with palm fronds.

Fig. 2. A 13th-century cottage in southern England, of timber-framed, clay-tiled, wattle-and-daub construction is heated by an open hearth that exhausts through the gable.

Human beings are much more durable and adaptable than the providers of goods and services suppose. Indeed, a degree of physical challenge and adversity appears to be desirable for a healthy life. Perhaps what is perceived as comfort is actually a component of disease. An understanding of this may inform the lifestyles of some religious groups, such as the Shakers and the Amish.

From the earliest times, buildings have also served symbolic purposes—demonstrating power, identity, or spiritual values. These "monumental" characteristics are usually achieved at some otherwise unjustified expense and inconvenience to the practical performance of the building. Unfortunately, in contemporary design and construction, the monumental has overshadowed the empirical, with the situation further complicated by the introduction of codes and standards. Such regulations attempt to capture the multifaceted performance of buildings by setting criteria for discrete aspects of construction, such as structural strength, fire resistance, thermal characteristics, ventilation, lighting, and security.[2] These factors are further defined by testing methods that are limited by available laboratory procedures. Such aspects of performance are inherent in vernacular construction, which resulted from the evolutionary optimization of the overall design. Important to both birds and humankind is the health of the occupants, which is fundamental to survival. In monumental buildings, this concern is accorded lesser significance (except perhaps in a spiritual sense).

Novelty and innovation in building are admired and promoted. Often the overall design is unique, with little basis in experience acquired over many projects or generations. Craftsmanship develops into a style, the appearance of which becomes sought after and an end in itself. The forms in nature, sometimes originally of constructional significance, are reproduced in a stylized way to become decoration, as in classical Greek architecture. Appearance then subsumes function. This sophistication replaces evolved experience and self-reliance based on knowledge of local materials and the environment. Construction now depends on specialisms and the mass production of components, complex assemblies, and ultimately entire buildings. This development has been accompanied by the division of related skills and theoretical studies into further specialization in production technology and the related professions. Such an approach to design results in a building that is remote from its users and their concerns—the choices the occupant is offered are increasingly superficial. The incorporation of imagery and structural components from monumental building into the mainstream has led to the dominance of conceptual design over the utilitarian product of evolution and the local environment.

The compartmentalization of construction for design, production, study, and evaluation obscures its essential intricacy. Lacunae develop at the interfaces between disciplines, knowledge, materials, data, components, and systems. Condensation (a trigger for mold growth) is a consequence of such unforeseen incompatibility. Architects and engineers—each with a visual or structural idea to promote, but without an overarching experience of the interactions between the diverse needs and performance requirements—fail to communicate effectively. They, in turn, serve investors, building owners, and users who have little personal comprehension of the issues and, therefore, rely heavily on professional advice.

The requirements for building performance are now less likely to be achieved with simple construction materials. Instead, complex components are used in assemblies and are bought as catalog items to achieve theoretically defined objectives such as weather resistance, soundproofing, and thermal insulation. The performance of individual items may be known from testing, but the consequences for the whole assembly—and particularly the interfaces—may not be fully understood, defined, or even measurable. Consequently, failures occur because of unanticipated effects or events, at interfaces between materials or sections, or as a result of environmental stress. Such failures often occur in voids inaccessible to inspection.

Caves or trees provide the most primitive shelter. Buildings were developed as either heavyweight structures (often of masonry), or lightweight assemblies of frames and infilling panels (normally of timber). Each has advantages for particular climates and occupancies. A masonry building (Fig. 3) has a higher thermal capacity and lower thermal response than a frame building, which will be easier to light and ventilate but may also contain more cavities. Modern buildings are generally of frame-and-fill construction, even if faced with masonry, as modern construction favors prefabrication and tries to limit the activities on site. Typically, such buildings have suspended ceilings, raised floors, and paneled or cavity wall constructions, thus creating voids that are not easily accessible and can be contaminated and difficult to clean. Builders' dirt and personal rubbish, which may be organic, are frequently left in places that now also accommodate engineering plant and distribution services, pipes, and ductwork.

Since the early 19th century, the exploitation of fossil fuels has provided an inexpensive and readily available source of thermal energy for the manufacture of products such as glass, metals, cement, and bricks, as well as the power needed for cutting, forming, handling, and distribution of such products. Fossil fuels also provide raw materials for plastics, fibers, dyes, paints, and adhesives used in construction, building services, furniture, soft furnishings, household utensils, and clothing, as well as for a host of lifestyle support products such as cleaning agents, polishes, insecticides, and perfumes. The processes used affect the environment and generate persistent wastes, and the products themselves may have undesirable emissions and toxic effects.[3]

Fig. 3. A fisherman's cottage in Scotland is constructed of roughly squared quarry stone in lime mortar (thermal energy is used to produce the lime). The chimneys improve ventilation and enhance thermal storage in the heavyweight building.

Problems caused by the resulting pollution of the atmosphere, land drainage, and the seas do not have political boundaries and may transcend topographical ones. The major use for fossil fuel is to power environmental services in buildings (e.g., heating, ventilation, air conditioning, elevators, and escalators), without which much of the world would be virtually uninhabitable at present densities. Management of such services is not precise, and fuel is inexpensive and generally available, consequently there is little incentive to economize (environmental services consume some 60% of all fuel). In historic terms, this relatively short-term, low-cost resource has changed lifestyles such that the thermal importance of clothing has declined, the size of individual spaces has increased, and the provision of natural light and ventilation has been reduced. The easy availability of hot water, and the consequent dispersal of activities such as bathing, laundering, and cooking, have resulted in higher levels of indoor humidity. Where once this would have been dealt with by natural ventilation and the passive qualities of the building enclosure,[4] the engineering now required to provide for such conditions has become increasingly complex (Fig. 4). It occupies an increasing proportion of available space and requires regular and diligent maintenance to achieve its theoretical performance. This is expensive, and its effectiveness cannot always be guaranteed, particularly in older buildings and those with less affluent occupants.

Interior environmental conditions are set at norms equivalent to a mild summer day in Europe or the United States, regardless of actual location, climate, or season. With divergent indoor and outdoor climates, people do not acclimatize and are faced with frequent step-changes in indoor temperature, humidity, and air quality, and when moving from the inside to outside environments. Urban life in many communities is now totally sheltered, with the wearing of clothing largely governed by convention, sexual display, and status. Garments make extensive use of synthetic materials, which hinder the evaporation of sweat; indeed, perspiration, as a natural cooling and excretory process, is considered embarrassing.

The artificial interior environment is as much, or more, a fashion statement than a necessary response to

Fig. 4. The apparently unbroken wall, floor, and ceiling surfaces of this building conceal a mass of distribution ducts, pipes, and cables that provide services to the plant and facility.

the exterior climate and atmospheric pollution. Humankind is well adapted to live in a wide range of climates and has developed appropriate clothing. However, local chemical and particulate air pollutants are additional insults to human tissue and produce conditions for which evolution has not prepared us. These conditions may result in direct damage and opportunities for organisms, such as fungi, to flourish. Condensation (distilled water) aggressively facilitates other forms of deterioration (e.g., bimetallic corrosion).

The environmental standards achieved by engineering services inside buildings—particularly if the buildings are sealed—subjects the exterior building envelope to greater stresses to provide the primary barrier to exterior conditions. The envelope must accommodate thermal movement, increased vapor pressure, and rates of heat transfer, while providing shelter and weather resistance. Under some conditions condensation will occur, and this may be within the structure. Heating and air-conditioning control systems operate reactively; their on-off switching creates step-changes in temperature and humidity, with lags in the rate of response that also can result in condensation.

Diurnal temperature reversal can lead to condensation on the outside of air-conditioned buildings. Condensation on the inside can result from activities that increase humidity, or can be the consequence of external/internal temperature reversals arising from processes or the climate. Mechanical plant pipes and ductwork may produce moisture or attract condensation. Activities such as cooking, bathing, and laundering—as well as the occupants themselves—give off moisture. Condensation may occur at some distance from the source of the moisture, normally on wall surfaces, in cavities, or interstitially within building materials. This intermittent moisture tends to accumulate, and is quite sufficient to initiate mold growth on surfaces and cause fungal decay of organic materials. In older conventional buildings, condensation is likely to occur on exterior walls and in roof spaces, but in mechanically ventilated and air-conditioned buildings, condensation may be found on air-handling equipment, ducts, pipes, and fan/coil units. Some plants, such as wet cooling towers, may emit aerosols and, if contaminated, can be a source of illness (e.g., Legionnaires Disease). Refrigerators and vending machines may be sources of liquid water, either through leakage or defrosting.

Reversal of stresses in the building envelope occur both seasonally and diurnally, and also with the movement of the sun and changes in cloud cover and wind direction. Achieving a sealed envelope under these conditions is a challenge which, even if successful, results in a building that cannot lose moisture through its skin, any failure of which will cause high local water penetration. Such a sealed envelope rarely has windows that open, or ventilators, and control of the interior environment by the occupants is limited.

The nature of frame-and-fill buildings favors the use of space heating, rather than structural heating that exploits high thermal capacity. The latter is characterized by slow adjustment to ambient conditions, which can normally be expected to give comfort with lower air temperatures, and thus is less likely to produce condensation. Space heating (or cooling) depends on processing large volumes of air, much of which is recirculated to reduce fuel consumption. The air, which carries the waste products of respiration and indoor activities, must be processed, filtered, and mixed with "fresh" air for reuse. Such a system is susceptible to contamination and condensation at a number of points, even if filters are routinely changed and well maintained. Biological materials form some of the contaminants and, with appropriate humidity, fungal growth can be anticipated. Chemical "air freshening" may be used, which only adds to the problem. The complete volume of air in such a system should be purged regularly to avoid the accumulation of contaminants.

As a consequence of building use and standards of housekeeping, various unobserved voids may become home to insects and other pests. These pests—encouraged by the ambient conditions and sustained by human litter and food droppings—may cause additional contamination and damage to the structure and furnishings. Particular activities, such as sleeping and cooking, bring particular infestations (e.g., house mites and cockroaches), adding to an ecology with potentially negative consequences for human health. With high levels of moisture, major destruction can result from fungi (e.g., wet and dry rots) and wood-boring insects, all of which are indicators of unhealthy environments and poor maintenance/housekeeping.

Fungi have evolved to exploit dead or morbid material as a nutrient and break it down into its chemical components, to be recycled by other organisms. In buildings, fungi attack dead organic material, which in nature would fall to the forest floor and be broken down as part of the nitrogen cycle. The natural environment for fungal success is damp, poorly lit, and poorly ventilated. Fungi cause direct decay, particularly of timber, and aid (by symbiosis) other agents of decay such as wood-boring insects. They also attract predators, such as viruses, bacteria, and wood lice. It is even possible that fungi may, to some extent, be capable of managing resource allocation and spatial networks. Mathematical modeling is being used to examine this concept.[5]

The conditions that favor the growth of molds and fungi in buildings usually support the growth of indigenous fungal organisms, which are spread by spores found in the soil or wind-blown from fruiting bodies. However, some fungi are "imports," incapable of surviving in the wild of their new homes (e.g., *Serpula lacrymans* originates in the Himalayan forest). These imports rely on the conditions required for germination being provided by accident within the building. This can occur as the result of domestic activities or a defect that permits rainwater leakage, penetrating dampness, or condensation. Poor ventilation is a common feature that promotes mold growth and structural decay in buildings.

On the basis of such understanding, fungal and insect infestations can be managed—without chemical treatment or the destruction of undamaged materials—by reversing the conditions that cause decay and ensuring adequate ventilation (Fig. 5). The following protocol has been used by our company for 20 yr[6] (e.g., at Windsor Castle[7]):

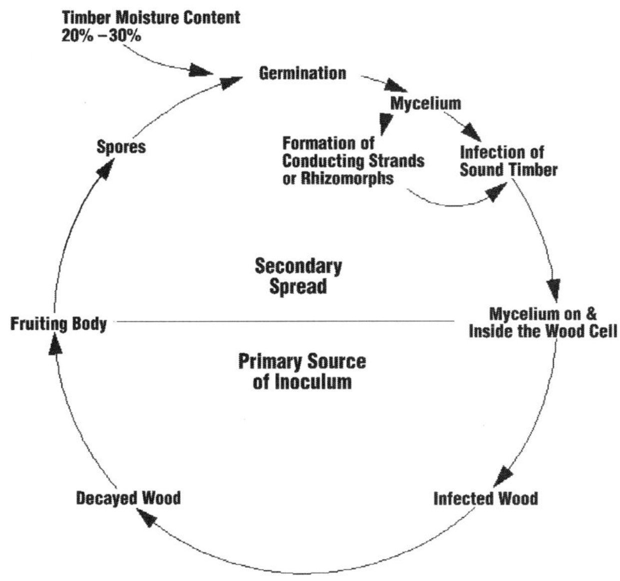

Fig. 5. To disrupt the life cycle of decay fungi, decisive interventions can be made in the material, environment, or organism, without the need for fungicides.

1. Investigation
   a. Trace the extent of decay, as an indicator of moisture presence/source.
   b. Plot moisture distribution.
   c. Check ventilation.
   d. Check for evidence of previous treatment(s).
   e. Check for evidence of physical damage.
   f. Identify the fungi involved.
2. Analysis
   a. Trace sources of moisture.
   b. Identify construction, service, and equipment faults.
   c. Establish environmental factors.
   d. Determine the effects of human activities or lifestyle.
   e. Value the construction at risk.
   f. Consider the impact on the health of occupants.
3. Remediation
   a. Reverse the conditions that favor fungal growth.
   b. Correct defects in construction and services.
   c. Reduce or isolate available nutrients.
   d. Monitor the building to ensure that conditions favoring fungal growth do not reoccur.

The conditions that favor the growth of fungi are also detrimental to human health and, for this reason alone, should be avoided. Because molds and fungi propagate by means of airborne spores, they can cause respiratory problems. As the air outside the building is frequently better than that inside, designers should again consider the use of systems based on cross-ventilation and the installation of windows that can be opened and closed.

Very heavily serviced buildings, such as hospitals, can become hygienic liabilities. The continuing cost of maintenance, and the difficulty of decontamination, may make simpler, disposable accommodation feasible. Ultimately, buildings are often demolished for reasons of fashion or return on investment; perhaps contamination and hazards to health should be factors in such decisions as well. Building is probably the single greatest consumer of energy, both in the form of capital incorporated in products and construction, and in electrical current consumed in operation and maintenance. The cost of building services can be more than half the total capital cost of a hospital, the operating cost of which may equal the capital cost in 3 yr. Clearly, some cost-benefit analysis is required.

Buildings of historic, technical, literary, or national interest should be preserved. These buildings help us understand how people adapted to the climate and context in which they lived during the thousands of years before the profligate use of energy. They give clues as to how we may have to live again. Old buildings often suffer from fungal decay and insect damage due to neglect, misuse, or the introduction of modern heating and air conditioning, new materials, and occupant lifestyles; they rarely suffer because of the work of the original builders.

Medicine, like building construction, has evolved. The shaman, witch doctor, wise woman, apothecary, and anatomist have knowledge that has continuing validity. These earlier practitioners were likely to see disease in both its social and physical contexts. The way in which a building is constructed and operated is an important part of this context and epidemiology.

Self-remediation seems to have been followed by dependence on the reasoning of others (e.g., sympathetic magic, the relation of "similars" [cure with likes and opposites], remedy by experiment and observed effect). This approach gradually gave way to surgical, physical, and spiritual intervention, followed by the application of scientific specialties such as chemistry, microbiology, electromagnetic effects, and psychotherapies, each with greater focus on a smaller part of the overall picture.

People, animals, and plants exist and survive, consuming similarly altered foods, in environments polluted by chemicals, byproducts, and wastes; humans occupy increasingly sophisticated buildings and vehicles made of synthetic materials. This is made possible by the use of energy-intensive processes, operated at the expense of fossil fuels, which are the only means by which many parts of the world can be inhabited at high density by humans and domesticated species. This system is not so much the result of evolutionary pressure as it is a short-term exploitation of available fossil fuels. Alternative sources of energy (other than nuclear and possibly geothermal) are not only expensive to implement, they also have environmental implications. For example, less than 100 megawatts of electricity is believed to be producible by ocean wave power, at present. The World Energy Council considers the maximum potential output from wave power to be 2 terawatts (i.e., 10% of global electricity consumption), at possibly twice the cost of the same output using fossil fuel.[8] Public resistance to large installations for wind generation is already evident. Neither wind or waves provide a consistent energy supply.

The viability of the ecological niche we occupy, using current technology and management methods, depends on the rate at which the cost of energy rises and whether the dispersion of pollutants in air and water can be kept to a tolerable level. Radical changes in building design, and lifestyle, will be necessary to achieve economies, particularly as the demand for energy from developing countries will increase.

Human and environmental health must be made a central concern. Examination of building pathology requires the exploration of construction as an aspect of life, health, welfare, material, and processes in a seamless relationship of observation and experiment, regardless of scientific discipline or profession. Humans and other life forms have evolved in a world occupied by fungi. To avoid circumstances that give rise to illness, humans have developed natural resistance to fungal attack (as have trees) and have adopted lifestyles, clothing, and food storage methods that discourage decay organisms while, at the same time, exploiting edible fungi. It appears that confidence and knowledge have been lost, which now leads to building decay and the increasing implication of molds in the ill health of occupants.

Meanwhile, the smaller descendants of the original wildlife continue, with minor adaptations, to preserve their ways of life and, in some cases, to exploit new features of the current environment. Storks are now building nests on power line poles, and have been so successful that they are often regarded as pests.

Life has evolved over eons, as a process of trial and error, and by selection against environmental stresses. Redundancy is part of the process, but waste is not; everything in nature is recycled. Evolution implies a process of reaction to the environment of which the organism is part, and change is not anticipated. Changes cannot be provided for beyond the limits of the redundancy of the system, for circumstances not previously encountered. Among these are the chemicals now in the environment (and in body fat), originating from biocides, cleaners, decorative finishes, adhesives, food additives, and preservatives, and their byproducts and residues. Evolution reacts to events by incremental change. It cannot anticipate or respond rapidly to new environmental circumstances. Evolutionary forces continue to operate, but on a time scale that favors simple organisms rather than higher animals, which are becoming increasingly reliant on technology for survival. The following equations may help us to retrace our evolutionary way until we regain the sustainable path to health:

Data + context = information.
Information + experience = knowledge.
Knowledge + observation = understanding.
Understanding + ethics = wisdom.

Whether directly or indirectly, the components of action are data, information, experience, knowledge, and wisdom.[9] We must actively seek the data and understanding necessary to reintegrate building and biology—our future depends on it.

* * * * * * * * * *

Submitted for publication September 20, 2003; revised; accepted for publication October 30, 2003.

Requests for reprints should be sent to Geoffrey Hewland Hutton, Hutton + Rostron, Netley House, Gomshall, Guildford, Surrey, GU5 9QA U.K.
E-mail: admin@handr.co.uk

* * * * * * * * * *

References

1. Von Frisch K, Von Frisch O. Animal Architecture. New York: Harcourt Brace Jovanovich, 1974.
2. The Royal Commission. General Report of the Committee Appointed for Improving the Sanitary Conditions of the Barracks and Hospitals. London: Parliamentary Papers (House of Commons), 1861.
3. Purdey M. Does an infrasonic acoustic shock wave resonance of the manganese 3+ loaded/copper depleted prion protein initiate the pathogenesis of TSE? Med Hypotheses 2003; 60b:797–820.
4. Douet J. British Barracks 1600–1914: Their Architecture and Role in Society. London: Stationery Office Books, 1998.
5. Watkinson SC, Boddy L, Burton K, et al. New approaches to investigating the function of mycelial networks. The Royal Society, London, 2003. (in press for *The Mycologist*).
6. Hutton + Rostron. <www.handr.co.uk>
7. Nicholson A. Restoration: The Rebuilding of Windsor Castle. London: Michael Joseph, 1997; pp 138, 141–143.
8. Knott M. Power from the waves. New Scientist 2003; pp 33–35.
9. The Bible. Ecclesiastes 7:(in general, and esp. verses 10–13).

Proactive Approaches for Mold-Free Interior Environments

KATHERINE WARSCO
PATRICIA F. LINDSEY
Department of Interior Design and Merchandising
College of Human Ecology
East Carolina University
Greenville, North Carolina

ABSTRACT. Interior design education and practice can contribute to the prevention of mold growth in indoor environments. The authors provide an overview of current thinking within the interior design educational and professional communities regarding proactive approaches to achieving mold-free building interiors, including identification of current best practices for the prevention of mold problems in buildings. They also discuss the development of certification programs. A review of recent literature points to the need for interior designers to be educated to specify the use of ecologically sound materials that support the health of building occupants. The authors present trade-offs between best practices for designing mold-free indoor environments (including considerations of cost, maintenance, and operation) and occupant comfort, health, and well-being.
<Key words: chemical sensitivity, indoor air, interior design, mold>

FEATURE ARTICLES in the news media regarding toxic mold in homes, schools, and public buildings have generated a groundswell of consumer demand for information on healthful indoor environments. In the midst of this publicity, the health-conscious homeowner more frequently requests residential designs that address a number of health concerns.[1] Design practitioners demand knowledge regarding air quality in indoor environments—an issue not addressed by traditional curricula of interior design educational programs.[2] This article presents an overview of current thinking in the interior design educational and professional communities regarding proactive approaches to achieving mold-free building interiors. We focus on the rationale for practicing "benign design," and identify current best practices of a general and conceptual nature for the prevention of mold problems in homes, and in institutional and commercial facilities. Philosophical positions within the interior design community, among educators, practitioners, and agricultural cooperative extension specialists are compared, and information is presented on organizations and consultancies that support designers, and their clients, in efforts to create mold-free interiors.

Mold-free interior design is especially important for individuals with Multiple Chemical Sensitivity. Those who receive a major exposure to a biological or chemical irritant may become sensitized to that irritant. Subsequent multiple low-dose exposures to the irritant can then trigger a response (e.g., dizziness, eye and throat irritation, chest tightness, nasal congestion[3]) in these sensitized individuals.[4]

A New Curriculum for Designers

The U.S. Environmental Protection Agency (EPA), and the faculty and staff at East Carolina University, Greenville, North Carolina (ECU), have supported the development of an environmental education curriculum that promotes interest in design for healthy living. The curriculum was developed and disseminated to assist educators and students of interior design in exploring links between microclimate, architectural shell, interior space planning, mechanical systems, materials specifications, indoor air quality, and environmental illness.[5,6] With expanded knowledge of design for the improvement of indoor air quality, coupled with critical thinking and problem-solving skills to address complex environmental design issues, educators can contribute to a balanced perspective in interior design curricula that gives equal attention to considerations of the natural environment and of the health and well-being of building occupants.

Health issues. Pilatowicz[7] states that "It is the designer's responsibility to address the special needs of the client, which may include dangers of prolonged exposure or sensitivity to chemical substances" within building interiors. The most defenseless members of our population—the elderly, young children, and individuals who are ill—are the ones most exposed to indoor pollution. These groups are physically weak and spend more time inside than the general population. Pilatowicz goes on to say that a growing number of individuals are sensitive to even low concentrations of a range of some chemicals present in indoor air. This sensitivity, called Multiple Chemical Sensitivity, is recognized as a disability by the U.S. Social Security Administration and the U.S. Department of Housing and Urban Development. Human health, safety, well-being, and productivity can be affected by the choices designers make.[7] Informed selection of interior materials, careful planning for airflow, the placement of furnishings, and fenestration can be important in the prevention of indoor pollution.

Regional focus. In this article, we concentrate on environmental illness in the coastal southeastern region of the United States. Factors in this region, such as demographics, building stock, geography, regional customs, and microclimate, make indoor air quality a critical component of the building design process.[8] The population of this region exceeds the national average for elderly residents (65 yr of age and older), and for the percentages of elderly in poverty and in poor health.[9] Elderly and infirm persons spend more time indoors, increasing their long-term exposure to indoor air pollution. Persons in poor health are also more susceptible to indoor air contaminants.[10]

The low-income elderly tend to live in older dwellings that contain hazardous building products (e.g., lead pipes, lead-based paint, and asbestos) and faulty building systems (e.g., leaking furnace ductwork, sewer pipes, and plumbing). Architectural shells are often in poor repair, with leaky roofs and inadequate air and vapor barriers.[8]

In addition, the southeastern region has a high proportion of regular cigarette smokers, with the least stringent restrictions on the smoking of tobacco in public spaces.[11] The EPA estimates that 3,000 lung cancer deaths occur annually in nonsmokers in the United States as a result of exposure to environmental tobacco smoke.[12]

Hot and humid climatic conditions can increase the levels of contaminants in indoor air.[8] Increases in biological and chemical activity can incite toxic and allergic reactions in hypersensitive individuals.[13,14] For the entire southeastern region, the average morning relative humidity exceeds the comfort range (20–80 °F) for at least 8 mo of the year.[15] Outdoor temperatures for coastal states in this region are uncomfortably hot during 42–69% of the year.[15] High moisture content and high temperatures of indoor air contribute to microbial growth and accelerate the release of volatile organic compounds (VOCs) in indoor living environments.[16] Thus, for households without adequate ventilation, the combined conditions of high relative humidity and high ambient temperatures can precipitate environmental illnesses.

It is essential that professionals engaged in residential design have access to information about indoor air quality that is relevant to the climate in which they practice.[8] Conventional thinking about design solutions for air quality associated with heating seasons in cold, arid climates does not fit indoor air considerations for hot, humid climates.[17–19] The interior design professional needs information about the interactions among indoor pollutants, physical conditions within a building, and the susceptibility of building occupants.[20] A study of 30 infants with acute pulmonary hemorrhage in Cleveland, Ohio, illustrated these associations.[21] Black mold was present in older urban residences as a result of chronic plumbing leaks or basement flooding. Exposure to toxigenic spores, which circulated from unfiltered gravity-fed air heating systems, combined with environmental tobacco smoke, was linked with acute pulmonary hemorrhage among the infants.[22]

Best Practices for a Mold-Free Environment

From space planning of the building interior, to the specification of interior finishes, furnishings, fabrics, and equipment, interior design decisions can affect the occurrence of moisture and nutrient matter that support microbial growth. Strategies for removing, isolating, and diluting moisture and nutrient matter in building interiors follow 2 schools of thought. These environmental alternatives typically focus on either the natural environment (i.e., "healthy planet") or on the interaction of the building with its occupants (i.e., "healthy people"). Criteria for the former include the judicious use of natural resources and preservation of the natural environment through sustainable design strategies that focus on the use of building materials and interior products with recycled content; reduction of fossil fuel consumption in the making, using, and discarding of building and interior products; and specification of renewable resources. Criteria for the "healthy people" approach include attention to the interaction of the building and its occupants, addressing physical health as well as mental and emotional well-being.

Molds require air, moisture, and a source of nutrients

(e.g., cellulose) for sustenance. Also required is protection from the ultraviolet (UV) rays in sunlight. A mold-free interior environment can be achieved through 4 strategies to reduce indoor air pollution: (1) source control, (2) separation, (3) filtration, and (4) ventilation.[16,23] These strategies are not always compatible with strategies promoting chemical-free environments. Our review of design recommendations discusses both the benefits and detriments for the chemically sensitive individual. Figures 1–3 are excerpted from student submissions to a design competition that was part of an environmental education project, and serve to illustrate interior design for a mold-free environment.[8]

Source control. Control of molds, and their toxic metabolites (mycotoxins), at the source is the most effective strategy for achieving a mold-free interior environment (Table 1); however, such controls can have a negative effect on the chemically sensitive individual. Eliminating the contaminant at its source minimizes risk of exposure.[23] It also reduces building costs associated with construction techniques and mechanical systems that isolate or exhaust contaminants from living areas. Because source control is the most effective, reliable, and cost-effective solution, it should be the 1st step taken to improve the quality of indoor air.[16,18] This strategy relies on identifying sources of indoor pollutants and eliminating them, or mitigating them through the use of benign materials.

An important approach to preventing the growth of molds, and their release of mycotoxins, is to remove conditions hospitable to microbial growth. This includes eliminating sources of moisture and nutrient matter. Designs that minimize cracks and crevices in interior millwork and casework, and that specify interior surface materials with a nonporous and nonfleecy texture prevent the entrapment of moisture and nutrients that sustain the growth of mold. In addition, UV light (daylighting and/or mechanical lighting) may be used to prevent microbial growth. Source control must be addressed in 3 areas: outdoors at the building perimeter, within the cavities of the architectural shell, and inside the building.

Four methods are typically used to control sources of mold. A primary method is removal or reduction of

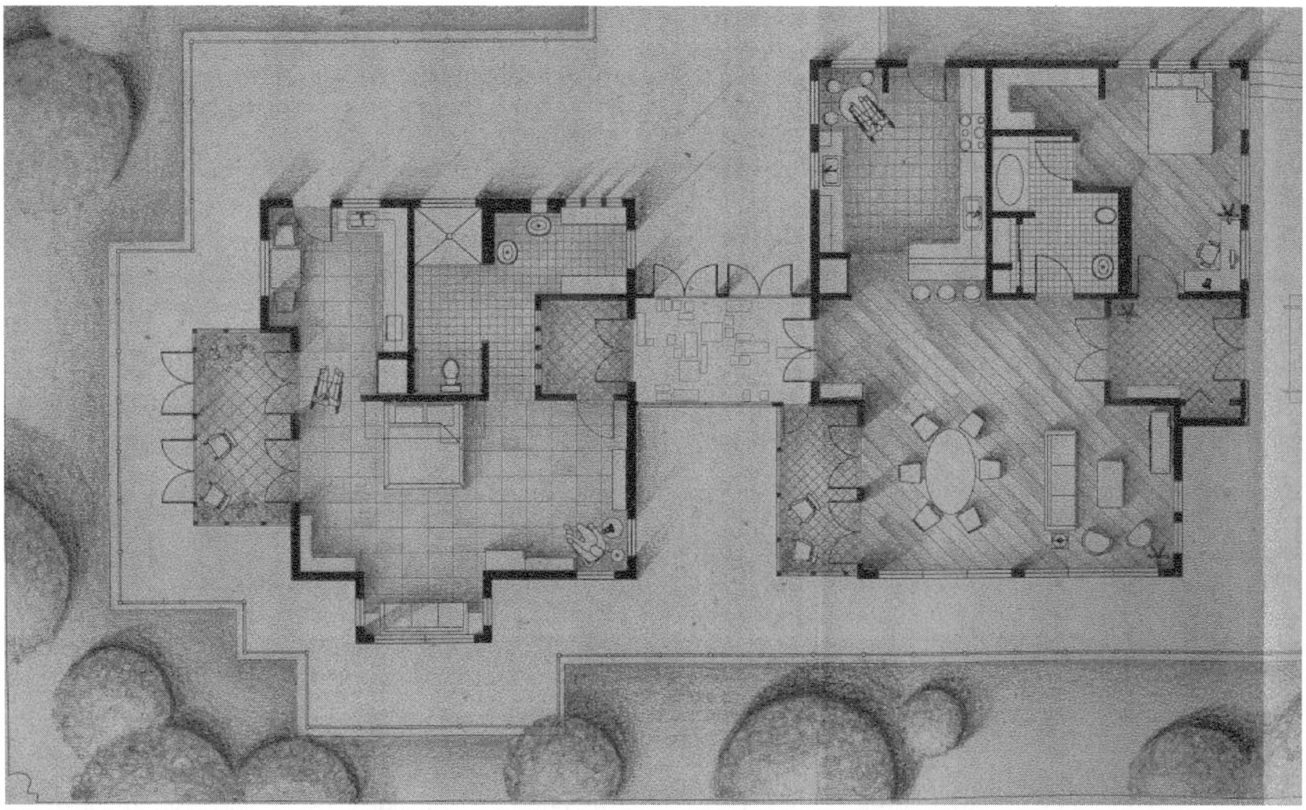

Fig. 1. Coastal cottage design for a chemically sensitive resident, demonstrating 2 methods of source control. (1) *Ventilation*: furnishings in the living/dining area are arranged to allow air to flow unobstructed around the perimeter, and closets in bedrooms are open to allow air to circulate around stored items of apparel (caveat: open closets should be separated from bathrooms to protect from moisture-laden air from bathing activities). (2) *Separation*: a front entrance serves as an air lock, isolating ambient, unfiltered air from the indoor living areas and providing a transitional space for shedding of outdoor garments and footwear prior to entering indoor living areas; a breezeway serves as intermediate buffer zone between living areas and sleeping quarters. *(Student: Lucy A. Fry; faculty: Janine King and Mary Joyce Hasell; University of Florida, Gainesville.)*

Fig. 2. Sitting area design for a chemically sensitive resident, demonstrating methods of source removal. Built-in metal storage cabinets enclosed with glass-fronted doors prevent the build-up of dust on stored items. A continuous soffit extending from the ceiling to the top of the storage unit eliminates a horizontal surface that would collect dust. Miniblinds encased within the window units further reduce surface area for collection of dust. Concrete flooring with mineral colorant is low-textured to eliminate reservoirs for moisture and nutrient matter. Chair upholstery formed from cotton canvas webbing can be laundered. All other low-textured, hard surfaces exposed to the breathing zone can be wiped clean. *(Student: Virginia Meza; faculty: Mary Lamb; Kansas State University, Manhattan.)*

Fig. 3. Bedroom design for a chemically sensitive resident, demonstrating methods of ventilation. The bed is designed with a dedicated air-exchange unit. Balanced air pressure is maintained, and contaminants from the bedding and from reading materials stored in the headboard are exhausted before they enter the breathing zone. A ceiling fan augments controlled ventilation to keep interior surfaces dry. *(Student: Jason Woynar; faculty: Kathleen Sullivan; LaRoche College, Pittsburgh, Pennsylvania.)*

Table 1.—Source Control Design Strategies to Promote Mold-Free Environments

Environment	Strategy and effect
Building site	• Landscaping to reduce standing water and maximize exposure to sunlight will retard microbial growth. • Maximizing natural daylighting, and supplementing with mechanical lighting in the interior, will retard microbial growth.
Architectural shell	• Maintaining an airtight architectural shell will reduce leaking and infiltration of unfiltered and/or contaminated air. • Specifying materials for perimeter walls so that permeability increases from exterior cladding to interior finish will allow moisture within the wall cavity to dissipate.
Mechanical systems for heating, ventilation, and air conditioning (HVAC)	• Properly installing and maintaining plumbing systems will reduce the risk of condensation and leaking of water from pipes. • Properly designing, sizing, installing, and maintaining air conditioning systems will reduce risk of condensation and standing water. • Properly maintaining dehumidifier coils and drainage pans will reduce risk of microbial growth.
Interior space and furniture plan	• Planning for adequate space to ensure airflow around furnishings and stored items will prevent the build-up of moisture in cellulose materials.
Interior materials specification	• Specifying nonporous finishes such as ceramic tile for interior walls, with low temperature difference between rooms, will avoid entrapment of moisture and organic materials resulting from activities of the building occupants. • Specifying noncellulose materials such as concrete, ceramic, glass, and metal will decrease hospitable habitats for microbial growth. • Specifying nonporous, continuous work surfaces will avoid entrapment of moisture and organic materials. • Specifying nonfleecy interior surfaces (e.g., hardwood flooring) and designing built-in furniture will reduce surface-to-volume ratio for collection of dust and the sink effect of volatile organic compounds (VOCs). • Specifying nonfleecy fabrics and designing built-in window treatments will reduce surface-to-volume ratio for collection of dust and sink effect of VOCs. • Specifying finishes and fabrics such as untreated cotton barrier cloth that is easily cleaned (laundered or wiped clean) will prevent build-up of contaminants. • Specifying upholstery with open construction will allow for air movement and dissipation of body heat.

molds outdoors, at the perimeter of the architectural shell. Although mold is ubiquitous in the natural environment, steps can be taken to minimize conditions that support the growth of mold on the building site. Control of moisture outside the building includes grading of the terrain to eliminate standing water, landscape design that eliminates the need for irrigation[1,23–25] (e.g., through the use of drought-tolerant plants[16,19]), providing adequate drainage for precipitation runoff,[16,19,25,26] and avoiding landscape designs that include fountains or stagnant pools.[1] Installing plantings at a distance from the building allows exposure to sunlight, thereby retarding microbial growth around the building.[1] The chemically sensitive individual benefits from a design that incorporates drought-tolerant plants, which usually require no chemical fertilizers or herbicides, thus reducing the risk of exposure to these toxins.[1]

A 2nd method of mold source control deals with control of moisture and nutrient matter within cavities of the architectural shell, and includes airtight construction that eliminates infiltration of spores from building cavities and exfiltration of moisture through cavities.[16,19,23] Vapor diffusion, which allows cavities to breathe, addresses the removal of moisture resulting from temperature differences between surfaces.[16,17,19]

The specification of materials for perimeter walls such that permeability increases from exterior cladding to interior finish allows moisture within wall cavities to dissipate. Examples of interior materials to avoid are vinyl and foil wall coverings, which create an unwanted vapor barrier on perimeter walls. Use of noncellulose construction materials, which are inhospitable to microbial growth, reduces the surface materials that serve as a nutrient source for molds.[16,17,19] Examples include the substitution of metal for wood in framing systems, and the exchange of plaster for gypsum wall board. For the chemically sensitive occupant, the benefits of this strategy include the use of airtight outside cladding for exterior walls, to reduce the risk of infiltration into interior living areas by toxins from outdoors and within wall cavities.[16,17]

A 3rd source control approach involves controlling moisture given off by plumbing and mechanical systems.[17,27] Proper installation and maintenance of plumbing systems reduces the risk of condensation and leakage from pipes. Proper sizing, installation, and maintenance of air conditioning systems reduces the risk of condensation and standing water. Proper maintenance of dehumidifier coils and drainage pans reduces the risk of microbial growth.

A 4th method of source control eliminates molds from inside the home. Because light discourages mildew growth,[28] this approach includes the provision of abundant natural daylight—with supplementation by mechanical lighting—in indoor areas likely to harbor mold. Control of moisture and nutrient matter inside the home includes dehumidifying and circulating indoor air to eliminate conditions conducive to microbial growth. Planning adequate, ventilated storage for cellulose-based items (e.g., books and textiles) allows airflow to reduce the buildup of moisture.[27,29] Open closets (Fig. 1), louvered doors, and ventilated shelving promote airflow in small areas that contain stored items susceptible to microbial growth. Arranging furniture away from perimeter walls allows airflow to decrease the risk of condensation on walls and furnishings (Fig. 1). The moisture content in indoor air is reduced through the use of desiccants, air conditioning, and dehumidifiers.[16,27] Circulation of air over damp surfaces will accelerate drying of those surfaces, if the moisture content of the air is low. Ventilation from openings in the architectural shell, ceiling fans, exhaust fans, and whole-house air-exchange systems also serves to dry interior surfaces and products.

A preventive approach to curbing the presence of mold nutrients inside the home includes specification of nonporous and nonfleecy, low-texture materials that cannot harbor dust, moisture, or nutrients.[29–31] Examples of mold-resistant materials for interior environments include glass, metal, concrete (Fig. 2), and ceramic.[23] These materials, as well as wicker and cane, have low moisture absorption capability. The specification of continuous surfaces reduces crevices that entrap dust, moisture, and nutrients. Continuous surfaces include soffits built over wall cabinetry (Fig. 2) and tight-fitting countertops with backsplashes.[1,23] Enclosing objects that have a high surface-to-volume ratio reduces the surface area that can gather dust, moisture, and nutrients. Means of enclosure include glass-fronted bookcases (Fig. 2) and etageres for holding personal effects, windows with movable blinds sealed between panes of glass (Fig. 2), and cornices that cover the tops of window treatments.[32,33] Removing sources of moisture and nutrient matter by cleaning surfaces and laundering textile products also will prevent microbial growth.[34,35] Effective means for cleaning surfaces include wiping bathroom walls, laundering slipcovers, and using bedding that can have all layers laundered.[28,30,32,33] Upholstery with open construction allows for air movement and dissipation of body heat (Fig. 2).

Unfortunately, source control best practices can adversely affect the chemically sensitive building occupant. Desiccants are useful for absorbing moisture from cellulose materials (e.g., books, textiles) stored in small, confined places that are difficult to ventilate.[27] However, desiccants contain chemical agents that may be toxic to children, pets, and the chemically sensitive. Although elimination of fleecy, highly textured, and porous surfaces in the interior environment prevents the entrapment of volatile organic compounds (VOCs) (i.e., the "sink effect") and sources of allergens other than molds (e.g., animal dander and dust mites[1,23]), some materials that are valued for their low porosity or low texture may produce allergic or toxic responses in the chemically sensitive individual. For example, terpenes in soft woods; curing agents added to concrete; phthalates in soft plastics; fungicides applied to wicker, reeds, and grasses; and fungicides added to grout preclude use of these products in indoor environments.[1,36,37] Fleecy materials, such as textile products, are desirable for floor and window covering because they absorb transmission of sound; however, products used should be capable of withstanding frequent and thorough cleaning and drying to prevent the collection of dust, nutrient matter, and moisture.

Open closets expose the chemically sensitive person to chemicals found in fabric finishes, residues of cleaning solvents, and contaminants adhering to footwear and outdoor apparel. If closets are not enclosed, dressing areas should be isolated from living areas and, in particular, from rooms with moisture-laden air (e.g., bathrooms and laundry rooms). Dedicated exhaust ventilation can be used to expel contaminants before they enter the breathing zones of indoor living areas.

Separation and filtration. The growth of molds, and the release of mycotoxins into indoor air, can also be controlled by separating molds and their spores from the occupancy areas in a home.[16] The use of barriers is one method of separation. Barriers may take the form of air pressure differences that prevent the infiltration of ambient pollutants, interior rooms that serve as buffers between outdoor pollutants and indoor living areas, and filters added to mechanical systems (Table 2).

Both molds and their airborne spores can exist anywhere outside a structure where there is moisture and nutrient matter for growth, and protection from the UV rays of the sun. Pressurization of the interior with a central heating, ventilation, and air conditioning (HVAC) system can provide a barrier to the infiltration of ambient airborne pollutants and dormant spores within cavities of the architectural shell.[16,17] Mechanical ventilation systems must run continuously to provide positive pressure and to balance negative pressure resulting from the operation of local ventilation used to expel indoor moisture and pollutants.

A 2nd method of separation is to provide transitional spaces that isolate incoming unfiltered air from primary living areas (e.g., a vestibule, air-lock entrance, mudroom, or breezeway) (Fig. 1).[29] In the manner of the traditional farmhouse mudroom, these spaces provide a transitional zone for shedding contaminants that have attached to shoes and clothing.[1,37]

A 3rd method of separation combines filters with a ventilation system to clean the outdoor air before a controlled ventilation system brings the air indoors. Some general ventilation systems have built-in filters.[16] In addition, portable room-sized filters can be used to clean the air after it is brought indoors. Filters come in a vari-

Table 2.—Separation and Filtration Design Strategies to Promote Mold-Free Environments

Environment	Strategy and effect
Mechanical systems for heating, ventilation, and air conditioning (HVAC)	• Pressurizing the interior with central air systems will prevent infiltration of ambient pollutants and dormant spores within cavities of the architectural shell. • Providing filters in HVAC systems will remove bioaerosols from indoor air.
Interior space and furniture plan	• Providing transitional zones (e.g., vestibules, air lock entries, mudrooms, or breezeways) to shed contaminants from outer garments and footwear will reduce amounts of airborne contaminants entering living areas.

Table 3.—Ventilation Design Strategies to Promote Mold-Free Environments

Environment	Strategy and effect
Building site and architectural shell	• Arranging fenestration and interior partitions to facilitate natural ventilation will dry moisture-laden areas and exhaust indoor contaminants. • Arranging air inlets and outlets to facilitate cross-ventilation will dry moisture-laden stored items and exhaust indoor contaminants.
Mechanical systems for heating, ventilation, and air conditioning (HVAC)	• Specifying fans will exhaust moisture-laden air prevalent in rooms such as kitchens, bathrooms, hobby rooms, and utility rooms. • Specifying air-to-air exchanger units will filter incoming air for interior living spaces and exhaust contaminants at their source.

ety of types, depending on the kind of pollutant to be removed. High-performance filters in HVAC systems can remove solids (e.g., bioaerosols) as well as gaseous pollutants from indoor air.[17,29] Impingement filters remove solids of different sizes; absorber filters remove gases or vapors. Filters vary in their ability to remove varying sizes of solids.[16,17]

Again, separation and filtration best practices may adversely affect the chemically sensitive individual. The necessary degree of separation from environmental irritants is relative to the susceptibility of the occupant and the duration of exposure. In spaces such as sleeping areas, in which building occupants experience many hours of exposure to indoor air, it is critical that biological and chemical irritants be filtered from indoor air. Many filters are not sufficiently effective to address the level of susceptibility of occupants who are chemically sensitized. It is desirable to isolate sources of contamination far away from indoor spaces that are occupied for extended periods of time.[1,29,37]

Ventilation. In addition to providing every indoor living space with a constant supply of fresh air, general ventilation—both natural and mechanical—serves to exhaust airborne mold spores and other contaminants from inside the house (Table 3). Not all moisture and nutrient matter can be controlled through source control, separation, and filtration.[16,18] The byproducts of metabolism (e.g., breathing and perspiration) and activities of building occupants (e.g., bathing, cooking, laundering) provide a constant source of moisture and organic matter to attach to interior surfaces and become airborne in interior living spaces. Ventilation strategies provide for controlled movement of air between the outdoors and indoors, which influences the movement of mold spores and the drying of moisture-laden surfaces. Strategies include architectural decisions that result in passive air movement, and engineering decisions that rely on mechanical fans and blowers to supply and circulate fresh air and exhaust polluted air.

"Natural ventilation" involves the manipulation of principles of passive convection to facilitate air movement inside a building, without the use of fossil-fuel–driven mechanical equipment.[16] Orientation of a building to prevailing winds, and alignment of exterior wall openings with interior wall openings, can facilitate cross-ventilation and negative air pressure, thus creating an exchange of air with the outdoors.[15,38,39] This form of uncontrolled ventilation includes the use of vents in the roof, closets, and storage cabinets to enable passive air movement to exhaust pollutants from the indoor breathing zone.[1] If ambient air contains low levels of mold spores and moisture, and the architectural shell is airtight, designing the building to take advantage of passive convection is an energy-efficient approach to expelling mold spores found in indoor air.

Mechanical ventilation systems provide both uncontrolled and controlled circulation of indoor air.[16,17] Uncontrolled mechanical ventilation includes the use of portable fans and stationary window and ceiling fans to circulate air within a room (Fig. 3). This form of ventilation facilitates air movement to retard microbial growth; however, it does not exhaust mold spores from indoor living areas. Controlled ventilation is necessary to expel indoor pollutants, moisture, and odors to the outdoors, and to provide a source of fresh air.[40] It involves the manipulation of air pressure through mechanical systems specifically designed to ventilate houses by using controlled pressures.[16] Controlled ventilation includes mechanical systems that provide positive air-pressure general ventilation, negative air-pressure exhaust ventilation, and neutral air-pressure balanced ventilation via exchange of air inside the building. Central HVAC systems can exhaust contaminants and moisture-laden air from all rooms. Fans to exhaust moisture-laden air prevalent in rooms such as kitchens, bathrooms, hobby rooms, and utility rooms remove contaminants at their source. Air-to-air exchanger units can be used to filter incoming air for interior living spaces and to exhaust contaminants at their source (Fig. 3).

The chemically sensitive individual may benefit from the isolation of molds found in the ambient environment. Such measures prevent additional biological and chemical irritants from infiltrating interior living areas; however, depending on the air-tightness of the architectural shell, some pollution is unavoidable. A combination of source control to prevent the build-up of most pollutants, and filters to clean up minor pollutants, is also helpful to building occupants with chemical sensitivities. Ideally, ventilation controls should be coupled with separation strategies to exhaust the build-up of all biological and toxic irritants found within the residential environment.

Once again, some ventilation best practices can pose problems for the chemically sensitive occupant. Exchange of fresh air to replace exhausted air is critical; however, decreases in air pressure inside the home can cause outside air to be pulled into the home through leaks and other openings, exposing occupants to allergens and toxins found within the architectural shell. In airtight homes, this depressurization can also result in backdrafting of the furnace and other combustion appliances, thereby exposing building occupants to carbon monoxide and other dangerous gases that are pulled back in to the house rather than being expelled up the chimney.[40] An optimal ventilation system for the chemically sensitive resident provides a continuous supply of fresh, filtered air to replace exhausted air, while maintaining neutral air pressure.[16,29]

Perspectives of the Interior Design Community

Interior design education and interior design practice contribute to both the prevention and remediation of mold growth in indoor environments. Our review of current literature points to 3 prevailing philosophical perspectives that influence approaches to achieving mold-free indoor environments: (1) the traditional materials philosophy, (2) the ecologically sound materials (eco-materials) philosophy, and (3) the human ecology philosophy.

Traditional materials philosophy. A traditional materials specialist's philosophy—which is indicative of mainstream interior design practice—values materials for their appearance, durability, availability, and adherence to fire and safety codes.[41] Designers who subscribe to this philosophy focus on client considerations and preferences, sometimes to the exclusion of ecological considerations. If cost is a consideration, the expectation is that designing with ecologically sensitive products will increase the cost of the project.[42] Historically, the knowledge underlying interior design decisions has been held by an elite group and applied on behalf of an elite clientele.

Eco-materials philosophy. This view, which is of growing interest in the "green design" and "sustainable design" movements, values materials for their environmental sensitivity over availability, convenience, or appearance.[42] This philosophy embraces an ecology-centered approach that expands traditional measures of building performance (e.g., cost per square foot, minimum standards for air quality) to include issues of occupant comfort, satisfaction, and overall well-being. Health factors affected by material selections include sensitivity to VOC emissions and susceptibility to allergens and toxins. Long-term environmental impact is considered in tandem with up-front building costs.[43,44] Knowledge based on product quality can lead to changes in the manufacture and composition of interior products, and in the specification of products by interior designers.[45,46] An intermediate level of understanding explores creative solutions for the development of new processes and products of the material world that give equal value to preservation of the natural environment and to human health.[46] The highest level of understanding explores a preventive health model as an evolutionary step in material culture and interior design—empowering the client to monitor their well-being relative to design decisions that affect both the natural environment and human health.

Human ecology philosophy. A human ecology philosophy is a "cooperative extension" approach to design that values materials for their durability and for wise use of natural resources, within a tradition of supporting health and wellness within the family. This philosophy embraces an holistic approach to design decisions that considers both the environment and human needs. Knowledge based on science, and on best practices developed by federal and state governments (e.g., programs addressing disaster relief for flooding), is translated to layman's language to empower the consumer.

Certification Programs and Consultancies

Certification programs and consultancies have emerged in the fields of building science and industrial hygiene. Current efforts to develop certification programs that rate building materials and interior products include the Leadership in Energy and Environmental Design (LEED) Program and the GREENGUARD Certification Program. The LEED Program is a voluntary certification program that analyzes the content of building materials used in new facilities.[47,48] From its beginning with the evaluation of exterior materials, the program is currently focusing on interior materials for commercial buildings, and eventually will develop a materials rating system for the remodeling of older buildings. Criteria for this program include recycled content, energy efficiency, and minimization of VOC emissions. The GREENGUARD Certification Program (Air Quality Sciences, Inc. [AQS] [Marietta, Georgia]) rates building materials and furnishings in accordance with performance-based standards that define goods with low chemical and particulate emissions for use indoors.[49] In addition, AQS provides the following services to address the causes and heath effects of indoor air quality problems related to molds and mycotoxins: emissions testing, building investigations, laboratory services, and consulting for design and construction. McDonough Braungart Design Chemistry (Charlottesville, Virginia) and the Environmental Protection Encouragement Agency (Hamburg, Germany) provide consulting services on the chemical composition of interior products from the perspective of a "cradle-to-cradle" impact on the natural environment and the wellness of building occupants.[45,46] Product certification programs and consultancies provide critical resources for the design profession on building projects that require mold remediation, and for clients who have biological and chemical sensitivities.

Conclusions

Our search of current literature pertaining to interior design education and practice points to a need for interior designers to be educated from an eco-materialism perspective, which links the selection of indoor materials with building occupant health through concern for maintaining good air quality in the indoor environment.[2] Professionals will benefit from the development of a design model that reflects a fusion of values indicative of an aesthetic tradition (beautiful design), a science of product quality evolving from the environmental movement (contaminant-free design), and the cooperative extension tradition of empowering individuals and families (user-driven design). To support this model, a more comprehensive methodology for the evaluation of building performance is needed—one that includes measures of occupant health and well-being, as well as lifecycle analyses of interior products that measure their long-term impact on the natural environment.

The implications of this review of literature for interior design education and practice include the need to address the following: (a) an increased interest in the human health effects of mold and mycotoxin infiltration into the interior architectural shell; (b) an increased interest in methods of mold prevention by interior design educators, by those in professional practice, and by agricultural extension specialists; (c) increased demand for information on the damaging effects of the failure to address design and material issues pertaining to molds and mycotoxins in building interiors and the environment in general; (d) increased demand for a bibliographical source for information on mold and mycotoxin issues; (e) a need for information on legislation involving mold and mycotoxin issues; (f) increased demand for the development of a process to empower clientele with knowledge of the medical and environmental implications of interior design decisions; (g) increased demand for information on sources for interior materials certified in regard to their impact on the environment and human health; (h) increased demand for the development of a method for evaluating building performance; and (i) development of a design philosophy that merges aesthetic comfort with human and environmental health concerns.

* * * * * * * * * *

This document does not express East Carolina University (ECU) or U.S. Environmental Protection Agency (EPA) recommendations. It is intended to raise general points for consideration, not to dictate a prescribed course of action. All actual applications must include consideration of individual project conditions, specific requirements, and limitations. In the end, designing for good indoor air quality and mold-free environments entails using professional judgment to determine which issues are important and how each issue should be pursued.

Views expressed by the authors do not necessarily reflect ECU or EPA policy. ECU and EPA in no way warrant the accuracy or reliability of the resources mentioned. Users of this information accept all responsibility for examining the validity and appropriateness of these resources for their own purposes.

Katherine Warsco, Ph.D., IIDA, IDEC, NKBA, AFCS, was project director for the environmental education project sponsored by East Carolina University, Greenville, North Carolina, and the U.S. EPA Environmental Education Grants Program, Washington, D.C., entitled Designing for Good Indoor Air Quality in Hot Humid Climates: Student Design Competition as a Teaching Tool. Dr. Warsco received financial support for this project during 1996–1998 from the EPA and ECU. Excerpts from student competition submissions were used in this article to illustrate design strategies for mold-free residential environments.

The following consultants provided content guidance in the development of the student design competition and teaching guide. In addition, they served as jurors for the review of student design competition submissions: Delores Ginthner, IDEC, IES, University of Minnesota, St. Paul, Minnesota; Robert Kobet, ASHRAE, AIA, Carnegie Melon University, Pittsburgh, Pennsylvania; William Rea, M.D., Environmental Health Center–Dallas, Dallas, Texas; and Henry Slack, PE, EPA Region IV, Office of Radiation, Air, Pesticides, and Toxics Management Division, Atlanta, Georgia.

The research assistant for this environmental education project was Michelle Puckett (now Michelle Puckett Jenkins), East Carolina University, Greenville, North Carolina. Ms. Puckett Jenkins provided, for inclusion in the teaching guide, examples of programming documents and schematic designs to illustrate principles of design for good indoor air quality.

Submitted for publication September 20, 2003; revised; accepted for publication October 30, 2003.

Requests for reprints should be sent to Katherine Warsco, 152 Rivers Building, East Carolina University, Greenville, NC 28858-4353.

E-mail: warscok@mail.ecu.edu

* * * * * * * * * *

References

1. Rea WJ. Optimum Environments for Optimum Heath and Creativity. Dallas, TX: Crown Press, 2002.
2. Hasell MJ, Scott SC. Interior design visionaries' explorations of emerging trends. J Interior Design 1996; 22(2):1–14.
3. Gist GL. Multiple chemical sensitivity—the role of environmental health professionals. J Environ Health 1999; 61 (6):4.
4. Aerias. Glossary of Indoor Air Quality Terms. <http://www.aerias.org/glossary.asp> Accessed 2003 Jul 8.
5. Warsco K. Designing for Good Indoor Air Quality in a Hot, Humid Climate: Student Design Competition [unpublished competition materials]. Prepared for the U.S. Environmental Protection Agency, Environmental Education Grants Program, Washington, DC, 1997.
6. Warsco K. Teaching supplement to the student design competition: Designing for Good Indoor Air Quality in a Hot, Humid Climate [unpublished supplement to competition materials]. Prepared for the U.S. Environmental Protection Agency, Environmental Education Grants Program, Washington, DC, 1997.
7. Pilatowicz G. Eco-interiors: A Guide to Environmentally Conscious Interior Design. New York: Wiley, 1995.
8. Warsco K. Designing for good indoor air quality in hot, humid climates: development of a student design competition. J Housing and Society 1999; 26:41–53.
9. Inman M, Shea J. Housing problems for older adult households in the southeast. Proceedings of the Annual Meeting of the American Association of Housing Educators; 1993 Oct; Columbus, OH. College Station, TX: Texas A&M University College of Architecture, 1993.
10. Zummos SM, Karol MH. Indoor air pollution: acute adverse health effects and host susceptibility. Environ Health 1996; 58(Jan–Feb):25–29.
11. U.S. Environmental Protection Agency, Consumer Product Safety Commission, American Medical Association. Trends in cigarette smoking. MMWR 1998; 47(43):1.
12. Browner CM. Environmental tobacco smoke: EPA's report. EPA Journal 1993; 4:18–22.
13. Kloeppel JE. Beware the fungus among us: emissions from mold and fungus may be culprits in indoor air problems. Georgia Tech News 1993; Apr:1–4.
14. Peart V. Indoor air quality in Florida: formaldehyde. Gainesville, FL: University of Florida Cooperative Extension Service, 1993; Fact Sheet He 3205.
15. Lechner N. Heating, Cooling, Lighting: Design Methods for Architects. New York: Wiley, 1991.
16. Bower J. Understanding Ventilation. Bloomington, IN: The Healthy House Institute, 1995.
17. Odom JD, DuBose G. Preventing indoor air quality problems in hot, humid climates: design and construction guidelines [unpublished manuscript]. Orlando, FL: CH2M Hill, Inc., Disney Development Co., 1996.
18. Lstiburek J, Carmody J. Moisture Control Handbook: Principles for Residential and Small Commercial Buildings. New York: Van Nostrand Reinhold, 1993.
19. Spaul WA. Building-related factors to consider in indoor air quality evaluations. J Allergy Clin Immunol 1994; 94: 385–89.
20. Danko S, Eshelman P, Hedge A. A taxonomy of health, safety, and welfare implications of interior design decisions. J Interior Design Education and Research 1990; 16 (2):19–30.
21. Dearborn DG, Smith PG, Dahms BB, et al. Clinical profile of 30 infants with acute pulmonary hemorrhage in Cleveland. Pediatrics 2002; 110:627–37.
22. Etzel RA, Balk SJ, Bearer CF, et al. Toxic effects of indoor molds. Pediatrics 1998; 101:712–14.
23. Bower J. Healthy House Building for the New Millennium. Bloomington, IN: The Healthy House Institute, 2000.
24. Speigel R, Meadows D. Green Building Materials: A Guide to Product Selection and Specification. New York: Wiley, 2003.
25. Institute of Food and Agricultural Sciences (IFAS). A guide to environmentally friendly landscaping: Florida yards and neighborhood handbook. Gainesville, FL: IFAS, Cooperative Extension Service, University of Florida, November 1994; Bulletin 295, 56 pp.
26. U.S. Environmental Protection Agency. A brief guide to mold, moisture, and your home. <http://www.epa.gov/iaq/pubs> Accessed 2002 Nov 11.
27. Minnesota Department of Commerce, Energy Information Center. Home moisture. <http://www.state.mn.us/mn/externalDocs/Home_Moisture_110802041415_Home Moisture05-03.pdf> Accessed 2002 Nov 11.
28. Bode M, Munson D. Controlling mold growth in the home [guidance document]. Kansas State University Agricultural Experiment Station and Cooperative Extension Service, 1995. <http://www.oznet.ksu.edu/library/hous2/mf2141.pdf> Accessed 2002 Nov 11.
29. Rousseau D, Wasley J. Healthy by Design: Building and Remodeling Solutions for Creating Healthy Homes. Point Roberts, WA: Hartley & Marks, 1999.
30. Aerias. Carpet, a haven for unwanted guests. <http://www.aerias.org/c_Doc_124.htm> Accessed 2002 Oct 11.
31. Lewis RD, Breysse PN, Lees PSJ, et al. Factors affecting the retention of dust mite allergen on carpet. Amer Ind Hyg Assoc J 1998; 59(9):606–13.
32. Aerias. Dust mites, the unwanted houseguests. <http://www.aerias.org/c_Doc_36.htm> Accessed 2002 Oct 11.
33. Fernández-Caldas E, Trudeau WL, Ledford DK. Environmental control of indoor biologic agents. J Allergy Clin Immunol 1994; 94(2):404–12.
34. Aerias. Textiles, upholstered furniture, and other fleecy materials. <http://www.aerias.org/c_Doc_104.htm> Accessed 11 Oct 2002.
35. Frank D, Northern C, Black M. Furnishing and the indoor environment. J Textile Institute 1994; 85(4):496–504.
36. Fuston A, Nadel KP. Creating nontoxic, health-enhancing environments. In: Marberry SO (Ed). Healthcare Design. New York: Wiley, 1997; pp 232–54.
37. Rousseau D, Rea WJ, Enwright J. Your Home, Your Health and Well-Being. Berkeley, CA: Ten Speed Press, 1990.
38. Haase RW. Classic Cracker: Florida's Wood-Frame Vernacular Architecture. Sarasota, FL: Pineapple Press, 1992.
39. Kibert CJE. Sustainable construction code—residential: for the Abacoa development [unpublished manuscript]. Gainesville, FL: University of Florida Center for Construction and Environment, 1996.
40. Minnesota Department of Commerce, Energy Information Center. Home Energy Guide: Indoor ventilation. <http://www.state.mn.us/mn/externalDocs/Indoor_Ventilation_110802042008_IndoorVent05-03.pdf> Accessed 2002 Nov 11.
41. Nielson K, Taylor D. Interiors: An Introduction. Boston: McGraw Hill, 2002.

42. Beecher MA, Davies B. Shades of green: the philosophical challenges of ecological responsibility in interior design education and practice. Proceedings of the Annual Meeting of Interior Design Educators Council: Mesas and the Mysteries: On the Edge of Imagination/Green Design; 19–24 Mar 2002; Santa Fe, NM.
43. Moussatche H, King J, Rogers TS. Material selection in interior design practice. Proceedings of the Annual Meeting of the Interior Design Educators Council: Mesas and the Mysteries: On the Edge of Imagination/Green Design; 19–24 Mar 2002; Santa Fe, NM; pp 26–27.
44. Moussatche H, Languel J. Life cycle costing of interior materials for Florida's schools. J Interior Design 2002; 28(2):37–48.
45. McDonough W, Braungart M. Cradle to Cradle. New York: North Point Press, 2002.
46. McDonough W, Braungart M. Redefining green: a new definition of quality empowers the next wave of design. Perspective 2003; Spring:20–25.
47. U.S. Green Building Council. Leadership in energy and environmental design. <http://www.usgbc.org/LEED/publications.asp> Accessed 2002 Nov 11.
48. Mendler S. LEED: a roadmap for added value. Perspective 2002; Winter:42–49.
49. GREENGUARD Registry of Healthy Indoor Materials. Air Quality Sciences, Inc., Marietta, Georgia. <http://www.aerias.org/GreenGrd.pdf> Accessed 2002 Oct 11.

Creating Mold-Free Buildings: A Key to Avoiding Health Effects of Indoor Molds

BRUCE M. SMALL
Foundation for Independent Research on Technology and Health
Georgetown, Ontario, Canada

ABSTRACT. In view of the high costs of building diagnostics and repair subsequent to water damage—as well as the large medical diagnostic and healthcare costs associated with mold growth in buildings—commitment to a philosophy of proactive preventive maintenance for home, apartment, school, and commercial buildings could result in considerable cost savings and avoidance of major health problems among building occupants. The author identifies common causes of mold growth in buildings and summarizes key building design and construction principles essential for preventing mold contamination indoors. Physicians and healthcare workers must be made aware of conditions within buildings that can give rise to mold growth, and of resulting health problems. Timely advice provided to patients already sensitized by exposure to molds could save these individuals, and their families, from further exposures as a result of inadequate building maintenance or an inappropriate choice of replacement housing.

<Key words: asthma, buildings, condensation, dampness, moisture transport, molds, health, water damage>

READERS of the *Archives of Environmental Health* have been alerted to the possibility of human illness syndromes related to indoor mold growth in buildings.[1] While further research is under way to fill the many gaps in the science connecting indoor mold exposure with ill health, it is important for architects, engineers, builders, and others associated with the built environment today to apply appropriate building science to prevent and mitigate mold growth of all kinds in buildings. The National Association of Home Builders has stated that growth of any type of mold in a home is never acceptable.[2] Stephen J. Redd of the U.S. Centers for Disease Control and Prevention (CDC) has stated:

> *While there remain many unresolved scientific questions, we do know that exposure to high levels of molds causes some illnesses in susceptible people. Because molds can be harmful, it is important to maintain buildings, prevent water damage and mold growth, and clean up moldy materials. . . . Because molds can be harmful, CDC concurs with the general recommendations of agencies such as EPA [U.S. Environmental Protection Agency] and FEMA [Federal Emergency Management Agency], which offer information on preventing and cleaning up mold growth in indoor environments.*[3]

The discussion that follows identifies common causes of mold growth in buildings and summarizes key building principles that are essential to preventing mold contamination indoors.

Dampness

By far the most common risk for deterioration of housing is that of damage by water and water vapor. Surveys in Scotland and England have estimated that between one-quarter and one-third of homes may be affected to some degree by condensation and dampness.[4] In a study of 30 Canadian communities, which focused on the association between the respiratory health of young children and home dampness and molds, flooding was reported in 24.1% of the homes surveyed, molds were reported in 32.4%, and moisture in 14.1%, with an overall prevalence of home dampness or molds of 37.8%.[5]

Dampness in housing has been correlated with various adverse health symptoms, particularly of the respiratory system, including (a) exacerbation of asthma, (b) increased incidence of wheezing and other respiratory symptoms, (c) increased incidence of upper respiratory infections, and (d) inflammation of the airways. Conditions of dampness appear to worsen respiratory conditions both directly (e.g., the exacerbation of existing asthma) and indirectly (e.g., increased incidence of upper respiratory infections, which in turn can exacerbate existing asthma).[6] It is unclear which mechanisms involved in damp housing produce these health effects. Some evidence suggests that the growth of mold indoors—and subsequent human exposure to spores and metabolic gases, as well as to mold cell wall proteins (e.g., (1→3)-β-D-glucan) and mycotoxins—may be involved in the production of adverse health effects.[6]

A review of 9 population-based studies found 1 or more positive associations between mold levels and health outcomes, as measured by respiratory disease symptoms, in 7 of the 9 studies. The authors concluded that molds do contribute to allergic disease, and that the extent of their involvement is likely greater than is indicated by available clinical and epidemiological studies.[7]

Mold Growth

Molds (i.e., microfungi) are most common in situations where conditions indoors are very moist, either from condensation of airborne water vapor on building surfaces (e.g., on or within uninsulated walls), or from leakage of water as a result of rain or runoff entering the building, or from pipes and taps inside the building. Awareness of the damaging potential of mold growth has recently reached crisis proportions in the United States because of a proliferation of thousands of lawsuits related to moisture damage in buildings, mold growth, and subsequent adverse health effects in occupants. Many instances of mold growth in buildings have included the following sequence of events:

- Building failure (e.g., deterioration of caulking around a chimney flashing as a result of inadequate maintenance)
- Moisture penetration into the building (e.g., intrusion of rain through unsealed roof cracks)
- Moisture damage (e.g., rotting of structural wood members)
- Mold growth (e.g., on the paper layer of a drywall ceiling)
- Awareness of occupant illness
- Diagnosis of mold-related illness by a physician
- Submission of insurance claims to cover building damage and healthcare costs
- Hiring of an indoor air quality consultant to assess moisture and mold damage
- Hiring of a mold remediation firm to remove moldy materials
- Hiring of a clearance testing contractor to confirm adequate mold removal
- Hiring of a renovation firm to complete repairs and rebuilding.

In some cases, the building is rebuilt in the same manner that led to its failure in the first place, and mold growth recurs. In other cases, delays in insurance assessment have resulted in increased mold growth during the waiting period, thereby escalating costs and health effects. In cases where insurers have not provided adequate payment to cover damages, 2 more steps have occurred that have led to the current awareness of a "mold crisis": (1) the occupant launches a lawsuit against all parties previously involved, and (2) the occupant obtains press coverage in an attempt to force an adequate out-of-court settlement.

Prevention of Mold in Buildings

It is certainly possible to design a building that will not be susceptible to mold growth. Well-known building consultant Joe Lstiburek has emphasized: "Controlling moisture is the only effective strategy for preventing mold growth in our homes."[8] It is also important to make the building sufficiently moisture-tolerant and capable of drying in the event of a mishap, so that damage is minimal and mold does not grow. The key is to incorporate into the building design a practical "risk management strategy" for every possible way that moisture could enter the building and accumulate to damaging levels.

External flooding. Proper location, grading, and drainage of a home can minimize the chances that flash storms, overflowing rivers, hurricanes, and other acts of nature will force water into the home. The use of a sump pit and automatic pump is advisable in basements that are subject to a high water table.

Internal spillage. High-quality plumbing fixtures, materials, and installation practices, combined with regular and competent inspection and maintenance, can minimize the chance of water spilling from broken pipes, malfunctioning washing machines, or overflowing sinks, or splashing past shower curtains. One example is the use of high-strength, steel-reinforced water supply hoses on washing machines. Such hoses are designed to withstand water pressure if the water is left on continuously. (Older style rubber hoses are notorious for bursting, particularly while the homeowner is away for the weekend.)

Rain leakage. Meticulous design and installation of roofing, windows, and siding, as well as adequate gutter drainage, site contouring, and foundation drainage, are all effective risk management strategies and good building practices to prevent rainwater from penetrating the building. Roofs must be installed with proper flashing and appropriate slope. Effective drainage and waterproofing are also essential in those areas in which basements are used. A properly designed and installed "drainage plane" behind the exterior building cladding ensures that any water forced into the wall from outside can find a path downward and outward, without penetrating to the interior of the building.

Capillary action. Water can be drawn into a building by capillary action (e.g., through brick in a driving rain). Again, the use of a proper drainage plane, including weep holes and ventilation behind exterior surfaces, can help ensure that any water that does penetrate the skin of the building is redirected out again.

Vapor transport. One of the strongest mechanisms for moisture transport is simply water vapor entering or leaving a building on an air draft, through leaks in the building envelope. Easily 100 times the amount of moisture that can penetrate a wall by diffusion can be blown back and forth across a wall by wind pressure and stack effect. Low-emission, soft foam insulation, combined with appropriate caulking (e.g., under wall base-plates), can be used as an effective air barrier to prevent moisture transport and avoid condensation and mold growth within the building envelope (Fig. 1).[9]

Diffusion. Although far less important than bulk moisture transport on air leakage, diffusion of water molecules directly through building materials still must be accounted for in building design. Proper placement of vapor-retardant materials—on the inside of a wall in a predominantly cold climate, and on the outside of a wall in a predominantly warm and humid climate—is essential in order to avoid condensation within walls when the vapor pressure is persistent in one direction. Moisture diffuses outward in climates in which heating is used, and inward in climates that require air conditioning. If the proper sequence of vapor-permeable materials is installed, vapor that does diffuse into a wall can continue through safely to the drying side (e.g., to the air-conditioned interior in hot climates, and to the outside winter air in cold climates). In mixed climates (those with cool winters and warm summers), it may be preferable to omit the vapor retarder, as long as the wall is well sealed.

Fig. 1. Low-emission insulation can serve as an effective barrier to prevent moisture transport via wall air leaks.

Standing water. Moisture can also accumulate in specific places within a home. For example, improperly drained drip pans on air conditioning units, and standing water in humidifiers and dehumidifiers, can grow bacteria or mold. Open earth crawl spaces are notorious as sources of standing moisture that can introduce excessive moisture into houses. Use of proper slopes, provision for continuous drainage, covering of crawl-space floors, and frequent inspection and maintenance are all important in these situations.

Surface condensation. Any moisture in a home will evaporate into the air. As long as the relative humidity remains in a tolerable range (30%–50%), few problems occur. However, if the humidity exceeds 60% conditions that support mold growth may develop in some materials. If the humid air contacts a colder surface (e.g., a poorly insulated area at the top of an outside wall), it can release its moisture as condensation on the surface. If this condition persists, mold can begin to grow. Sometimes, air leakage around clothes in a cupboard, into a wall through leaky outlets, or through wall insulation can bring evaporated moisture into contact with very cold surfaces, on which condensation and mold growth can form. Consistent insulation, air sealing of walls and ceilings, and appropriate whole-house ventilation and moisture exhaust can reduce or eliminate these risks.

Building Design Matched to Climate

It has become clear that building design must be specific to the climatic conditions of a given region. Examples abound of "imported" building styles that do not function well in new climates. For example, ventilated attics are used in many cooler-climate areas of North America to permit easy conduct to the outside of water vapor that has leaked from the interior of a building. Cold, dry, winter air flows into the attic area from eave venting and out through vents spaced along the roof. When the same ventilated attic design is used in warmer-climate areas, such as southern Florida, condensation and mold can result because the ventilation air is warm and very humid.

Combine the above factors with the tendency to install uninsulated air conditioning equipment and ducting in Florida attics, and the result is a built-in mechanism for mold growth. The warm, humid outside air drawn into the attic reaches the cool, metal surfaces of the air conditioning ducts and condenses. Exposure of uninsulated ducting to humid air, and the resultant condensation, renders the air conditioning highly inefficient. In addition, the accumulated water drips onto the ceiling insulation between the attic and the living space below, feeding mold growth on the insulation and the drywall beneath it (Fig. 2).

In response to an increased occurrence of moldy attics in Florida, a new technique is now being adopted that involves the "cathedralization" of such attic spaces. This effectively converts the attic into air-conditioned space by placing air-sealing insulation on the bottom side of the roof deck, between the roof rafters, instead of on the floor of the attic. Once this is done, the attic requires no outside ventilation and can be fully finished for storage or human use. The efficiency of the air conditioning system is increased, and the source of moisture and condensation—and resulting mold growth—is effectively removed.

Fig. 2. Mold often forms on the cellulose layer in moist drywall.

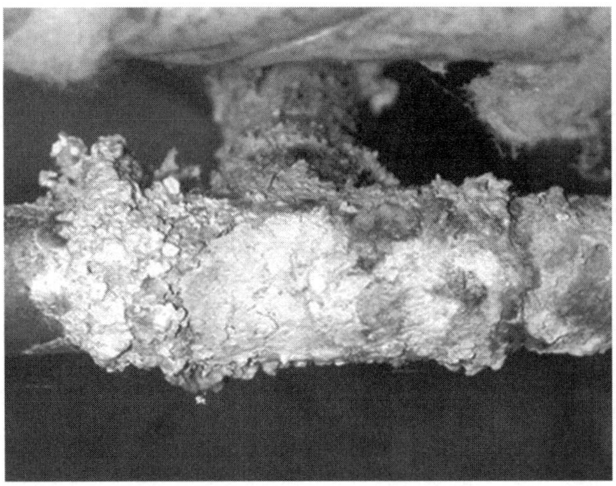

Fig. 3. Many homeowners rely on "just-in-time" repairs like this, which can lead to more serious deterioration and costly repairs.

"Just-in-Time" Maintenance

Homeowners have, for various reasons, developed a prevailing philosophy of "just-in-time" maintenance, which is based on the hope that there will still be time to repair something that has been obviously failing, before irreparable damage is caused. For example, a corroded pipe with a slow leak will be left unrepaired as long as a simple bucket below the leak keeps the problem under control. Unfortunately, it is common for this kind of deterioration to lead to more major water leaks before repairs are mounted (Fig. 3).

In view of the devastating costs of building diagnostics and repair—as well as the potentially large medical diagnostic and healthcare costs associated with mold growth in buildings—commitment to a philosophy of proactive preventive maintenance for home, apartment, school, and commercial buildings could result in considerable cost savings and avoid major health problems among building occupants.

Conclusions

Physicians and healthcare workers must be made aware of conditions within buildings that can give rise to mold growth, and of resulting health problems. Timely advice provided to patients already sensitized by exposure to molds could save these individuals, and their family members, from further exposures as a result of inadequate building maintenance or an inappropriate choice of replacement housing.

* * * * * * * * * *

Submitted for publication September 20, 2003; revised; accepted for publication October 30, 2003.

Requests for reprints should be sent to Bruce M. Small, Executive Director, Foundation for Independent Research on Technology and Health, 100 Rexway Drive, Georgetown, Ontario, Canada L7G 1R5.

E-mail: enquiries@envirodesic.com

* * * * * * * * * *

References

1. Kilburn KH. Janus revisited, molds again. Arch Environ Health 2002; 57(1):7–8.
2. National Association of Home Builders. 2002. Get the facts on mold. <http://store.builderbooks.com/cgi-bin/builderbooks/398?id=hDJECdBm&mv_pc=101>
3. Redd SC. 2002 Jul 18. Summary: state of the science on molds and human health. Statement for the Record before the Committee on Financial Services, U.S. House of Representatives. <http://www.cdc.gov/nceh/airpollution/images/moldsci.pdf>
4. Strachan DP, Sanders CH. Damp housing and childhood asthma; respiratory effects of indoor air temperature and relative humidity. J Epidemiol Community Health 1989; 43:7–14.
5. Dales RE, Zwanenburg H, Burnett R, et al. Respiratory health effects of home dampness and molds among

Canadian children. Am J Epidemiol 1991; 134(2): 196–203.
6. Ontario Lung Association. 2002 Apr. Indoor air pollutants in residential settings: respiratory health effects and remedial measures to minimize exposure. <http://www.on.lung.ca/cando/IAPRS_contents.html>
7. Verhoeff AP, Burge HA. Health risk assessment of fungi in home environments. Ann Allergy Asthma Immunol 1997; 78(6):544–56.
8. Lstiburek J. 2002 Jun (NWReporter). The National Green Building Conference, Seattle, Washington, 24–26 March 2002. <http://www.nahbrc.org/tertiaryR.asp?DocumentID=3621&CategoryID=1676>
9. Small BM. 2001 Nov/Dec. Creating airtight and healthy homes. Home Energy Magazine Online. <http://www.envirodesic.com/sfl/icynene1.htm>

Lipoic Acid as a Potential First Agent for Protection from Mycotoxins and Treatment of Mycotoxicosis

SHERRY A. ROGERS
Northeast Center for Environmental Medicine
Sarasota, Florida

ABSTRACT. Mycotoxins—toxic substances produced by fungi or molds—are ubiquitous in the environment and are capable of damaging multiple biochemical mechanisms, resulting in a variety of human symptoms referred to collectively as "mycotoxicosis." In fact, mycotoxins mimic multiple xenobiotics, not only with respect to their ultimate damage, but also in their routes of detoxification. This suggests potential therapeutic options for the challenging treatment of mycotoxicosis. In this brief review, the author examines the use of lipoic acid as an example of an inexpensive and available nutrient that has been shown to protect against, or reverse, the adverse health effects of mycotoxins.
<Key words: lipoic acid, mycotoxicosis, thioctic acid, xenobiotic detoxification>

MYCOTOXINS are the potentially toxic and immunogenic byproducts of the life cycles of certain fungi or molds. Exposure to mycotoxins is a relatively unsuspected cause of a wide range of symptoms that, in humans, are collectively termed "mycotoxicosis." Given that lipoic acid has a unique record for countering various forms of mycotoxins when all else in medicine has failed, a review of its actions can set the stage for increasing our knowledge of its effectiveness against this ever-expanding aspect of environmentally caused diseases.

Ubiquitous, unavoidable airborne molds cause many symptoms and diseases—from common allergies of the respiratory tract to a variety of less-well-defined syndromes (e.g., Sick Building Syndrome).[1–3] We can sometimes differentiate symptoms caused by ambient molds from those caused by environmental chemicals by means of an environmental history, or with environmental (home and office) mold testing[4] and intradermal testing.[5,6] However, fungi can also produce dangerous mycotoxins that we do not test for intradermally. Also prevalent is dietary exposure to a wide range of mycotoxins, which is heavily influenced by improper harvesting and storage of foodstuffs.[7] Mycotoxins can even be passed to man through, for example, the consumption of meat and milk from animals that have consumed unsuspected mycotoxin-contaminated feed.[8]

In addition, fungi grow in the human body as a secondary effect of medications such as antibiotics and immunosuppressives. For example, the widespread use of inhaled steroids for simple nasal allergies, to complex asthma, fosters the growth of *Candida*.[9] Given that mycotoxins damage human immune cells in culture[10,11] and are also potent carcinogens,[12–18] there is a growing need to explore the use of inexpensive and readily available nutrients that will block the effects of unsuspected mycotoxins from multiple environmental sources (i.e., air and food).

As knowledge of the effects of mold byproducts continues to emerge, it is apparent that mycotoxins can damage any target organ in the human body and cause not only cancer, but several symptoms and diseases via multiple mechanisms.[10,19–22] Mycotoxins can bind to and damage deoxyribonucleic acid (DNA),[12,17] form epoxides,[7] mutate the cancer-protecting wild-type p53 gene to a cancer-promoting gene,[13] and act as carcinogens.[14,23] Furthermore, mycotoxins can do everything that pesticides do: they can (a) inhibit adenosine triphosphatase[10] and protein synthesis[14]; (b) trigger osteoporosis[24]; (c) be immunosuppressive[10]; and (d) damage mitochondria, genes, nerves, the heart, or the liver.[14] Mycotoxins can also be estrogenic[25] or cardiotoxic,[10] and they alter the barrier function of the intestinal epithelium.[10] In essence, the multiple mechanisms of potential damage from mycotoxins are not unlike those of many xenobiotics.[14] With this in mind, the inactivation of mycotoxins in the body should parallel aspects of xenobiotic detoxification.

Clearly, we must begin to build a foundation on which to prevent the carcinogenic cascade and arrest the symptoms of mycotoxicosis. A logical beginning would be to examine agents that historically have (a) reversed life-threatening mycotoxicosis, (b) inhibited aflatoxin-induced cancers, (c) inhibited the potentially fatal cardiotoxic side effects of mold-derived chemotherapeutic agents (e.g., adriamycin [doxorubicin is the generic]), (d) boosted xenobiotic detoxification, and (e) broadly inhibited a diverse range of reactive oxygen species. Lipoic acid (1,2-Dithiolane-3-pentanoic acid [also known as thioctic acid]) can do all of these.

Lipoic Acid

Amanita mushroom poisoning—1 form of mycotoxicosis—has a mortality rate of 50–90% from total liver failure within 48 hr of exposure.[26,27] Yet, more than 30 yr ago, files from the U.S. National Institutes of Health and the U.S. Food and Drug Administration recorded cures of otherwise fatal *Amanita* poisoning with the use of (nonpatentable) lipoic acid.[28] To this day, nothing in the modern emergency room is comparable to lipoic acid for saving the lives of entire families and picnickers who have mistakenly ingested poisonous mushrooms.[28–40] With the addition of other antioxidants, lipoic acid has also been used successfully in the treatment of end-stage hepatitis C in patients seeking liver transplantation. Interferon and antivirals have less than a 30% positive response rate. All 3 patients in 1 study using lipoic acid improved so remarkably that liver transplantation was no longer needed.[41]

In addition to being an outstanding antioxidant[42–45] and correcting mitochondrial damage in rats,[46] lipoic acid chelates heavy metals.[47–49] In culture, it inhibits human immunodeficiency virus replication[50,51] and enhances adenosine 5′-triphosphate production and overall protein synthesis[52]; in animals, it slows the complications of diabetes.[53] Furthermore, lipoic acid protects animal livers from aldehydes and other alcohol metabolites[54] and improves hepatic insufficiency.[55] Environmental industrial acrylamides and hexacarbon intoxications cause primary axon diseases, yet lipoic acid protects animals from hydrocarbon-induced neuropathies,[56,57] nephrotoxicity,[58] antibiotic-induced hearing loss,[59] and autoimmune hepatitis.[60]

There are many types of free radicals, and all have specific detoxifiers. For example, the singlet oxygen radical is quenched by beta carotene, the peroxyl radical is quenched by vitamin E, the hydroxy radical is quenched by glutathione, and the oxy radical can be quenched by coenzyme Q10. But, as the "universal antioxidant," lipoic acid—in its 2 forms: lipoic acid and reduced dihydrolipoate—is the only antioxidant that neutralizes all types of free radicals.[44] Although the body makes this antioxidant, it is not known exactly how this process occurs.[44] An age-related increase in peroxidation, paralleled by a depletion of protective detoxification enzymes, has been observed in rats.[61] Lipoic acid catalyzes the decarboxylation of α-keto acids in the Krebs cycle.[62] In animals, lipoic acid penetrates cell membranes and reaches high intracellular concentrations within 30 sec of administration.[63] It reaches all areas of the central and peripheral nervous systems within 30 min.[64] When given orally to rats, lipoic acid has decreased lipid peroxidation in brain tissue by 50%[64] and significantly restored conduction velocity in rats with experimentally induced diabetic neuropathy.[65]

Because lipoic acid cycles between the oxidized and reduced forms,[66,67] it also restores vitamins C and α-tocopherol, plus glutathione, to their reduced forms.[44,68] In addition, glutathione plays a vital role in the reduction of dehydroascorbate to ascorbate,[69] and ascorbate regenerates α-tocopherol from its oxidized form, as a result of which tocopherol continues to scavenge free radicals.[70] Given that lipoic acid protects against vitamin E deficiency[71] and increases intracellular glutathione levels,[72] it is unsurpassed in its perpetuation of antioxidant capacity. No other nutrient recycles so many antioxidants. In experimental animals, lipoic acid protects against bone marrow damage after radiotherapy[73] and reduces liver injury from radiation.[74] In Russian children, following the 1986 Chernobyl nuclear reactor disaster, lipoic acid markedly reduced radiation health effects on liver and kidney function, even though it was used for only 28 days.[75] Lipoic acid's main mode of action in rats appears to be protection of microsomal[76] and nuclear lipid peroxidation,[58,77–79] as well as being a powerful multipurpose antioxidant[49,80] and a recycler of numerous other key antioxidants (e.g., glutathione and vitamins C and E).[49,68,81]

Although animal data elucidate many of lipoic acid's mechanisms, the fact that lipoic acid has saved the lives of entire families, when all else failed, from rapidly lethal mycotoxins is impressive.[7,10,44] For example, lipoic acid infusions were used in the treatment of *Amanita* mushroom poisoning in 75 patients during the period between 1974 and 1978. Sixty-seven patients recovered, compared with the expected rate of 10–50%,[39] and other protocols have been reported.[37,38,82] The author has used 300 mg lipoic acid, administered orally 1–2 times daily, for more than 2 decades in well over 1,000 patients as part of a "detox cocktail," which also contained ascorbic acid and glutathione,[83] for the documented reduction of heavy-metal toxicity and for the reversal of chemically induced toxic encephalopathy (as clinical examples). A major drawback to statistical analysis of the use of lipoic acid is that it is always used in concert with other therapeutic agents.

In terms of safety, during more than 3 decades of scientific research and clinical use of lipoic acid, no serious adverse effects have been reported. The LD_{50} (i.e., low dose, 50% fatality) of lipoic acid is approximately 400–500 mg/kg oral dose in dogs.[44] In 1 study of rats, fatal toxicity was prevented by prior administration of thiamine.[84] In another rodent study, the authors reported that lipoic acid inhibited glycine conjugations of benzoic acid[85]; however, no assay of deficient detoxification nutrients was performed. In clinical practice, we tend to assay and correct nutrient deficiencies before instituting therapies, which may further account for beneficial therapeutic results accompanied by no adverse effects.[83]

A case can also be made for other diverse agents that support xenobiotic detoxification and also, for example, protect against aflatoxin-induced tumors.[15] These other nutrients and phytochemicals include, but are not limited to, glutathione,[16,86] *Silybum marianum* (called milk thistle or silymarin),[87] *Brassica* indoles,[88,89] aged garlic extract,[17,90] α-tocopherol,[10,91] and selenium.[10,18,83,92] In view of its (a) extensive antioxidant capability, (b) recycling of other antioxidants, (c) inhibition of chemically and mycotoxin-induced liver damage, and (d) salvage of humans from otherwise fatal *Amanita* mushroom poisoning, lipoic acid appears to head the list of indispensable and inexpensive nonprescription nutrients for mitigating mycotoxicosis.

* * * * * * * * * * *

Submitted for publication September 26, 2003; revised; accepted for publication October 30, 2003.

Requests for reprints should be sent to Sherry A. Rogers, M.D., Northeast Center for Environmental Medicine, P.O. Box 2716, 2800 West Genesee Street, Syracuse, NY 13220.

* * * * * * * * * * *

References

1. Rogers SA. Indoor fungi as part of the cause of recalcitrant symptoms of the tight building syndrome. Environment Int 1991; 17(4):271–76.
2. Savilahti R, Uitti J, Roto P, et al. Increased prevalence of atopy among children exposed to mold in a school building. Allergy 2001; 56:175–79.
3. Dziadzio L, Bush RK. Assessment and control of fungal allergens. Curr Allergy Asthma Rep 2001; 1:455–60.
4. Rogers SA. A 13-month work, leisure, sleep environmental fungal survey. Ann Allergy 1984; 52:338–41.
5. Rogers SA. Diagnosing the tight building syndrome. Environ Health Perspect 1987; 76:195–98.
6. Santilli J Jr, Rockwell WJ, Collins RP. Individual patterns of immediate skin reactivity to mold extracts. Ann Allergy 1990; 65:454–58.
7. Hayes JR, Campbell TC. Food additives and contaminants. In: Klaasen CD, Amdur MO, Doull J (Eds). Casarett and Doull's Toxicology. The Basic Science of Poisons, 3rd ed. New York: Macmillan, 1986; pp 794–800.
8. Watson DH. Toxic fungal metabolites in food. Crit Rev Food Sci Nutr 1985; 22(3):177–98.
9. Fukushima C, Matsuse H, Tomari S, et al. Oral candidiasis associated with inhaled corticosteroid use: comparison of fluticasone and beclomethasone. Ann Allergy Asthma Immunol 2003; 90:646–51.
10. Beasley VR. Trichothecene Mycotoxicosis: Pathophysiologic Effects, Vol 2. Boca Raton, FL: CRC Press, 1989.
11. Yarom R, Sherman Y, Yagen B, et al. T-2 toxin effect on bacterial infection and leukocyte functions. Toxicol Appl Pharmacol 1985; 75:60–65.
12. Wang JS, Groopman JD. DNA damage by mycotoxins. Mutat Res 1999; 424:167–81.
13. Aguilar F, Hussain SP, Cerutti P. Aflatoxin B_1 induces the transversion of G–>T in codon 249 of the p53 tumor suppressor gene in human hepatocytes. Proc Natl Acad Sci USA 1993; 90:8586–90.
14. Newberne PM. Chemical carcinogenesis: mycotoxins and other chemicals to which humans are exposed. Semin Liver Dis 1984; 4(2):122–35.
15. Nyandieka HS, Wakhisi J, Kilonzo MM. Association of reduction of AFB1-induced liver tumors by antioxidants with increased activity of microsomal enzymes. Indian J Med Res 1990; 92:332–36.
16. Novi A. Regression of aflatoxin B_1-induced carcinomas by reduced glutathione. Science 1981; 212(4494):541–42.
17. Tadi PP, Teel RW, Lau BHS, et al. Organosulfur compounds of garlic modulate mutagenesis, metabolism and DNA binding of aflatoxin B_1. Nutr Cancer 1991; 15: 87–95.
18. Lei DN, Wang LQ, Ruebner BH, et al. Effect of selenium on aflatoxin hepatocarcinogenesis in the rat. Biomed Environ Sci 1990; 3(1):65–80.
19. Ueno Y. The toxicology of mycotoxins. Crit Rev Toxicol 1983; 14(2):99–132.
20. Uragashi K. Mycotoxic origin of cardiac beriberi. J Stored Prod Res 1969; 5:227.
21. Kroft WA, Jarvis BB, Yatawara CS. Airborne outbreak of trichothecene toxicosis. Atmos Environ 1986; 20(3): 549–52.
22. Etzel RE, Montana E, Sorenson W, et al. Acute hemorrhage in infants associated with exposure to *Stachybotrys atra* and other fungi. Arch Pediatr Adolesc Med 1998; 152:757–62.
23. Leszczyszyn J. Growth of transplantable aflatoxin B_1-induced mammary cancer in Wistar rats during treatment with BCG vaccine. Patol Pol 1986; 37(3):222–36.
24. Nassar AY, Galal AF, Mohamed MA, et al. The effect of aflatoxin B1 on the utilization of serum calcium. Mycopathologia 1985; 91(2):127–31.
25. Lindsay DG. Zeranol—a "nature-identical" oestrogen? Food Chem Toxicol 1985; 23(8):767–74.
26. Parish RC, Doering PL. Treatment of *Amanita* mushroom poisoning: a review. Vet Hum Toxicol 1986; 28(4): 318–22.
27. Wieland TH, Faulstich H. Amatoxin, phallotoxins, phallolysin and antaminide: the biologically active components of poisonous *Amanita* mushrooms, Crit Rev Biochem 1978; 5:185–260.
28. Culliton BJ. The destroying angel: a story of a search for an antidote. Science 1974; 185:600–01.
29. Berkson BM. Thioctic acid in treatment of hepatotoxic mushroom poisoning. N Engl J Med 1979; 300:371.
30. Berkson B. Treatment of four patients with thioctic acid. In: Faulstich H, Kommerell B, Wieland T (Eds). *Amanita* Toxins and Poisonings. Baden Baden, Germany: Verlag Gerhard Witzstrock, 1980; pp 203–07.
31. Teutsch C, Brennan RW. *Amanita* mushroom poisoning with recovery from coma, a case report. Ann Neurol 1978; 3:171–79.

32. Vesconi S, Langer M. Therapy of cytotoxic mushroom intoxication. Crit Care Med 1985; 13:402–06.
33. Roldan EJA, Lloret AP. Thioctic acid in *Amanita* poisoning. Crit Care Med 1986; 14:753.
34. Zulik R. Death cap poisoning. Lancet 1972; 11:288.
35. Becker CE. Diagnosis and treatment of *Amanita phylloides*-type mushroom poisoning: use of thioctic acid. West J Med 1976; 125:100–09.
36. Piering WF, Bratanow N. Role of the clinical laboratory in guiding treatment of *Amanita virosa* mushroom poisoning: report of two cases. Clin Chem 1990; 36(3):571–74.
37. Nagy I, Pogatsa-Murray G, Zalanyi S Jr, et al. *Amanita* poisoning during the second trimester of pregnancy. A case report and a review of the literature. Clin Investig 1994; 72:794–98.
38. Sabeel AI, Kurkus J, Lindholm T. Intensive hemodialysis and hemoperfusion treatment of *Amanita* mushroom poisoning. Mycopathologia 1995; 131:107–14.
39. Bartter FC, Berkson B, Gallelli J. Thioctic acid in the treatment of poisoning with alpha-amanitin. In: Faulstich H, Kommerell B, Wieland T (Eds). *Amanita* Toxins and Poisoning. Baden Baden, Germany: Verlag Gerhard Witzstrock, 1980; pp 197–202.
40. Steyn DG. The treatment of cases of *Amanita phylloides* and *Amanita capensis* poisoning. S Afr Med J 1966; 40: 405–06.
41. Berkson BM. A conservative triple antioxidant approach to the treatment of hepatitis C. Combination of alpha lipoic acid (thioctic acid), silymarin, and selenium: three case histories. Med Klin 1994; 94:S84–89.
42. Suzuki Y, Tsuchiiya M, Packer L. Thioctic acid and dihydrolipoic acid are novel antioxidants which interact with reactive oxygen species. Free Rad Res Commun 1991; 15:255–63.
43. Scott B, Aruoma O. Lipoic and dihydrolipoic acid as antioxidants. Free Radic Res 1994; 20:119–33.
44. Packer L, Witt EH, Tritschler HJ. Alpha-lipoic acid as a biological antioxidant. Free Radic Biol Med 1995; 19: 227–50.
45. Dikalov S, Khramtsov C, Zimmer G. Reaction of lipoate and its derivatives with reactive oxygen species. In: Fuchs J, Packer L, Zimmer G (Eds). Lipoic Acid in Health and Disease. New York: Marcel Dekker, 1997; pp 47–66.
46. Hagen TM, Ingersoll RT, Lykkesfeldt J, et al. (*R*)- α-lipoic acid-supplemented old rats have improved mitochondrial function, decreased oxidative damage, and increased metabolic rate. FASEB J 1999; 13:411–18.
47. Ou P, Tritschler HJ, Wolff SP. Thioctic (lipoic) acid: a therapeutic metal-chelating antioxidant? Biochem Pharmacol 1995; 50:123–26.
48. Biewenga GP, Haenen GR, Bast A. The pharmacology of the antioxidant lipoic acid. Gen Pharmacol 1997; 29:315–31.
49. Muller L, Menzel H. Studies on the efficacy of lipoate and dihydrolipoate in the alteration of cadmium toxicity in isolated hepatocytes. Biochem Biophys Acta 1990; 1052:386–91.
50. Bauer A, Harrer T, Peukert M, et al. Alpha-lipoic acid is an effective inhibitor of human immunodeficiency virus replication. Wien Klin Wochenschr 1991; 69:722–24.
51. Grieb G. Alpha lipoic acid inhibits HIV replication. Med Monatsschr Pharma 1992; 15(8):243–44.
52. Zimmer G, Manika L, Kruger E. Dihydrolipoic acid activates oligomycin-sensitive thiol groups and increases ATP synthesis in mitochondria. Arch Biochem Biophys 1991; 288:609–13.
53. Packer L. Antioxidant properties of lipoic acid and its therapeutic effect in prevention of diabetes complications and cataracts. Ann NY Acad Sci 1994; 734:257–64.
54. O'Neill CA, Hallowell B. Aldehyde-induced protein modifications in human plasma. Protection by glutathione and dihydrolipoic acid. J Lab Clin Med 1994; 124(3):359–70.
55. Thompson CM. Addition of thioctic acid to a plan for management of hepatic insufficiency. Am J Med 1956; 21:131–32.
56. Kemplay S, Martin P, Wilson S. The effects of thioctic acid on motor nerve terminals in acrylamide-poisoned rats. Neuropathol Appl Neurobiol 1988; 14:275–88.
57. Altenkirch H, Stoltenberg-Didinger H, Wagner H, et al. Effect of lipoic acid in carbon-induced neuropathy. Neurotoxicol Teratol 1990; 12:619–22.
58. Sandhya P, Varalakshmi P. Effect of lipoic acid administration lipid peroxidation in rats. J Appl Toxicol 1997; 17: 405–08.
59. Conlon BJ, Aran JM, Erre JP, et al. Attenuation of amino glycosides-induced cochlear damage with the metabolic antioxidant alpha-lipoic acid. Hear Res 1999; 128(1–2): 40–44.
60. Gut J, Christen U, Frey N, et al. Molecular mimicry in halothane hepatitis: biochemical and structural characterization of lipoylated autoantigens. Toxicology 1995; 97 (1–3):199–224.
61. Arivazhagen P, Juliet P, Panneerselvam C. Effect of DL-alpha lipoic acid on the status of lipid peroxidation and antioxidants in aged rats. Pharmacol Res 2000; 41: 299–303.
62. Halliwell B, Gutteriedge JMC. Free Radicals in Biology and Medicine, 3rd ed. New York: Oxford University Press, 1999; p 194.
63. Peinado J, Sies H, Akerboom TP. Hepatic lipoate uptake. Arch Biochem Biophys 1989; 273:389–95.
64. Packer L, Tritschler HJ, Wessel K. Neuroprotection by the metabolic antioxidant alpha-lipoic acid. Free Radic Biol Med 1997; 22:359–78.
65. Nagamatsu M, Nickander KK, Low P, et al. α-lipoic acid improves nerve blood flow, reduces oxidative stress, and improves distal nerve conduction in experimental diabetic neuropathy. Diabetes Care 1995, 18(8):1160–67.
66. Nohl H, Gille L. Evaluation of the antioxidant capacity of ubiquinol and dihydrolipoic acid. J Biosci 1998; 53: 250–53.
67. Loffelhardt S, Bonaventura C, Locher M, et al. Interaction of α-lipoic acid in enantiomers and homologues with the enzyme components of the mammalian pyruvate dehydrogenase complex. Biochem Pharmacol 1995; 15(5):637–46.
68. Han D, Handleman G, Marcxocci L, et al. Lipoic acid increases the de novo synthesis of cellular glutathione by improving cystine utilization. Biofactors 1997; 6:321–38.
69. Wells WW, Xu DP, Yang Y, et al. Mammalian thiol transferase (glutaredoxin) and protein disulfide isomerase have dehydroascorbate reductase activity. J Biol Chem 1990; 265:15361–64.
70. Hansen DE, Black G, Levine M. Ascorbic acid: biological functions and relation to cancer. J Natl Cancer Inst 1991; 83:547–50.
71. Podda M, Tritschler HJ, Ulrich H, et al. Alpha-lipoic acid supplementation prevents symptoms of vitamin E deficiency. Biochem Biophys Res Commun 1994; 204: 98–104.
72. Han D, Tritscherl HJ, Packer L. Alpha-lipoic acid increases intracellular glutathione in human T-lymphocyte Jurkat cell line. Biochem Biophys Res Commun 1995; 207: 258–64.
73. Ramkrishnan N, Wolff WW. Radioprotection of haemo-

topoietic tissues in mice by lipoic acid. Radiat Res 1992; 130:360–65.
74. Kropachova K, Mishurova E. Flavobion and thiocic acid lessen radiation-induced latent injury of the liver. Biull Eksp Biol Med 1992; 113(5):547–49.
75. Korkina LG, Afanasef IB, Diplock AT. Antioxidant therapy in children affected by a radiation from the Chernobyl nuclear accident. Biochem Soc Trans 1993; 21:S314.
76. Bast A, Haenen GRMM. Interplay between lipoic acid and glutathione in the protection against microsomal lipid peroxidation. Biochim Biophys Acta 1988; 963: 558–61.
77. Devasagayam TPA, Subramanian M, Pradhan DS, et al. Prevention of singlet oxygen-induced DNA damage by lipoate. Chem Biol Interact 1993; 86:79–92.
78. Mizuno M, Packer L. Effects of alpha lipoic acid and dihydrolipoic acid on expression of proto-oncogene c-fos. Biochem Biophys Res Commun 1994; 200(2):1136–42.
79. Sumathi R, Devi VK, Varalakshmi P. DL α-lipoic acid protection against cadmium-induced tissue lipid peroxidation. Med Sci Res 1994; 22:23–25.
80. Kagan VE, Shvedova A, Servinova E, et al. Dihydrolipoic acid: a universal antioxidant both in the membrane and in the aqueous phase. Biochem Pharmacol 1992; 44: 1637–49.
81. Scholich H, Murphy ME, Eies H. Antioxidant activity of dihydrolipoate against microsomal lipid peroxidation and its dependence on α-tocopherol. Biochim Biophys Acta 1989; 1001:256–61.
82. Berkson BM. Alpha-lipoic acid (thioctic acid): my experience with this outstanding therapeutic agent. J Orthomol Med 1998; 13(1):44–48.
83. Rogers SA. Detoxify or Die. Syracuse, NY: Prestige, 2002. Available from <www.prestigepublishing.com>
84. Gal EM. Reversal of selective toxicity of (-)-α-lipoic acid by thiamine in thiamine-deficient rats. Nature 1965; 207:535.
85. Gregus Z, Fekete T, Klaassen CD, et al. Lipoic acid impairs glycine conjugation of benzoic acid and renal excretion of benzoylglycine. Drug Metab Dispos 1996; 245: 682–88.
86. Trickler D, Shklar G, Schwartz J. Inhibition of oral carcinogenesis by glutathione. Nutr Cancer 1993; 20(2):139–44.
87. Vogel G, Tuchweber B, Trost W, et al. Protection by silibinin against *Amanita phylloides* intoxication in beagles. Toxicol Appl Pharmacol 1984; 73:355–62.
88. Boyd JN, Babish JG, Stoewsand GS. Modification by beet and cabbage diets of aflatoxin B_1-induced rat plasma α-foetoprotein elevation, hepatic tumorigenesis, and mutagenicity of urine. Food Chem Toxicol 1982; 20:47–52.
89. Fong AT, Swanson HI, Dashwood RH, et al. Mechanisms of anti-carcinogenesis by indole-3-carbinol. Studies of the enzyme induction, electrophile-scavenging, and inhibition of aflatoxin B_1 activation. Biochem Pharmacol 1990; 39(1):19–26.
90. Kojima R, Toyama Y, Ohnishi ST. Protective effects of an aged garlic extract on doxorubicin-induced cardiotoxicity in the mouse. Nutr Cancer 1994; 22:163–73.
91. Geetha A, Sankar R, Marar T, et al. Alpha tocopherol reduces doxorubicin-induced toxicity in rats: histological and biochemical evidences. Indian J Physiol Pharmacol 1990; 34:94–100.
92. Chen J, Goetchius MP, Campbell TC, et al. Effects of dietary selenium and vitamin E on hepatic mixed-function oxidase activities and in vivo covalent binding of aflatoxin B_1 in rats. J Nutr 1982; 112(2):324–31.

Treatment of Cancer with Mushroom Products

JEAN A. MONRO
Medical Director
Breakspear Hospital
Hemel Hempstead
Herts, United Kingdom

ABSTRACT. Cancer has been attributed to 3 causes: pollution, infection, and poor nutrition. Conventional treatments include surgery, chemotherapy, and radiotherapy. The author proposes that immunotherapy also be considered. Among other environmental influences, dietary deficiencies and carcinogenic viral infections must be investigated and treated wherever possible. It has been suggested that mushrooms, in particular, have a structure that is immunomodulatory because it resembles the proteoglycan structure in the human extracellular matrix, and both are metabolically active. Inasmuch as mitochondria have a bacterial origin, proteoglycans may have a mushroom origin. The author describes a study which shows that natural killer cells can double in number with 8 wk of treatment with *Coriolus versicolor*. Also described is an epidemiological survey of cancer deaths among *Flammulina velutipes* farmers in Japan, which found that the mushroom farmers had lower rates of cancer deaths than controls who were not involved in mushroom farming.
<Key words: cancer, *Coriolus versicolor*, immunotherapy, matrix, mushrooms>

THE GLOBAL INCIDENCE OF CANCER has been attributed to 3 principal causes. Of 10 million cancer cases diagnosed each year worldwide, 3 million are attributed to tobacco smoking, 3 million to poor diet (inadequate or incorrect nutrition), and 1.5 million to infection, in particular by viruses. The remaining 2.5 million cases of cancer have no identifiable cause.[1]

Pollutant-caused cancers are predominantly attributable to the use of tobacco products, and involve the lung, oropharynx, larynx, bladder, and kidney. Causes related to diet involve the stomach, colon, esophagus, breast, liver, oropharynx, and prostate. Infectious causes affect the stomach; cervix; liver; nasopharynx; neural, hematological, and lymphatic systems; and bladder.

Current treatments for cancer consist principally of surgery for eradication or bypass of a diagnosed mass, and chemotherapy and radiotherapy directed at ablation of the abnormal growth of cells by direct effects on the development of cancerous tissue. In the future, cancer treatment may include these modalities in combination with immunotherapy alone, or with targeted immunotherapy, in which chemotherapeutic agents are linked both chemically and structurally to an antibody that targets cancerous tissue, which allows direct transport of the agent to the site of primary or secondary tumors.

Treatment of cancer has focused on fighting the *established* cancer; however, treatment of the 3 principal causes is not managed well. Inasmuch as biological systems are integrated to ensure long-term survival, the clinician should address both treatment of a cancer *and* its causes, to protect against conditions favorable for initiation or recurrence of the disease. For example, cancer of the lung is more likely to recur in a persistent smoker—without strong encouragement and effort to stop smoking, the provocant remains. Similarly, because asbestos and other pollutants act synergistically with tobacco smoke, appropriate avoidance strategies must be considered.

Among other environmental influences, poor diet or incorrect nutrition must also be addressed, and carcinogenic viral infections[2–4] must be investigated and treated wherever possible. Ironically, chemotherapy causes immune suppression, which allows infectious causes to flourish and possibly increase the likelihood of recurrence or secondary spread.[5] The relationship between diet and cancer is complex. We have to eat, but our food, its preparation, and interactions between dietary components during digestion and assimilation can cause significant changes in levels of hormones and other cellular messengers. Pollutant contaminants (e.g., polychlorinated biphenyls [PCBs], dioxins, and pesticides) can be estrogen mimics.[6,7]

Certain diets—such as high-fiber, low-fat diets with a high content of fresh fruit and vegetables—protect against cancers because many phytonutrients enhance immunity and are free-radical scavengers. Conversely, the low-fiber, high-fat (or its contaminants) diets common in Northern Europe and the United States are associated with increased cancer risk.[8]

Infection is also important as a cause of cancer. Papillomavirus infection induces cervical cancer—the 5th most common cancer in humans and the 2nd most common cause of cancer death in women. At least 90% of all cervical carcinomas are thought to be related to human papillomavirus infection (HPV),[2] with the HPV-16 strain accounting for more than 50% of cases worldwide, and the HPV-18 strain for about 14%.[3] Hepatitis viruses are associated with increased risk of liver cancer and hepatoma,[9] and Epstein-Barr virus (EBV)[4,10] is associated with increased risk of lymphoma. It may be possible to immunize against such infections; for example, researchers in Taiwan have shown that immunization against hepatitis reduced the incidence of childhood hepatoma by 50%.[9]

Pollution causes cancer by promoting the production of abnormal fragments of deoxyribonucleic acid (DNA). Chemical irritants such as tobacco smoke, physical agents such as ionizing radiation, and infectious agents such as viruses can cause cancer many years after initial exposure. The time delay may be associated with a cytokine shift from a Th1 to Th2 pattern, which weakens the Th1 immune response (i.e., the antiviral, antibacterial cell-mediated immune response). If a Th1-to-Th2 shift occurs over a period of time, proinflammatory cytokines may be induced and may maintain a heightened state of humoral response, during which triggers (individually or in combination) continue to weaken the immune system. As stated previously, nutrition can exert a major influence on hormonal systems. For example, it has been shown that folic acid deficiency causes patchy loss in colonic mucosa, which is a precursor to colon cancer; estrogen and its mimics (e.g., pesticide pollutants, phthalates from plastics, or PCBs) can deplete folic acid.[11]

Cancer prevention and treatment should, therefore, include reversal of the cytokine Th1-to-Th2 shift, allowing for the periodic restoration of Th1 activity through the use of biological agents that restore or support Th1 immune response activity. Personal pollution (e.g., tobacco smoking, exposure to estrogen mimics in the diet or wider environment) should be reduced or eliminated. Knowledge of disease causation frequently results in effective prevention and treatment, and avoidance of cause may prevent recurrence.

Coriolus versicolor and Chronic Fatigue

Very few agents are effective for both treatment and prevention of disease, but our research has shown that certain mushrooms can be used in this way. The fungus *Coriolus versicolor* is found almost worldwide; however, its bioactivity varies depending on the habitat in which it grows. To eliminate these variations, Gourmet Mushrooms, Inc. (Sebastopol, California) established the CV-OH1 strain approximately 15 yr ago. This strain demonstrates rapid and aggressive colonization and far outmatches other isolates in bioactivity and vigor. A mother culture of strain CV-OH1 has been developed and maintained to produce spawn. In-house checks confirm that the mother culture does not alter from the original isolate. The production process involves the inoculation of sterile organic edible grain with spawn from the mother culture. The fungus is allowed to completely colonize the growth medium aseptically. At the correct stage of development—at maximum bioavailability—the living biomass is aseptically air-dried, granulated, and tested microbiologically.

This reproducible technique produces a raw material that is sterile, contains no pesticides or heavy metals, and is free from foreign matter. The material is shipped to a U.K. GMP (Good Manufacturing Practice) pharmaceutical facility (Mycology Research Laboratories Ltd (MRL) [East Yorkshire, U.K.]), where it is manufactured into 500-mg tablets (Coriolus-MRL®) with the addition of cellulose, silica, a granulating agent, and a tablet press lubricant. The finished tablet complies with the British Pharmacopoeia requirements for tablets. The long-term stability of this product is being established with total viable count and polymerase chain reaction (PCR) tests. A PCR/electrophoresis method is being developed to determine the DNA of this product, as an additional means of quality control.

We studied 30 patients who fulfilled international criteria for the diagnosis of chronic fatigue.[12,13] The patients ranged in age from 17 to 83 yr, with a female-to-male ratio of 2:1. We measured the following 8 immunological parameters in each patient: (1) T lymphocyte subsets, (2) natural killer cells CD3–CD16+ CD56+,[14] (3) EBV viral capsid antibody immunoglobulins IgG and IgM, (4) EBV early antigen antibody diffuse IgG, (5) EBV nuclear antigen antibody, (6) EBV nuclear antigen IgM, (7) human herpes virus 6 (HHV6) IgG, and (8) cytomegalovirus (CMV) IgG and IgM. Patients were then given 6 Coriolus-MRL® tablets daily for 15 days, followed by 3 tablets daily for 45 days.

Before treatment, all patients were found to have combinations of high antibody levels to EBV and/or

HHV6 or CMV. The results of our study revealed that T cells (percent CD3+CD26) showed increased activation in two-thirds of the patients initially and, following review of the patients after 8 wk, T cell depression in 22% of patients. The T cell level was unchanged in 11% of the patients.

The normal range of natural killer cells in humans is 5%–20% of the lymphocyte population; the normal cell count is 75–1,800/mm^3. Natural killer cells were low in our patients prior to treatment (average = 129.64/mm^3). Following treatment with Coriolus, the average cell count was 175/mm^3—*an increase of 35%*.[15] Thus, we have demonstrated that natural killer cell production can be induced by use of a mushroom product. These findings indicate an increased probability of immunological improvement in patients suffering from chronic fatigue.

Flammulina velutipes and Cancer

It has been well documented that tumor growth can be inhibited through functional modulation of the following components and processes: cytotoxic macrophage activity, monocytes, neutrophils, natural killer cells, dendritic cells, cytokines (e.g., interleukins, interferons, and colony-stimulating factors), induction of gene expression, cytokine receptors, T cells (i.e., cell-mediated cytotoxicity), and B cells (i.e., antibody production). Each of these functions can be affected by mushroom products.[16]

In Japan, Korea, and China, Phase I, II, and III clinical trials are under way on the use of mushrooms as adjuncts to chemotherapy. Various immune-modulating mushroom polysaccharides have been identified (e.g., lentinan from *Lentinula edodes*; schizophyllum from *Schizophyllum commune*; PSK Krestin [polysaccharide-Krestin] and PSP [polysaccharide-peptide], components of *Coriolus*; and grifon D from *Grifola frondosa*). Lentinan and schizophyllum increase numbers of T helper cells and macrophages, which are involved in altering acute-phase proteins and colony-stimulating factors; they also alter activation of macrophages, neutrophils, and complement. PSK Krestin and PSP are immunostimulators that increase total white blood cell counts, gamma interferon, and interleukin 2.[16]

The National Cancer Center Research Institute of Japan, founded in 1966, has established that basidiomycetes (the fungal group with freestanding fruiting bodies, often used as an edible mushroom) can have anticancer activity. Protective properties of mushrooms were demonstrated in a 15-yr epidemiological study (1972–1986) in the Nagano area of Japan, where farmers grow the edible mushroom *Flammulina velutipes*. In that study of 174,505 individuals, investigators compared the cancer death rates of farmers who produced this edible mushroom with death rates of controls, with the assumption that the farmers would eat some of the mushroom species they farmed. Results showed a cancer death rate in controls of 160.1 per 100,000, compared with 97.1 per 100,000 among the mushroom farmers (Fig. 1).[17]

Discussion

It is thought that eukaryotic organisms evolved through incorporation of bacteria into cells as mitochondria.[18] Thus, it could be further postulated that the proteoglycans in mushrooms formed the evolutionary background for the matrix in human connective tissue.[19] Polysaccharides are important mushroom components. Monosaccharides with glycosidic links can produce branched molecules with a huge variation in linkages. There are as many as 35,560 permutations for 4 monosaccharides, providing enormous diversity and flexibility of linkages. In the matrix of the connective tissue between human cells, the principal linked polysaccharides are attached to a protein core in much the same way as seen in mushroom polysaccharides (Fig 2).[18] Mushroom proteoglycans may represent an information transmission system, inasmuch as the diversity and flexibility of glycosidic linkages of the carbohydrate saccharides permit rapid transmission of information. It

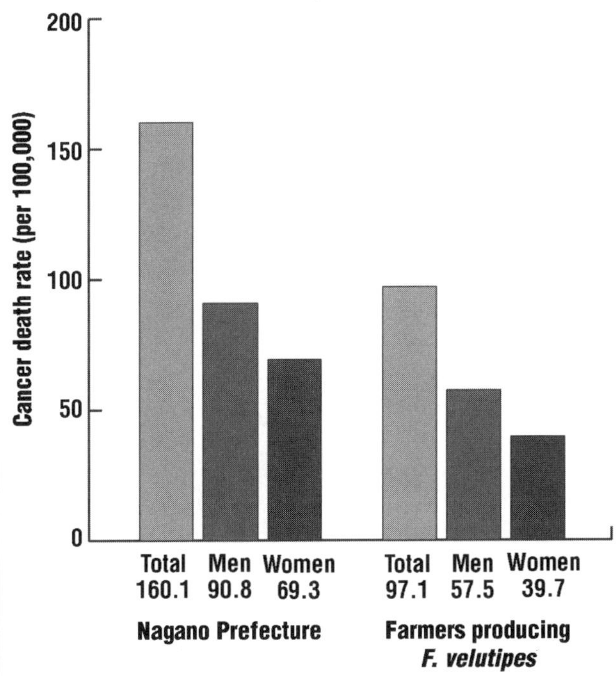

Fig. 1. Comparison of cancer death rates among farmers producing *Flammulina velutipes* mushrooms, vs. controls, in Nagano Prefecture, Japan, 1972–1986. *Source: Dr. Tetsuro Ikekawa: from his slide presentation, "Beneficial Effects of Mushrooms, Edible and Medicinal, on Health Care in Japan," given at the Global Holistic Health Summit, Bangalore, India, January 2003. Notes:* $p < 0.01$. Age-adjusted standard population. Total objective population analyzed = 174,505.

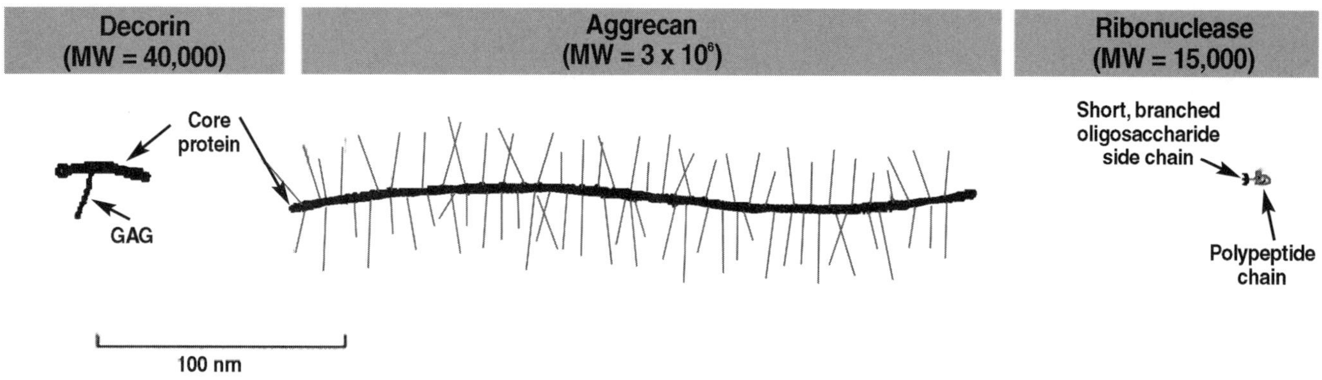

Fig. 2. Similar polysaccharide linkages are found in the extracellular matrices of human cell connective tissue and in mushroom cells. ***Source: Reproduced from Reference 18, p 976, by permission of the publisher. Notes:*** **Aggrecan is a high-molecular-weight proteoglycan, and decorin is a low-molecular-weight proteoglycan. GAG = glycoaminoglycan, and MW = molecular weight.**

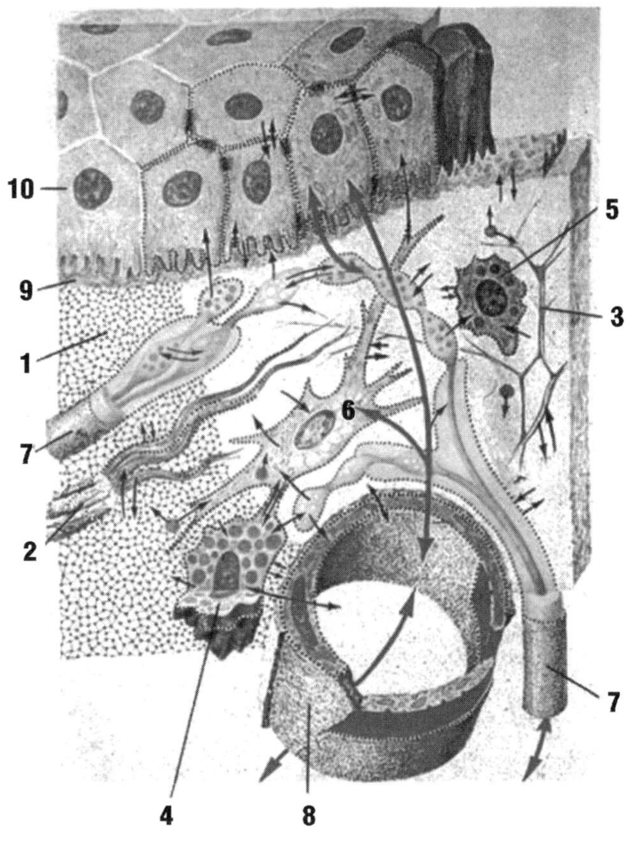

Fig. 3. Reciprocal relationships (arrows) between capillaries (8), ground substance (matrix; proteoglycans and structural glycoproteins [1], collagen [2], and elastin [3]), connective tissue cells (mast cell [4], defense cell [5], and fibrocyte [6]), terminal autonomic axons (7), and organic parenchymal cells (10); basement membrane (9). The fibrocyte (6) is the regulatory center of the ground substance. Only this type of cell is able to synthesize extracellular and nervous components. The main mediators and filters of information are the proteoglycans, structural glycoproteins, and the cell surface sugar film (glycocalyx [dotted line on all cells], collagen, and elastin). ***Source: Reproduced from Reference 19, p 21, by permission of the publisher.***

could be that evolutionary development from unicellular to multicellular organisms requires information transmission; this transmission through the extracellular matrix of higher organisms can occur, it is said, at the speed of sound (Fig 3).[16,19] Perhaps, in mushrooms, similar information transmission is part of the structural function of proteoglycans (Fig 2).[18]

Natural killer cells recognize high-molecular–weight glycoproteins on virally infected cells. Perforin, apoptosis, tumor necrosis factor, proteases, natural killer cytotoxic factors, and ionized adenosine triphosphate apoptosis are all involved in natural killer cell effects. Both virally infected cells and cancer cells are killed by cytotoxic T cells and antibody-dependent, cell-mediated cytotoxicity.[20]

Our research has demonstrated that mushrooms may be useful in protecting against virally induced cancers through enhancement of natural killer cells, and may also play a role in the prevention of cancers induced by diet and poor lifestyle choices. Thus mushrooms may have a significant role in cancer treatment.

* * * * * * * * * *

Submitted for publication September 20, 2003; revised; accepted for publication October 30, 2003.

Requests for reprints should be sent to Jean A. Monro, M.B., B.S., M.R.C.S., L.R.C.P., F.A.A.E.M., D.I.B.E.M., M.A.C.O.E.M., Medical Director, Breakspear Hospital, Hertfordshire House, Wood Lane, Paradise Estate, Hemel Hempstead, Herts, HP2 4FD, U.K.

E-mail: jmonro@breakspearmedical.com

* * * * * * * * * *

References

1. Sikora K. The future of cancer care. Pharmacia Lecture at The Royal Society, London, April 2001. Available from Prof.

Karol Sikora, 79 Harley Street, London W1G 8PZ, UK.

2. Eiben GL, Da Silva DM, Fausch SC, et al. Cervical cancer vaccines: recent advances in HPV research. Viral Immunol 2003; 16:111–21.
3. Bosch FX, Manos MM, Muñoz N, et al. Prevalence of human papillomavirus in cervical cancer: a worldwide perspective. J Natl Cancer Inst 1995; 87:796–802.
4. de-The G, Geser A, Day NE, et al. Epidemiological evidence for causal relationship between Epstein-Barr virus and Burkitt's lymphoma from Ugandan prospective study. Nature 1978; 274:756–61.
5. British National Formulary. 45th ed. Sec 8: Malignant disease and immunosuppression. London: British Medical Association and Royal Pharmaceutical Society of Great Britain, 2003; pp 407–38.
6. Cheek AO, Vonier PM, Oberdorster E, et al. Environmental signaling: a biological context for endocrine disruption. Environ Health Perspect 1998; 106(Suppl 1):5–10.
7. Crisp TM, Clegg ED, Cooper RL, et al. Environmental endocrine disruption: an effects assessment and analysis. Environ Health Perspect 1998; 106(Suppl 1):11–56.
8. World Health Organization (WHO) Study Group on Diet, Nutrition and Non-Communicable Diseases. Diet, Nutrition and the Prevention of Chronic Diseases. Geneva: WHO, 1990; pp 67, 69.
9. Alter MJ. Protecting future generations through immunization against hepatitis B. Ann Intern Med 2001; 135: 835–36.
10. Mueller N, Evans A, Harris NL, et al. Hodgkin's disease and Epstein-Barr virus: altered antibody pattern before diagnosis. N Engl J Med 1989; 16:689–95.
11. Bukin YV, Draudin-Krylenko VA, Levchuk AA, et al. The effect of high doses of folic acid on the overexpression of ornithine decarboxylase and S-adenosylmethionine content in human colon adenomatous polyps. Ann NY Acad Sci 2001; 952:175–76.
12. Holmes GP, Kaplan JE, Gantz NM, et al. Chronic fatigue syndrome: a working case definition. Ann Intern Med 1988; 108:387–89.
13. Fekuda K, Straus SE, Hickie I, et al. The chronic fatigue syndrome: a comprehensive approach to its definition and study. Ann Intern Med 1994; 121:953–59.
14. Vojdani A. Natural Killer Cytotoxicity, Apoptosis and Cell Cycle in Human Health and Disease. Beverly Hills, CA: Immunosciences Lab., Inc., 1997; p 15.
15. Monro JM. Coriolus: the use of the medicinal mushroom Coriolus MRL® as an immunotherapeutic agent in the treatment of patients with chronic fatigue syndrome. (Submitted for publication 2004 to *Journal of Integrative Medicine.*)
16. Smith JE, Rowan NK, Sullivan R. Medicinal Mushrooms: Their Therapeutic Properties and Current Medical Usage with Special Emphasis on Cancer Treatments. London: Cancer Research UK, 2000; p 220. <http://sci.cancer researchuk.org/labs/med_mush/med_mush.html>
17. Ikekawa T. Beneficial effects of edible and medicinal mushrooms on health care. Int J Med Mushrooms 2001; 3:291–98.
18. Alberts B, Bray D, Lewis J, et al. Molecular Biology of the Cell. 3rd ed. New York: Garland Publishing, 1994; p 17.
19. Pischinger A. Matrix and Matrix Regulation: Basis for a Holistic Theory in Medicine. Brussels, Belgium: Editions Haug International, 1991.
20. Roitt IM. Essential Immunology. 9th ed. Oxford: Blackwell Science, 1997; p 34.

Summary of the 5th International Conference on Bioaerosols, Fungi, Bacteria, Mycotoxins, and Human Health

Kaye H. Kilburn, M.D.

REPRESENTATIVES of the world scientific community met to consider the adverse health effects of contaminants from indoor air at the 5th International Conference on Bioaerosols, Fungi, Bacteria, Mycotoxins, and Human Health, held at Saratoga Springs, New York, September 9–12, 2003. At 5 times the size of the first Saratoga Springs conference, held in 1994, this gathering comprised 71 8-minute scientific papers, 14 summary papers, and a poster forum. Major topic areas were epidemiology; clinical case studies emphasizing neurological, nasal, and ocular effects of mold exposure; laboratory signs of these effects; and remediation. Presentations on cytotoxicity and immune issues ranged from studies of pulmonary surfactant, to mice and rat lung models. Environmental sampling studies of mold-contaminated buildings included factors such as moisture, relative humidity, and dust. Sampling methods varied for homes and their crawl-spaces, schools, offices, and high-rise buildings. Also considered were interactions of molds and bacteria with calcareous materials (especially gypsum), release of enzymes from mold in indoor dust, cross-reactivity of monoclonal antibodies to molds, and applications of the polymerase chain reaction (PCR) test for identification of fungal DNA.

The increase in problems attributed to mold in buildings is evidenced by the fact that the U.S. Occupational Health and Safety Administration (OSHA) had a 10-fold increase in requests for mold inspections from 1999 to 2003 <DGShendell@lbl.gov>. Adverse health effects from dampness and mold indoors are found around the world. They are indicated by cough, wheeze, and asthma, at odds ratios of 1.4 to 2.2. A 75% increase in asthma occurred between 1980 and 1993 in Finnish children. One mold product, β-1-3-D glucan—possibly the most important mold-related inflammatory agent—acts like the lipopolysaccharides from gram negative bacteria. Mold and bacteria in indoor air are troublesome in Jordan, Brazil, and South Africa, as well as in North America and Europe. Prevention, control, and remediation of molds to improve indoor air quality were discussed with regard to criteria established by the U.S. Environmental Protection Agency (EPA), the American Conference of Governmental Industrial Hygienists (ACGIH), and the city of New York. For analysis of the economic impact of mold remediation vs. the demolition of a contaminated building, the Brown-Gibson economic decision-making model appears useful.

Conference Sessions

Session 1: Epidemiological research. Thirty-one teachers in a Finnish school showed an asthma rate of 26%. Their hoarseness and other respiratory symptoms decreased after mold remediation of the building. At a 2-yr follow-up, their immunoglobulin (Ig)G antibodies to 10 molds associated with sinusitis had decreased <ritta.patovirta@ktl.fi>. A study in Kuopio, Finland, found that joint pain, stiffness, and morning swelling accompanied respiratory irritation and infections in long-term occupants of moldy buildings. Autoimmune disease markers were 8–10 times higher than in average Finnish citizens, and the CD4:CD8 antigen ratio was altered <tuula.husman@ktl.fi>.

Session 2: Toxicology and neurological effects. Some patients with Indoor Air Syndrome have abnormal laboratory and pulmonary function tests. Associations were found between their symptoms in musty, mold-infested buildings and abnormal visual contrast sensitivity performance, as well as deficiencies in melanocyte-stimulating hormone (MSH) and leptin levels. Treatment with cholestyramine, at doses sufficient to lower cholesterol, appeared to reduce both MSH and leptin <ritchieshomaker@msn.com>. Impaired visual contrast sensitivity has been associated with exposure to other biotoxins, such as that produced by *Pfiesteria*, but this observation has not been compared with other visual tests, such as color discrimination and visual field performance <hudnell.ken@epamail.epa.gov>.

Follow-up by Gordon in 8 of 31 mold-exposed subjects who underwent neurobehavioral testing showed persistent functional defects of attention, learning, executive functions, short- and long-term recall, and visual memory at 12–36 mo <Wayne.Gordon@mssm.edu>.

Observations in 65 mold-exposed patients, compared with 202 non-mold–exposed community controls, showed deficient balance, reaction time, blink reflex latency, color discrimination, visual field performance, and grip strength. These physiological impairments were accompanied by decreased problem-solving, recall memory, vocabulary, attention, concentration, and perceptual motor performance, as were found by Gordon. In 8 of 8 patients restudied after 1 yr, functions had worsened, again confirming Gordon. Functions in 23 children in the studied families also decreased <Kilburn@usc.edu>.

Session 3: Health assessment and case studies. Children in Finland who occupied a mold-contaminated school building suffered from asthma and/or allergy and had higher IgG antibodies than nonallergic children. Their lymphocytes exhibited less spontaneous proliferation, and their eosinophilic cationic protein response was stronger <atiilika@sun3.oulu.fi>.

A mouse model, subjected by tracheal aspiration to extracts or cultures of *Metarhizium anisopliae*, *Stachybotrys chartarum*, and *Penicillium chrysogenum* showed increased protein, leukocytes, interleukin (IL)-5, and lactate dehydrogenase in bronchopulmonary lavage fluid. Doses of 100 µg of *P. chrysogenum* were equal in effect to doses of 10 µg of *S. chartarum* and *M. anisopliae* <Ward.Marsha@epamail.epa.gov>.

One fraternal twin, from a family living in a house contaminated with *Stachybotrys chartarum* and *Aspergillus versicolor*, *A. niger*, and *A. fumigatus*, died at 18 months of respiratory failure. Autopsy showed hemosiderosis, with *Aspergillus* found by PCR matching of DNA in environmental sampling and in lung samples. Staining revealed antibodies to mycotoxin T-2 and aflatoxin in lung, liver, and brain. The surviving twin had mycotoxins in serum by enzyme-linked immunosorbent assay (ELISA) testing <drordog@earthlink.net>.

Respiratory illnesses and shortness of breath, difficulty concentrating, and memory loss were found in occupants of a Nevada building <jcraner@drcraner.com>. These symptoms were not explained by allergic rhinitis, asthma, hypersensitivity pneumonitis, or somatization disorder, but seemed building-specific. From 425 adults and 176 children referred for moisture-related symptoms, Johanning found low sensitivity, but high specificity, for 6 fungal IgG and 2 fungal IgE titers. Among the adults, 59% had lower CD3 and 51% had higher than normal CD4:CD8 ratios. Findings in children were 48% and 1.6%, respectively <eckardtjohanning@fungal researchgroup.com>.

Session 4: Experimental research. Inflammatory responses toward specific organic material in mold and bacteria include antibodies and cell activation (which produces cytokines, nitrous oxide, and hydrogen peroxide), and cytotoxicity, but responses lack specific connecting links. Eye and upper respiratory irritations are common symptoms from mold exposure, and molds may emit volatile organic compounds (VOCs) <maya-ritta.hirvonen@ktl.fi>. Nasal lavage fluid parameters did not differ among 62 healthy women exposed to alpha terpene and d-limonene at 150 ppb plus ozone for 3 hr, to clear air, and to the agents alone <lambach@eohsi.rutgers.edu>. Molhave found ocular effects in 36 other volunteers to be equally inconclusive <lm@mil.au.dk>.

In a different group exposed in 3 mold-contaminated indoor environments, nasal lavage showed increased concentrations of nitrite, IL-6, and tumor necrosis factor alpha (TNF-α) <lm@mil.au.dk>.

In mouse macrophages, exposures to metabolites of *Stachybotrys chartarum* (e.g., atranones B+E, satratoxin G), microbial toxins (i.e., sterigmatocystin, citrinin, and ochratoxin A), and the actinomycete *Streptomyces californicus* were synergistic for cytotoxicity. TNF-α and IL-6 were elevated, and *S. californicus* was bound to nuclear DNA <Kati.Huttunen@ktl.fi>.

Exposure of rabbit alveolar type II cells in culture to spores of 2 strains of *Stachybotrys chartarum* inhibited phospholipid synthesis via conversion of choline phosphate to CDP choline (cytidyldiphosphocholine) <thomas.rand@stmarys.ca>.

Rand et al. found, with intratracheal exposures of mice to spores of *Stachybotrys chartarum* strains JS 58-17 (high satratoxin) and JS58-06 (no satratoxin), and to *Cladosporium cladosporioides* at varied doses, that fewer than 30 spores of *S. chartarum* had an effect equal to > 3,000 spores of *C. cladosporioides*, although responses to JS 58-17 exposure via tracheostomy. The inflammatory effects of J58-06 spores were less than those of JS 58-17, but more severe than from other fungi such as *Penicillium* <ixy@po.cwru.edu>.

Glucan added to office dust increased the IL-8 and eosinophil cationic protein (ECP) (C-reactive protein) responses of 36 human subjects, with greater response noted in those hyperresponsive to histamine <ts@mil.au.dk>.

Session 5: Assessment methods and field findings. Viable fungi constitute only 1% of the fungal biomass in indoor air. Sampling should include both personal and stationary sampling to help characterize the particular "cloud" surrounding each individual. Comparison of mold counts in dust from 68 intact ("healthy") homes vs. 145 water-damaged homes showed a 7 times greater level of phylloplane molds, mainly *Cladosporium*, in the damaged homes <aino.nevalaien@ktl.fi>.

With *Mycobacterium tuberculosis* complex as the model, air samples obtained in gel traps mounted horizontally and vertically were combined with PCR and KY18 and KY75 primers specific for *Mycobacterium* spp., including *M. tuberculosis*, *M. bovis*, and *M. africanum*. Inasmuch as air, adhesive, and contact petri dishes were complementary as sampling methods, this technique can be recommended for the assessment of molds and *Legionella pneumophilia*. Another approach, in which home dust was evaluated as the catchment for microbials, showed correlation with the degree of water damage <microvital@qc.aira.com>.

In a Berlin study of 25 homes, 5 sampling methods were compared: (1) microbial growth from vacuumed dust samples collected over a 7-day period, (2) air sampling with a 2-stage Anderson cascade impaction device, (3) petri dishes for sedimentation samples, (4) contact samples, and (5) analysis of microbial VOCs (MVOCs) on thermal desorption tubes, by gas chromatography/mass spectrometry. Elevated levels of *Aspergillus* and *Penicillium* species characterized moldy vs. nonmoldy homes, with the best results being obtained from carpet dust sampling and other long-term collection methods. Most of the MVOCs were not specific molds <Hans.scheibinger@medizin.fu-berlin.de>.

Comparison of molds enumerated in indoor air, outdoor air, and in pulsed air in high-volume air conditioners detected indoor mold and revealed leaks, errors in cleaning, and other defects in heating, ventilation, and air conditioning (HVAC) systems. Other microbials yield similar results and, when graphed as hexagonal centriles of concentrations at 25°C and 37°C in outdoor, indoor, and pulsed air, aid in identifying the likelihood of human illness <bprezant@prezant.com> or <C.Chasseur@iph.fgov.be>. Another device—the fungal spore source strength tester (FSSST)—assists in analysis of vacuum-collected samples. A bioaerosol sampler connected to the FSSST outlet aerosolized spores from surfaces contaminated with mold <sivasus @ucmail.uc.edu>.

Comparisons of bacterial and fungal loads by culture, from samples collected with an impaction sampler, showed *Staphylococcus hemolyticus* and *Micrococcus lutens* as the most common in indoor air. *Penicillium* was the common indoor mold in winter, and *Cladosporium* in summer (in Italy) <carraro@mfn.unipssuc.it>. In Marseilles, a Protimeter Survey Master device was used to measure relative humidity in walls, for comparison with direct microscopic identification of molds. High humidity correlated with the identification of *Stachybotrys chartarum* <denis-andrecharpin@mail.ap-hm.fr>.

Baseline data for MVOCs in buildings without visible mold are being collected in Denmark and Germany, and are correlated with moisture, intense use, mold, construction, and air-handling equipment. Sampling methods—including spore counts, frequency, and microbial identification techniques—vary widely. Inasmuch as there is no consensus, it was advised that at least 2 sampling methods be used to characterize indoor air spaces <Reinhard.keller@hygiene.ukl.mc-lucbeck.de>. New sampling methods are under development. <Annal.Pasanen@uku.fi>.

Session 6: Field findings (cont.) and mycology. The EPA sponsored a nationwide study of 100 office buildings, from 1994 through 1998. This Building Assessment Survey and Evaluation (BASE) study has produced a Microsoft Excel database that compares physical conditions with sampling for chemical and biological agents <DGShendell@1b1.gov>. One analysis of these data revealed that *Cladosporium, Alternaria, Epicoccum* species (CAE) and *Penicillium/Aspergillus* were higher outdoors in BASE buildings, whereas in water-damaged buildings *Penicillium/Aspergillus* was 7 times higher indoors but CAE ratios were near baseline. In 10 North Carolina homes with central HVAC systems, crawl-spaces were important reservoirs of mold <thomas010@mc.duke.edu>.

An interactive model for spore equilibrium in a building considers infiltration, exfiltration, and entrainment, after assuming that all spores are either airborne or settled <jbanta@restcon.com>.

High-rise buildings are a special case in regard to water intrusions because they contain shafts for elevators, plumbing, and ventilation, as well as stairwells. Often, water runs into the elevator shaft. Negative pressures between the elevator shaft and the floors are stronger near the bottom of the shaft, gradually becoming positive up the stack of floors. Fungi found were mainly *Penicillium* and *Aspergillus* <etchrom@aol.com>.

Real-time measurements of particles aerosolized from cultures of *Aspergillus versicolor*—sampled at 1 wk and at 2 mo and grown in malt extract agar—showed bimodal distributions of particles 1.26–1.99 μm in diameter (representing spores) and 0.07–0.13 μm (representing fragments), with a minimal concentration at 0.32–0.50 μm. These distributions should aid in the development of better techniques for field sampling of fungal fragments <chosh@ucmail.uc.edu>.

Microparticles of fungi pass through slits in buildings, easily becoming airborne and remaining in the air, from which they are readily inhaled into the lung alveoli. Growth and aerosolization were important determinants of the liberation of protein-containing particles before mold growth was visible <amm@ami.dk>.

The bioactive substances cereulide and valinomycin (toxic ionophoric antibiotics), from streptomycetes and bacilli isolated from water-damaged buildings, were found to inhibit boar sperm motility, depolarize mitochondria in boar sperm cells, and damage plasma membranes. Fifty percent of bacterial isolates had such cytotoxic activity. Building materials from an indoor wall contaminated with *Streptomyces* contained 33 ng of valinomycin <mirja.salkinoja-solanen@helsinki.fi>.

Session 7: Assessment/mycology. Gypsum added to nutrient medium (2% plain malt extract agar) improves mold growth and enhances spore production. Relative rates of sporulation will be examined in the future <anna-mari.pessi@utu.fi>.

Streptomycetes grow at pH from 4 to 11.5. Isolates grown at pH 4 increased the lethality of spores. *Streptomyces californicus* and *Stachybotrys* sp. are frequently found growing together on moist inner wall surfaces. *S. californicus* stimulates production of nitrous oxide and cytokines, and causes death of mouse macrophages; *Stachybotrys* produces potent mycotoxins.

When cultured together, their biomass is increased (as estimated from fatty acid methyl ester content) and cytotoxicity and inflammatory triggering are enhanced <Piia.Penttinen@ktl.fi>.

Quantitative PCR analysis of molds produces accurate speciation and numbers and appears useful for finding masked species in the fungal biomass. Primer and probe sequences have been published, and the database is growing <haugland.rich@epamail.epa.gov>. The brown wood rot fungus, *Meruliporia incrassate,* can be identified quantitatively by using real-time PCR <klin@Stlinc.com>. Airborne *Cryptococcus neoformans* isolated from trees and soil of Vancouver Island, British Columbia, accounted for human, cat, and porpoise cases of fungal infection—a first for a temperate climate <kbarlett@interchange.ubc.edu>. IgG antibodies in people who lived near a composting site did not differ significantly from those of individuals remote from the site, and thus did not serve as a biomarker for actinomycetes <Caroline.herr@hygiene.med.unigiessen.de>.

Session 8: Assessment/mycology (cont.). Monoclonal antibodies against *Aspergillus versicolor* showed cross-reactivity to related *Aspergillus, Penicillium, Alternaria,* and *Stachybotrys* species, and demonstrated sharing of epitopes and different amounts of antigen per spore. Therefore, this method is not suggested for recognition of genera of molds <zvf9@cdc.gov>. A substantial effort to characterize 192 clones of fungi from moisture-damaged buildings by using PCR showed large proportions of chimeric products. Although PCR compliments other procedures for identification of fungi, the size, coverage, and reliability with the existing sequence databases mean that this procedure must wait on the development of large databases and more specific methods <mviikari@operoni.helsinki.fi>.

The fungal enzymes N-acetyl-β-D glucosaminidase and N-acetyl glucosidase are likely associated with lung damage from fungi. Both are found in dust, and differences may be the result of large particle sizes in dust that settles to the floor vs. airborne dust. Results from culture-based and PCR panel methods differed, with more species found by PCR, thus these methods did not correlate <czk4@cdc.gov>.

Session 9: Transitional countries. In Durban, South Africa, 30% to 40% of 6 hospital buildings were damp and had carbon dioxide levels exceeding 1,000 ppm in some indoor areas. Half of staff members were stressed, and 32.5% had allergies. Many fungi were found, including 8 *Aspergillus* species, *Cladosporium,* and *Fusarium.* Skin testing of staff showed that 18% reacted to dust mites and 10% to molds <Gqalenin@nu.ac.za>.

Microbial exposures judged by biochemical markers—ergosterol for fungal biomass, 3-hydroxy acids for lipopolysaccharide, and muramic acid for bacterial biomass—differed greatly in dust samples from schools in Sweden, Jordan, and Poland <loay.wady@mmb.lu.se>.

Session 10: Remediation. Remediation standards were discussed in regard to guidelines set forth by the city of New York, the EPA, and ACGIH for bioaerosol assessment and control. The EPA mold remediation guidelines for schools and commercial buildings were considered to be adequate for consumers, regulators, and the media, but not specific enough for remediation contractors. Items addressed were: (1) moisture surveys, (2) information on restorative drying, (3) employee training, (4) inspection procedures and cross-contamination, (5) cleaning procedures, (6) remediation/abatement levels and, (7) professional judgment. For example, phylloplane fungi, normally more abundant outdoors, should dominate the flora of remediated buildings. Construction dust may overwhelm and render unreliable the sampling of air for mold <mjuinni@nni.com>.

The "electronic nose" (Cyranose®320) was evaluated for use in indoor air quality assessment and was found not to be useful in detecting hidden mold in buildings <stewilso@ttacu.tty.edu>.

Fan coils of an HVAC system were the source of contamination in 1 condominium. Swab samples and a Reuters centrifugal sampler with Rose Bengal agar strip showed *Aspergillus versicolor* and helped in the selection of units for remediation <kwhite@jacqueswhitford.com>.

For those faced with the knotty problem of whether, and how, to remediate a mold-contaminated building—by removing and replacing the HVAC system, or by demolishing the building itself—modeling with the Brown-Gibson economic decision procedure is useful in determining a course of action <michael-larranga@ttu.edu>.

Session 11: Prevention and control. Trends were examined which showed a 10-fold increase in requests for inspections for mold from 1999 to 2003, and a 3-fold increase in building inspections performed. The buildings microbial database, developed by Forintek Canada Corp., includes more than 700 articles in tabular form <adnan@van.fornitek.ca>. Large high-rise buildings are complex structures and are often difficult to characterize. Ventilatory design has a significant impact on schools. For example, mold growth was enhanced in a New England school building when the central ring of air conditioned classrooms was surrounded by a ring without air conditioning, leading to condensation of moisture on inner walls <BBarry@eheinc.com>.

Summary

Risk assessment of bioaerosols indoors has been summarized in terms of human health effects (e.g., allergy, infection, and irritation), but neurotoxicity has been a topic of heated debate <fseif@esceng.com>. However, this debate has been resolved rather clearly by evidence presented at this conference (and at the 21st Annual International Symposium on Man and His Environment in Health and Disease, Dallas, Texas, June 2003) which showed that neurotoxicity, as well as pulmonary and immune dysfunction, can result from exposure to molds in the indoor environment. Toxicity is apparent, and confirming studies are being published.

For more information regarding the conference, visit <http://www.bioaerosol.org/index.html>. For details on the research summarized herein, please contact the presenting authors via their e-mail addresses. This reference system should aid in the dissemination of information that is current but not yet published.

ENDANGERED BRAINS

by Kaye H. Kilburn, M.D.

A "dirty dozen" of chemicals cause brain damage. Everyone should know how to recognize them. Most are familiar, many are found in homes, and they hurt people. The damage they do is measurable and progressive. It resembles that from trauma to the brain, but is more generalized and usually permanent. Nothing reverses the damage—it does not go away. The result is sudden aging, premature by 20 years or more. Therefore, one must prevent damage by avoiding exposure.

Over the years, the author has measured brain damage from chemicals and has seen deaths from problems many physicians dismissed as "It's all in your head." His search for the cause led to chemicals, to certain conveniences of modern life found at home, carried in insecticides, cleaning agents, room deodorizers, carpets, and drapes—chemicals that cause brain damage. This book is a collection of patients' stories that saddened but enlivened his journey, the steps he took, and thoughts he had about this problem. Patients' tears, fears, and fates are the substance of this account—freed of numbers and statistics.*

Endangered Brains
Published April 2004
$20 plus $5 shipping, prepaid.
Send check to: Kaye Kilburn M.D., P.O. Box 5374, Pasadena, CA 91117 USA

It is also time to consider that ill health appears to be man-made, and thus controllable. A book on this topic, *Avoid Chemicals To Prevent Disease*, is in press.

Currently, much attention and concern are focused on problems stemming from exposure to molds at home, in schools, and in the workplace—as a result of water damage to buildings. A book describing mold toxicity and its effects on human health is on the way.

*Data, statistics, and detailed information regarding brain damage resulting from chemical exposure are presented in *Chemical Brain Injury*, New York: John Wiley publisher, 1998 (obtainable from Dr. Kilburn for $110, including prepaid shipping).